DISCARDED

LINKING EXPERTISE AND NATURALISTIC DECISION MAKING

Expertise: Research and Applications

Robert R. Hoffman, Nancy J. Cooke, K. Anders Ericsson, Gary Klein, Eduardo Salas, Dean K. Simonton, Robert J. Sternberg, and Christopher D. Wickens, Series Editors

---

Hoc/Caccibue/Hollnagel (1995) • *Expertise and Technology: Issues in Cognition and Human–Computer Interaction*

Zsambok/Klein (1997) • *Naturalistic Decision Making*

Noice/Noice (1997) • *The Nature of Expertise in Professional Acting: A Cognitive View*

Schraagen/Chipman/Shalin (2000) • *Cognitive Task Analysis*

Salas/Klein (2001) • *Linking Expertise and Naturalistic Decision Making*

LINKING EXPERTISE AND NATURALISTIC DECISION MAKING

Edited by

Eduardo Salas
University of Central Florida

Gary Klein
Klein Associates, Inc.

LEA LAWRENCE ERLBAUM ASSOCIATES, PUBLISHERS
2001 Mahwah, New Jersey London

Lawrence Erlbaum Associates, Inc., Publishers
10 Industrial Avenue
Mahwah, NJ 07430

Cover design by Kathryn Houghtaling Lacey

Library of Congress Cataloging-in-Publication Data

Linking expertise and naturalistic decision making / edited by Eduardo Salas, Gary Klein.
p. cm.
Includes bibliographical references and index.
ISBN 0-8058-3538-5 (cloth : alk. paper) — 0-8058-3539-3 (pbk. : alk. paper).
1. Decision making. 2. Group decision making. 3. Expertise.
I. Salas, Eduardo. II Klein, Gary A.
BF448 .L56 2001
153.8'3—dc21

2001033160
CIP

Books published by Lawrence Erlbaum Associates are printed on acid-free paper, and their bindings are chosen for strength and durability.

Printed in the United States of America
10 9 8 7 6 5 4 3 2 1

To the firefighters, pilots, commanders, nurses, and other experts who have given us their cooperation and shared their experiences with us so that we could learn about naturalistic decision making

Contents

Series Editor's Preface

Robert R. Hoffman

In the Series Editor's Preface to the volume that stemmed from the 1994 Conference on Naturalistic Decision Making (Zsambock & Klein, 1997), I wrote:

> It would be safe to say that something of a revolution or paradigm shift has occurred ... [but] we still see in NDM something of a promissory note, understandable given the newness of the approach and the untouched problems and challenges. (pp. xi–xii)

The present volume, like its predecessor, includes forays into questions and paradigm shifting. Like its predecessor, this work shows how the NDM movement has a strong international base, offering integration among historically diverse traditions and methods. However, the present volume speaks most loudly to the promise of the NDM movement.

In one way or another, each chapter hints at some way in which the world (or at least a piece of it) might be changed for the better. This volume includes research having goals involving training, decision aiding, and the facilitation of teamwork. Appropriate to this series, this book includes reports on research in significant domains of expertise, and enriching discussions of the nature of expertise. The pay-off in terms of a broadened research base is matched by a broadening and refinement of theories and methods in NDM.

It would be safe to say that something of a maturation has occurred ...

REFERENCE

Zsambock, C. E., & Klein, G. (Eds). (1997). *Naturalistic decision making*. Mahwah, New Jersey: Lawrence Erlbaum Associates.

Preface

We are pleased to say that the field of Naturalistic Decision Making (NDM) continues to grow and mature as an applied psychological science. More than a decade has gone by since the birth of this movement, and it is clear that progress has been made in many ways. There are more theoretical frameworks, and new constructs are emerging. A wider range of studies are empirical, and better methods are being designed and tested. This is not to say that we are satisfied that the field has reached a mature level of methods and models. Nonetheless, there has been movement in the right direction. We are happy to document the progress in this book.

This book contains selected papers presented at the 1998 conference on NDM, held at the Airlie Center in Warrenton, Virginia, May 29 to 31. This was the fourth and largest international NDM conference to date, with more than 180 people attending during the 3 days. The first NDM conference was held in 1989 in Dayton, Ohio (see Klein, Orasanu, Calderwood, & Zsambok, 1993). The second NDM conference was also held in Dayton, Ohio, in 1994 (see Zsambok & Klein, 1997). The third NDM conference was held in Aberdeen, Scotland in 1996 (see Flin et al., 1997). By the time this book is printed, the fifth NDM conference will have been held in Stockholm (May, 2000).

There are additional forums where the NDM community continues to meet to exchange ideas and debate its principles. For example, a significant group of researchers meet every year as one of the technical groups (Cognitive Engineering and Decision-Making) of the Human Factors and Ergonomics Society. This technical group keeps growing and is very active in the Society's activities. It has sponsored a special issue of the *Human Factors* journal on decision making in complex environments (see Cannon-Bowers, Salas, & Pruitt 1996).

We would like to acknowledge the sponsors of the fourth NDM conference: The Federal Aviation Administration, The Naval Air Warfare Center Training

Systems Division (Orlando, FL), the Office of Naval Research, the Space and Naval Warfare System Center (San Diego, CA), the U.S. Army Research Institute, and the Air Force Research Laboratory/Human Effectiveness Directorate. In addition, the conference was endorsed by the American Psychological Association Human Factors and Ergonomics Society (Cognitive Engineering and Decision-Making Technical Group) and the Institute of Electrical and Electronic Engineers-Systems, Man and Cybernetics.

We appreciate their support, and we hope it continues. We think this support signifies the extent to which the field of NDM matters to applied psychology. We know it is not easy to conduct research in natural settings. A considerable amount of time is needed to learn the domains and to arrange for data collection while juggling schedules that are not under the researchers' control. The payoff is in making findings that may transfer more easily to other field settings and are easier to put into use.

The theme for the fourth NDM conference was to explore linkages between NDM and related research traditions. We wanted to make U.S. researchers more aware of NDM research being conducted abroad, particularly in Europe. We wanted to connect NDM research with work in management and industry and to stretch beyond the military and paramilitary focus. And, we wanted to make a more explicit connection between NDM and expertise. These objectives are reflected in the chapters of this volume.

Many people contributed to the success of the conference, and, of course, to this book. We thank Laura Militello for her leadership in organizing and running the fourth NDM conference. We also thank Paula Sydenstricker for managing the logistics of the conference. And we thank Robert Hoffman, our series editor at Erlbaum, for his unfailing encouragement and guidance. Finally, we express our gratitude to Anne Duffy for her constant support to the NDM community.

—Eduardo Salas
University of Central Florida

—Gary Klein
Klein Associates Inc.

REFERENCES

Cannon-Bowers, J. A., Salas, E., & Pruitt, J. S. (1996). Establishing the boundaries of a paradigm for decision-making research. *Human Factors, 38*(2) 193–205.

Flin, R., Salas, E., Strub, M., & Martin, L. (Eds.). (1997). *Decision making under stress: Emerging themes and applications*. Aldershot, UK: Ashgate Publishing Ltd.

Klein, G. A., Orasanu, J., Calderwood, R., & Zsambok, C. E. (Eds.). (1993). *Decision making in action: Models and methods*. Norwood, NJ: Ablex.

Zsambok, C., & Klein, G. (Eds.). (1997). *Naturalistic Decision Making*. Mahwah, NJ: Lawrence Erlbaum Associates.

I

Orientation

1

Expertise and Naturalistic Decision Making: An Overview

Eduardo Salas
University of Central Florida
Gary Klein
Klein Associates Inc.

We can describe Naturalistic Decision Making (NDM) as the effort to understand and improve decision making in field settings, particularly by helping people more quickly develop expertise and apply it to the challenges they face. One of the significant features of the field of NDM is that it seeks explicitly to understand how people handle complex tasks and environments. Instead of trying to reduce these to variables that can be studied at leisure, NDM examines the phenomena themselves in the context of the situations where they are found, and uses this understanding to develop useful types of tools, training, and supports.

Another important feature of NDM is that it has engaged a community of researchers working in different domains, studying different but related phenomena (see Cannon-Bowers & Salas, 1998; Zsambok & Klein, 1997). We have found that we cannot study decision making in isolation from other processes, such as situation awareness, problem solving, planning, uncertainty management, and the development of expertise. We have also found that studies in medical decision making, military command and control, aviation emergency responses, and nuclear power plant management, may all be relevant to each other.

Therefore, in this book we bring together chapters that draw on NDM assumptions and try to understand decision making in complex domains. It is our objective through this volume that all of those involved in NDM efforts can use each other's work to learn important lessons. We hope that the NDM applications will continue to illustrate theoretically driven but practical solutions to supporting skilled performance in natural environments.

OVERVIEW OF THE BOOK

This book is divided into five parts. The first part is an orientation. In this first chapter we provide a brief overview of the chapters that follow. Next is a commentary by Frank Yates (chap. 2) who discusses the contribution of NDM, and how it brought attention to the inadequacies of prior paradigms of traditional decision making. Yates believes that the inadequacies were more a matter of previous paradigms being incomplete rather than being wrong. He conducted an informal survey of traditional decision-making and judgment researchers and presented a number of challenges to the NDM paradigm. These are worthy of study and analysis. The challenges offered by Yates can create opportunities for the NDM community to improve the science and practice of NDM.

The second part describes tools for training and for system design. Thus, Pliske, McCloskey, and Klein (chap. 3) developed an approach to train recognition decision-making skills. The authors briefly reviewed the types of approaches that have been used previously to train decision-making skills (i.e., decision analysis) and described their training tools that focus on critical judgments and decisions. Finally, the authors discussed future directions for research on how to improve judgment and decision-making skills. Next, Fallesen and Pounds (chap. 4), employing a naturalistic decision-making approach, attempted to understand better the actual problem-solving strategies used by U. S. Army officers. Their study was conducted in two phases, first identifying the focus of cognitive skill training, and then conducting the training. The authors' results show that NDM-derived methods are useful for providing a basis for cognitive skill training.

DiBello (chap. 5) addressed the user/tool interface and the relationship between human cognition and information technology. The purpose was to train unskilled workers to employ sophisticated software packages for manufacturing and maintenance control. The author presented an updated field report based on her latest work and a "state of the theory" dispatch. The mechanisms of tool mastery and appropriation remain the same, even with the most advanced technologies. The author's issue of employee resistance to technology involves further study.

Miller (chap. 6) applied a cognitive task analysis to identify the processes underlying the judgments of air campaign planners. The author's research was to first capture the tactical and strategic concerns of air campaign planners, then incorporate this understanding into planning technology to assist with filtering out the unacceptable options as a plan was being developed. The cognitive task analysis formed the foundation of a software tool, the Bed-Down Critic, which highlights potential problem areas and vulnerable assumptions and summarizes aspects of quality as the plan is being developed.

Finally, Roth and her colleagues (chap. 7) described a case study of the design of a large wall-mounted group view display intended to support broad situation awareness of individuals and teams in a compact computerized control room for power plants. The new system illustrates the use of a cognitive work analysis to define the cognitive and collaborative activities and to establish the display content and organization that is required to support these activities. The results supported the importance of situation awareness of individual and teams.

The third part of the book discusses decision-making models. First, Baumann, Sniezek, and Buerkle (chap. 8) developed a model of how self-evaluation may influence performance under acute stress through motivation, anxiety, and self-regulatory processes. After elaborating on the model, the authors presented preliminary empirical support obtained using a simulation-based methodology. Next, Montgomery (chap. 9) presented a description of judgment and decision making labeled a perspective model and described how motivational and information processing factors interact in professionals' decision making. In addition, Montgomery showed how the perspective-motivational model allows a distinction between reflective and nonreflective decision making. Next, De Keyser and Nyssen (chap. 10) addressed the issue of temporal expertise in the anesthesia domain. The authors focused on three types of competence: dynamic diagnosis, anticipation and planification, and synchronization. All are potential sources of errors and accidents. The authors discussed how simulation can improve performance and reduce errors.

Kirschenbaum (chap. 11) asserted that a detailed analysis of the interaction between decision maker and information at the cognitive level could yield useful insights into the decision process. The author discussed the concept of levels of analysis and then described how the submarine problem and the task of the Submarine Approach Officer fit within the NDM framework. She then investigated their decision-making processes in detail, with an emphasis on information-gathering strategies.

The final chapter in this part, Orasanu, Martin, and Davison (chap. 12), addressed the concept of "decision error," specifically focusing on decisions in the aviation domain by re-examining National Transportation Safety Board (NTSB) accident reports. The authors' view was that the errors may be an inevitable consequence of experts behaving like experts, applying their knowledge while performing tasks, frequently following the principle of cognitive economy and efficiency. The authors concurred with previous literature that says we must move beyond trying to pin the blame for accidents on a culprit, seeking instead to understand the systematic causes underlying the outcomes.

Part IV focuses on expertise in several different fields. Shanteau (chap. 13) discussed what it means when experts disagree. He believes that disagreements mirror the way that experts think and work. He discussed 10 structural and functional factors relating to why experts disagree. The chapter then focused on domain differences, experts' previous views, and an alternative hypothesis, the Multiple Solution Model. Militello (chap. 14) focused on the types of representations offered by cognitive task analysis methods. The view of expertise is represented by the output of the cognitive task analysis, which is then available to designers and developers of training, expert systems, and decision support systems.

Schraagen and Leijenhorst (chap. 15) focused on knowledge and search strategies that are used by forensic scientists. They discussed how to formalize strategies and how to incorporate them into forensic procedures. They then discussed the trade-off between efficiency and effectiveness. This trade-off often makes it difficult to decide which strategies to formalize and incorporate, especially if both efficiency and effectiveness are important.

Sonnentag (chap. 16) described how expert software professionals accomplish tasks and handle the complex requirements that are typical for their work situation. The results of two empirical studies indicated that high performers put added emphasis on comprehension of problems, preparatory activities, local planning, and seeking feedback. High performers also engaged in more communication activities. The results also indicated that there is no relationship between the number of years of professional experience and moderate or high performance, which suggests that other aspects of experience may be more necessary to achieve higher performance. One implication is that we cannot rely on years of experience as a measure of skill.

Dominguez (chap. 17) addressed the issue of expertise in laparoscopic surgery, outlined the background and theory of laparoscopic surgery, and presented the results of two studies on anticipation and affordances. Results showed that surgeons must predict whether a procedure is safe or too risky, which determines whether to convert to an open procedure. Surgeons must continually assess the situation that they are facing.

H. Klein, Vincent, and Isaacson (chap. 18) focused on driving proficiency and the development of decision skills. They used an NDM approach to study how decisions are made while driving. They compared the results of long-time drivers to less-experienced drivers. They then provided recommendations for decision skills training, such as initially focusing on declarative knowledge and advancing the development of control.

Weick (chap. 19) addressed the issue of why firefighters hold on to their tools when trying to outrun fires and how this contributes to the number of firefighters who are killed by fires. He discussed several reasons as to why firefighters do not drop their tools when told to do so. Weick used this problem to examine fundamental aspects of NDM investigations and theorizing. Lipshitz (chap. 20) reacted to several of the "rough spots" in the "NDM mindset" that were discussed by Weick. Lipshitz addressed several questions: the meaning of NDM; the areas of NDM that should be studied; and ways in which to study NDM.

The final part of the book deals with teams. Carroll, Rudolph, Hatakenaka, Wiederhold, and Boldrini (chap. 21) discussed learning in the context of team diagnosis of incident investigations and organizational decisions that occurred at four nuclear power plants. The results of their study showed that decision makers treat symptoms more frequently than fundamental causes, leading to ineffective corrective actions and rare opportunities for organizational learning. Carroll et al. also addressed how team composition, team process, and receptivity for management affect team decisions and organizational capabilities.

Smith, McCoy, and Orasanu (chap. 22) examined distributed cooperative problem solving in the air traffic management system. Changes are being made in the U.S. air traffic management system in order to improve safety and efficiency. The authors addressed how these changes affect decision making and human performance. They showed how high-level decisions concerning locus of control and task decomposition can impact the way people interact to make decisions.

Patel and Arocha (chap. 23) addressed the nature of constraints on collaborative decision making in health care settings. The authors discussed how decision making is affected by cognitive, situational, organizational, and epistemological constraints. Two studies were presented with the results suggesting that decision making is dependent on these four constraints.

Worm (chap. 24) discussed a tactical mission analysis by means of NDM and cognitive system engineering. He focused on the development of theories and models that are intended to continue analysis, evaluation, and assessment of both emergency and military response units that are performing high stakes and complex tactical operations. Simulations showed how these concepts were tested and validated by exercises.

FINAL REMARKS

In the following chapters, it will be clear how decision making is linked to problem solving and situation awareness and to the development and sustainment of expertise. Some of the chapters aim at a high-level view of the cognitive challenges found in natural settings, but most chapters focus on single domains and types of decision makers. The investigations are about specific phenomena, usually involving time pressure, uncertainty, high stakes, ill-defined goals, and other features of field settings. The investigations are aimed at understanding these phenomena—in all their complexity and messiness. The NDM community is committed to maintaining research at the level of the phenomenon.

In some ways, this book can be seen as a report from the trenches. Researchers have come back from their explorations, taken off their hard hats (literally, in the case of DiBello, who worked with maintenance crews), described what they have learned about aviation accidents (Orasanu, Martin, & Davison), nuclear power plant emergencies (Carroll et al.), and wildland firefighting disasters (Weick). They explained the critical judgments facing submariners (Kirschenbaum), forensic scientists (Schraagen et al.), air campaign planners (Miller), laparoscopic surgeons (Dominguez), and software developers (Sonnentag). Not all of the locales were exotic. H. Klein, Vincent, and Isaacson described the terrors facing teenage drivers (and their parents). Some of the challenges are set in the future, such as new decision requirements for air traffic controllers. The researchers have described the different cognitive landscapes they encountered, hoping that we can all learn from their experiences and their interpretations.

REFERENCES

Cannon-Bowers, J. A., & Salas, E. (Eds.). (1998). *Making decisions under stress: Implications for individual and team training.* Washington, DC: American Psychological Association.

Zsambok, C., & Klein, G. (Eds.). (1997). *Naturalistic decision making.* Mahwah, NJ: Lawrence Erlbaum Associates.

2

"Outsider:" Impressions of Naturalistic Decision Making

J. Frank Yates
The University of Michigan

When I received the invitation to give this talk, I was both honored and apprehensive. That is because talks like this have a "history" at naturalistic decision making (NDM) conferences. That history was reflected in the light-hearted comment of a friend on inspecting the program: "Ummm ... , so you're the one giving the curmudgeon talk this year, huh?" You see, for each of the NDM conferences, a perceived outsider, such as myself, has been asked to provide a critique of the field. I was flattered because the invitation suggested that the organizers thought I might have something useful to say. I also felt honored because of the stature of previous "curmudgeons," Michael Doherty and William Howell—fast company indeed. My apprehension arose from the very nature of the assignment and is implicit in the "curmudgeon" characterization given to those who have accepted the assignment. Critics criticize. They tell us that, in at least some respects, something we are doing is not up to snuff. Even when such criticisms are true, they are no less unpleasant to hear. And who wants to be a pariah?

I agreed to the assignment because my desire to see the NDM community succeed outweighed my need to feel comfortable. My overarching professional aim (and that of the groups I represent, the University of Michigan's Judgment and Decision Laboratory and its Decision Consortium and Business School) is to contribute to the effort to understand and improve decision behavior under any and all circumstances. And, it seems clear to me that that aim would be advanced significantly if NDM were to flourish. Hence, the positive way I framed my mission.

As something of a consultant, I sought to prepare a review that would provide NDM researchers with answers to the following sorts of questions they might ask themselves:

- "How well are we achieving what we set out to do?"
- "Why haven't we gotten (even) closer to our goals than we have?"
- "What unintended effects—positive and negative—are we having, and why?"
- "What are some plausibly viable means for being (even) more successful than we have been so far?"

My planned method was to re-read carefully—and with a different, more critical eye—the previous NDM volumes (Klein, Orasanu, Calderwood, & Zsambok, 1993; Zsambok & Klein, 1997) as well as other NDM writings I could find. I would then try to interpret that work against the backdrop provided by broader literatures related to how people decide. Over the years, I have been fortunate (actually, forced) to acquire considerable familiarity with those literatures. That is because of varied research, writing, teaching, editorial, and review responsibilities, for example, in connection with such journals as the *Journal of Behavioral Decision Making*, *Organizational Behavior and Human Decision Processes*, *Psychological Review*, *Medical Decision Making*, the *Journal of Applied Psychology*, *Management Science*, the *Journal of Forecasting*, and *Risk Analysis*. My interpretations would also be cast within a new conceptual framework my colleagues and I have found useful for thinking through decision problems of all sorts, the "cardinal issue perspective" (e.g., Yates & Estin, 1998; Yates & Patalano, 1999).

As I got toward the end of the review process, I began having doubts like these: "Am I being comprehensive?" "Am I being fair?" "Am I being representative of the kinds of important opinions my NDM 'clients' need to know about and understand?" Thus, as a check on myself, I decided to conduct an informal electronic mail survey. I wrote to people I knew in the United States and Europe who are among the world's best known and most widely read and respected experts on various aspects of decision making, none of whom is commonly identified with NDM per se. I asked them essentially the following: "What are your 'off-the-top-of-your-head' impressions of the NDM paradigm? What positives and negatives come to mind?" More than 25 were able to reply in the limited time available. Their remarks coincided with many of the conclusions I had reached independently, but several surprised me. They were so compelling that I decided that my NDM clients' interests would be better served if those responses took a more prominent role than I had planned for them. Thus, I begin with what my correspondents had to say.

THE SURVEY

Statistical analyses are inappropriate in situations like this, irrespective of control and sampling considerations. The aim of the survey was not to establish anything like defensible estimates of population tendencies. Instead, the purpose was to identify the *kinds* of sentiments and concerns that capture the attention of one particular, selective community of observers of the NDM scene, views that might not be immediately obvious to NDM researchers themselves. My respondents said a variety of things, often repeatedly, and in several different ways. In what follows, I try to categorize their comments and capture each idea in a paraphrasing that conveys the substance of what the respondents said and how they said it. I first describe the positive comments, which I call "plaudits." I then present the negative remarks, which I also prefer to think of positively, called "challenges." After that, I offer suggestions for what NDM researchers might make of the respondents' opinions.

Plaudits

- *Plaudit 1—Attention to Topics:* "NDM has identified important, interesting areas of inquiry that were neglected before, for instance, particular aspects of process, dynamics, and field settings."
- *Plaudit 2—Attention to Limitations:* "NDM has brought attention to the limitations of previous work, particularly formal approaches."
- *Plaudit 3—New Concepts:* "NDM has introduced some useful new concepts and ideas, such as the role of recognition."
- *Plaudit 4—Methods:* "NDM has introduced new methods that can be helpful sometimes."
- *Plaudit 5—Practical Intent:* "NDM reflects an admirable aim of helping people with significant practical problems they face in real life."
- *Plaudit 6—Recruitment:* "NDM has brought lots of applied investigators into the field."
- *Plaudit 7—Applied Respectability:* "NDM has brought new respectability to applied work."

Challenges

Before I describe my respondents' challenges, as a point of reference, it is useful to consider the list of NDM criticisms Gary Klein has assembled over the years, as described in an address to the Decision Consortium at the University of Michigan in December, 1997. These are the highlights of that list:

- *Criticism 1:* "NDM is just a reaction to decision analysis."
- *Criticism 2:* "NDM cannot explain errors."
- *Criticism 3:* "There is no theory in NDM."
- *Criticism 4:* "You can't do research in decision making without normative theory."
- *Criticism 5:* "There is nothing new about NDM."
- *Criticism 6:* "NDM isn't about decision making."

As the next list shows, Klein's inventory did a good job anticipating several of the respondents' concerns. But, it did not foresee them all. This suggests that the respondents did indeed provide a service that others had not rendered already.

The challenges posed by the respondents were far more numerous than the plaudits, which is the case in virtually every review situation I have witnessed. There were, in fact, so many challenges that it made sense to organize them into categories, as follows:

Challenge Category 1: Presentation

- *Challenge 1—Self-Definition:* "NDM has described itself negatively, as what it is *not* (for example, not decision analysis and not 'classical decision theory') rather than what it *is*, adopting a mainly critical rather than constructive role."
- *Challenge 2—Tone:* "NDM has taken on a needlessly negative and combative tone that has created polarization instead of cooperation."
- *Challenge 3—Quality Control:* "NDM writings are unusually uneven in quality, with too many weak contributions making their way into print, suggesting deficient peer review."
- *Challenge 4—Impact:* "NDM has had minimal impact on the field."
- *Challenge 5—Progress:* "NDM has made (or at least reported) surprisingly little documented progress, 'surprising' given NDM's comparatively generous funding."

Challenge Category 2: Connections

- *Challenge 6—Originality:* "Numerous of the key ideas claimed by NDM actually have been around a long time, and this has been unrecognized or at least unacknowledged."
- *Challenge 7—Awareness:* "NDM researchers seem unaware of work in other areas (for instance, in learning) that deals with essentially the same phenomena and, hence, could be of use to them."
- *Challenge 8—Misunderstanding:* "NDM writers have tended to either misunderstand or misrepresent 'classical decision theory,' offering little more

than a caricature of it, as aiming to demonstrate human stupidity, for example."

Challenge Category 3: Substance—High Level

- *Challenge 9—Generality:* "NDM has trouble with generality: (a) Much of its work has been so narrowly focused (say, on a specific military problem) that it bores those outside the immediate domain; and (b), when attempts *have* been made to generalize, the claims have been overly extravagant, for instance, patently false sweeping claims about what (presumably all) 'real' decision making is like."
- *Challenge 10—Theory:* "NDM theorizing has been deficient—imprecise, untestable, unsophisticated, narrow, non-exploitable."
- *Challenge 11—Description:* "NDM has been almost exclusively and superficially descriptive, offering little in the way of deep explanation."
- *Challenge 12—Methods:* "NDM methods are often clearly flawed. Some examples: (a) NDM studies frequently employ tiny sample sizes, yet researchers offer broad generalizations from them; (b) the methods seldom test NDM's claims rigorously and definitively; (c) researchers take too seriously decision makers' utterances as accounts for what the decision makers actually did in particular episodes; and (d) researchers fail to appreciate the appropriate roles that a variety of techniques, including experimentation, can play in enlightening NDM questions."

Challenge Category 4: Substance—Details

- *Challenge 13—Decision Quality:* "NDM sidesteps the critical issue of what constitutes good decision making."
- *Challenge 14—Reverence for Naturalism:* "NDM embraces the false assumption that, if people naturally do something, then it must be good, thereby setting too low a standard."
- *Challenge 15—Reverence for Expertise:* "NDM neglects to consider the negative aspects or risks of experts' methods, for instance, the downsides of decision making via pattern recognition."

What to Make of the Comments

Why should NDM researchers not simply ignore the respondents' remarks, particularly the negative ones? I can imagine several reasons they might be inclined to do just that. These reasons are captured well in the voices of skeptics like those I have actually heard:

- *Validity: "Some of these things are simply untrue."* In some instances, the criticisms may indeed be untrue. But, it is in the interests of NDM researchers to confirm that they are untrue. Furthermore, if a claim is false yet is nevertheless widely believed to be true, it is still damaging to progress in the field. Thus, NDM researchers would do well to actively correct such false beliefs.
- *Ignorance: "These folks don't really know NDM."* Actually, several respondents readily acknowledged their (relative) ignorance of NDM. A couple said they had not been exposed to NDM work. Others said that they had been exposed to it but had been "turned off" by the experience and thus avoided it thereafter. Still others saw NDM as irrelevant to the problems that occupy their attention. The action implications of these various kinds of ignorance are all similar. The NDM community should do more to get its message out broadly and convincingly. For instance, it would make sense to determine the reasons particular NDM writings have turned off some readers and, when possible, reframe accounts to be responsive to those concerns.
- *Antagonism: "These people are biased against NDM because NDM directly challenges their own views."* There might be an element of truth to this in rare instances, although this would be hard to verify. Besides, it is usually more productive to simply assume people's good intentions.
- *Representativeness: "We have no idea whether these ideas are typical of those held by* any *group or are nothing more than oddities—outliers, as it were."* This may be true as well, given the acknowledged fact that the survey was uncontrolled and included only a small number of cases. But, several of the comments were made sufficiently often that we should suspect that they might well be held widely. And, the more serious criticisms are sufficiently important that the NDM community might find it worthwhile trying to determine how broadly they are in fact believed to be valid.
- *Irrelevance: "It really doesn't matter what these particular people think as long as we're meeting the needs of our true constituency."* There is an important practical reason NDM researchers should care not only what these *kinds* of respondents think but also about what these *particular* respondents think: their opinions are highly influential in the academy and in publishing. If these people are unconvinced by NDM claims or, even worse, think that those claims are flat-out wrong, they will not remain silent. And, when key parties such as funding agencies and editors are deliberating matters critical to the interests of the NDM effort, they will not ignore those negative assessments.

A HIGHER VIEW: THE CARDINAL ISSUE PERSPECTIVE

For years, I was deeply puzzled and dismayed by the obvious antipathy between the NDM and traditional decision research communities (e.g., in my role as president of the Society for Judgment and Decision Making). The misgivings of the latter about the former are evident in the challenges described earlier. Those of the former about the latter shine through in much of the NDM literature, especially the earliest (e.g., in the papers from the first NDM conference, Klein et al., 1993). I am still mystified to some degree, but I believe I have made significant progress in understanding the bases and origins of the antagonism. Unfortunately, if my analysis is essentially correct, the forces driving the ill will are still very much alive. On the other hand—again, if I am essentially correct—they are in principle easy to neutralize, for the good of all concerned. I first summarize my thesis in broad outline. I then try to fill in key details, including prescriptive suggestions.

A Preview

Both NDM and traditional decision research have aspired to understand and improve people's decisions. Yet, when NDM researchers have sought to apply major conclusions from traditional research, they have been profoundly disappointed with the results. The same story, only with the roles reversed, describes the experience of the many fewer traditional researchers who have tried to see what the NDM point of view could do for them. How could this be? The crucial oversight seems to be this: the particular decision problems that NDM and traditional researchers have chosen to study are significantly and *qualitatively* different from each other. To mutilate a metaphor badly, it is as if one group has chosen to study apples, the other oranges, each assuming the applicability of the other's methods and principles. Under the false presumption that they are studying basically the same thing, each walks away from the experience frustrated, embittered, and with reduced respect for the other party. ("How could they possibly think that's informative?") My expectation is that, if both parties come to recognize the true nature of the problems they are confronting, two things will happen. First, their misgivings will largely disappear, dissolving into respectful coexistence. Second, each will be able to proceed more expeditiously and effectively, unfettered by false expectations and antagonisms.

Sad Tales

For well over two decades, I have welcomed and actively encouraged any and all comers (e.g., students and laypersons) to confront me with decision problems that matter to them personally. My feisty challenge is, in effect: "Bring it on, give me your best shot." One aim has been to try to motivate interest in what our field has to offer. Another is to find and push the limits of what we (think we) know, to pressure us into getting better at what we do. To my recurring and embarrassing horror, early in this campaign, far more often than I expected and wanted to admit, I discovered that what we had to offer was virtually nothing. It was not so much that our wares failed to do what they promised. Instead, what they promised and were capable of delivering was simply and obviously irrelevant. A stark (and real) example: Picture a young, U.S.-bred woman of Asian Indian heritage. She is faced with a troubling personal dilemma. On the one hand, she can insist on a U.S.-style, free-choice marriage in which she dates and eventually chooses a husband for herself. On the other, she can accede to her parents' wishes that she follow Indian arranged marriage customs (performed fully within the requirements of U.S. law, of course). Clearly, recommending traditional standbys such as expected utility maximization or even multiattribute utility maximization would be absurd in such circumstances. The same could be said for anything in the NDM researcher's toolkit too, incidentally.

I have been told that NDM was inspired by similar sad stories in settings such as military operations, aviation, and industrial process control. Many of these stories have become a part of NDM's oral history, being repeated in conference talks and private conversations. At least a few have been described in NDM writings. Several excellent examples appear in the chapter on training by Means, Salas, Crandall, and Jacobs (1993). Means et al. vividly summarized developers' repeated failed attempts to improve decisions in operational situations by training people to avoid the kinds of biases (e.g., hindsight) that for so long have been a major focus of attention in decision research. They tell a similar story for efforts to get people to apply formal techniques such as multiattribute utility theory. Just as such tools proved totally inadequate for helping my Indian informant come to terms with her marriage dilemma, they were just as fruitless for pre-NDM decision trainers, only worse. That is because, in contrast to what happened in the marriage case, the trainers described by Means et al. actually tried to apply the tools at hand, and the attempts were disastrous. Not only did they fail to yield clearly better choices, but they also proved to be cumbersome, slow, and wasteful.

Whereas the experiences described by Means et al. (1993) led to the development of NDM, my sad stories had a different kind of ending. Repeatedly faced with (and defeated by) challenges from all quarters, I concluded that I must go back to basics. I had to ask what really is at the heart of decision making, to understand why our efforts had been so obviously far off the mark for so many people's needs. So I collected and analyzed scores of "decision conversations" and "decision stories" like that involving the woman in the arranged marriage quandary. I also read and re-read hundreds of formal writings on decision making from myriad disciplines and traditions. The result has been the "cardinal decision issue perspective," which addresses four things: (a) what decisions are; (b) what decision quality is; (c) what are the modes by which people decide; and

(d) what are the fundamental—cardinal—issues that must be resolved in virtually all practical decision problems. I have found that the perspective helps me see more clearly what is and is not in question in a given decision situation or a given discourse about decision making. My hunch is that it might prove similarly beneficial to NDM researchers as they continue their own struggle to serve the needs of their constituencies. There is insufficient space here to offer a full rendering of the cardinal issue perspective. But I can sketch several aspects of the perspective that seem especially pertinent to NDM at this juncture. (The chapters by Yates & Estin, 1998, and by Yates & Patalano, 1999, provide partial yet more extensive accounts.)

Decisions

Recall that one of the common criticisms of NDM included in the list Gary Klein accumulated is the assertion that NDM researchers do not actually study decision making. Such a complaint reflects a frequent conclusion that prototypical traditional decision researchers reach when learning more about NDM. Those researchers had thought that they and NDM researchers were both studying decisions. But, after a while, the stark differences between whatever they were investigating became inescapable. And, because the traditional researchers knew that what they were studying really *was* decision making, then NDM researchers must be studying something altogether different. As I argue, they are absolutely right: they are investigating markedly different things. After this is established, the only language issue that remains is who gets to claim the term decision making. In one sense, worrying about who owns the expression is the height of pettiness. All that ought to matter is that the phenomena of interest are understood, whether they are called *decision making* or *blixtzzy*. But it does matter when people use the same term for significantly different things without realizing that they are doing so. Confusion reigns, as has been the case, apparently, among decision researchers of various stripes. Researchers have tended to implicitly define *decision making* very narrowly as any activity closely associated with the particular variety or aspect of decision making that happens to interest them personally. But suppose one listened carefully to all kinds of people—including but not limited to researchers—talk about the things they consider to be "decisions." Then, a surprisingly coherent common denominator emerges. That common denominator is the definition articulated in the cardinal issue perspective: A *decision* is the commitment to an action whose aim is producing satisfying outcomes. And *decision making* is the process of solving a particular type of problem, arriving at a good decision.

The cardinal issue perspective recognizes three broad categories of decisions, one of which has a major variant of its own. The first is *choice*, the selection of one or more options from a specified larger pool of options (e.g., choosing a dinner party's entrées from a restaurant's complete menu). An important special case of choice consists of *accept/reject decisions*, whereby the decision maker either agrees to take the one particular option that has been proffered or maintains the status quo, for a while at least (e.g., rejecting a proposal of marriage or accepting an investment opportunity). The second major decision category consists of *evaluation decisions*, wherein the decision maker articulates the worth of some entity and is prepared to act on that statement (e.g., offering a bid on a house). *Construction decisions* define the third and final major variety of decision, instances in which the decision maker attempts to use available resources to assemble the most suitable option possible given existing constraints (e.g., creating the administrative structure of a new business).

Careful reflection makes it apparent that what most traditional decision researchers consider to be decisions are consistent with the cardinal issue definition, for example, choices among stock portfolios or minimum buying prices for options presented in a laboratory experiment. The same is true for the decisions commonly acknowledged by NDM researchers, for example, a decision to attack a fire in a particular way or to respond to an intrusion in military airspace in some manner. However, at the same time, certain other things sometimes uncritically called *decisions* are conceded to be better called something else, even when they are related to decisions. An especially prominent example is that of a judgment. In the cardinal issue perspective (see also Yates, 1990, chap. 1), a *judgment* is an opinion as to what was, is, or will be some decision-relevant aspect of the world. A good illustration of a judgment is a physician's belief that there is a 70% chance that Patient Smith has pulmonary fibrosis. This is usefully distinguished from the decision to treat Patient Smith as if he has that disease. That decision arguably ought to take into account not only the given diagnostic belief, but also such considerations as the consequences of treating Smith for pulmonary fibrosis when he actually has lung cancer. One reason for distinguishing decisions from supporting judgments is that judgment quality is much easier to characterize and to assess than is decision quality, as I discuss later (see, for instance, Yates, 1994).

What does the conception of decisions in the cardinal issue perspective imply for questions important to the NDM enterprise? First of all, it highlights the enormous breadth of activities routinely and legitimately called *decision making*. Making decisions is a fundamental and ubiquitous human activity. From our earliest years and in virtually every context, we must and do make decisions of all sorts. Thus, it would be surprising indeed if adequately addressing those diverse decisions did not demand measures that sometimes differ sharply from one another. Traditional decision scholarship was developed to deal with economic decision problems, whereas NDM has been driven by the needs of people faced with other sorts of decisions, mainly operational ones. Given the radically different characteristics of typical economic and operational situations, we should expect (rather than be troubled) when we observe great differences in the descriptive and prescriptive conclusions derived by traditional and NDM decision specialists.

The cardinal issue perspective on different decision types also suggests that some of the controversies that have consumed the NDM community perhaps should not have been considered particularly controversial in the first place. An example: Much has been made of demonstrations in the NDM literature that decision makers often entertain only a single alternative rather than make comparisons among two or more (e.g., Klein, 1989). But such single-option deliberations are a defining characteristic of the accept/reject decisions acknowledged in the cardinal issue perspective, decisions that are extremely common in real life. Indeed, in some real-world instances, such as that involving marriage proposals, entertaining more than one option at once would be considered tacky at best. And in others, over the years a variety of techniques have evolved for guiding the sensible handling of single options, for example, discounted cash flow algorithms used in deciding about prospective investment projects. Thus, rather than making too much of the mere existence of single-alternative deliberations in the situations of interest to most NDM investigators, it might be more fruitful to try to determine what can be generalized from analogous deliberations in arenas where they have been recognized all along.

Decision Quality

In the cardinal issue perspective, a *good decision* is one that has few serious deficiencies, and a *good decision process* is one that tends to produce good decisions. And what is a decision "deficiency?" Deficiencies come in five varieties:

- *Aim:* An *aim deficiency* occurs when a decision fails to meet the decision maker's explicitly formulated aim (e.g., a book purchase fails to meet the purchaser's aim of becoming informed about a certain topic).
- *Need:* A *need deficiency* occurs when a decision maker fails to meet the actual need (or needs) in the given situation, perhaps because no decision was made when there should have been a decision. In one form of need deficiency, the decision does not relieve adverse circumstances that are impending if not already present (e.g., a book purchase fails to relieve the boredom that inspired the purchaser to buy a book in the first place, even if the purchaser perhaps never recognized that boredom). Another form of need deficiency takes place when a decision (or non-decision) fails to exploit an opportunity for improving current circumstances (e.g., when a person suffers from ignorance about a topic but never realizes it and thus never even considers buying a book that would alleviate that ignorance).
- *Aggregate outcomes:* An *aggregate outcomes deficiency* occurs when, collectively, all the outcomes of a decision (beyond those associated with the aim and need) leave the decision maker worse off than some effective reference, such as the status quo (e.g., a new book is so engrossing that the purchaser has to stay up all night reading it, making the next day at work unbearable).

- *Competitor:* A *competitor deficiency* occurs when, in the aggregate, a decision is inferior to some competing alternative (e.g., it turns out that a book purchased for $10 could have been bought at the store next door for $5).
- *Process cost:* A *process cost deficiency* occurs when the costs of arriving at a decision (in money, time, effort, "aggravation," and so on) are inordinately high (e.g., a purchaser agonizes for 45 minutes in front of a bookshelf before finally picking one to buy). Note that this deficiency does not refer to the cost of executing a decision (e.g., the price of a book or the distance one must travel to fetch it).

Decision problems differ in their *deficiency risk profiles.* That is, they vary in the extent to which, if no decision were made, the decision maker would be likely to suffer from particular deficiencies and the severity of the impact of those deficiencies. Consider the personal finance decision of buying life insurance. There, aim deficiencies—failing to leave adequate income for one's family—are central concerns. But, a particular kind of process cost is prominent for many people in that situation, too, that is, the aggravation of even thinking about something as unpleasant as one's own death. For some people contemplating financial decisions, competitor deficiencies are a serious matter, too, as implied in research showing that anticipated regret is often a major consideration (e.g., Josephs, Larrick, Steele, & Nisbett, 1992). The regret theories guiding such research posit that the attractiveness of options is not defined solely by the returns those options would yield. It is also affected by how those returns compare to what the investor could have earned had some other option been selected (e.g., Loomes & Sugden, 1982). The kinds of operational decisions that are the subject of most NDM research also focus on aims, but ones that are very different in character, for example, achieving certain military objectives in the heat of a battle. Much more so than in financial planning, a different kind of process cost is often critical too—decision time. But, as is implicit in paper after paper in the NDM literature, competitor deficiencies are unimportant for most operational decisions. Within broad margins, who cares if there was a better way of getting a child out of a burning house if the method that was actually chosen did the job in the precious little time available? The key point illustrated here is that the quality concerns of investigators in the traditional and NDM paradigms ought to be different because their focal problems are different. It is, therefore, little wonder that such investigators often have little to say to one another or find little value in one another's conclusions.

Real-life decision makers (if my informants are representative of them) find the treatment of decision quality by both traditional and NDM researchers puzzling and inadequate. Decision quality concepts in traditional decision research emphasize internal consistency or coherence (see the discussion by Yates, 1990, pp. 262–263). Thus, for example, some writers assert that a good decision is one that maximizes expected utility (e.g., Edwards, Kiss, Majone, & Toda, 1984). The typical practical decision maker naturally then asks, "What's so good about that? How does that make me better off?" The response implicit and sometimes explicit in the traditional literature is that maximizing expected utility prevents the decision maker from contradicting him- or herself, for example, violating principles equivalent to the axioms underlying utility theory and which the decision maker agrees should be upheld. But our practical decision maker might persist: "That's all well and good, but what would be so bad—materially—if I contradicted myself?" A traditionalist might then respond that a self-contradictory decision maker can be exploited in Dutch books or material traps, which are cycles of transactions that are guaranteed to leave the decision maker no better off after the transactions are completed than before. Our practical decision maker persists even further: "But how do I know that there are people (or circumstances) out there that would, in fact, take advantage of me like that?" Our decision maker is actually asking about the kind of empirically based goodness implicit in the cardinal issue perspective's definition of decision process quality. The decision maker is asking about the chances and extent to which maximizing expected utility would yield decisions that enhance his or her circumstances. The answer to this question undoubtedly depends on the nature of the decision problem. But by its nature, as argued later, there is little reason to expect that the benefits of maximizing expected utility can be substantial in the big picture.

Curiously, NDM seems to largely ignore questions about decision quality. Thus, Klein (1989) emphasized that the recognition-primed decision (RPD) model applies to proficient decision making. But "proficiency," which connotes high-quality decision making, is never actually defined. This is problematic for several reasons; perhaps the main reason is that the overriding concern of NDM's constituents, for example, the sponsors of NDM research, is to enhance proficiency. How, concretely, can they tell whether and when this has been done without a well-conceived specification of what proficiency is? The quality notions of the cardinal issue perspective might provide an avenue for achieving that specification.

Decision Modes

The cardinal issue perspective acknowledges several different *decision modes*, which are qualitatively distinct ways that people solve decision problems. The decision modes' interrelationships can be illustrated like so:

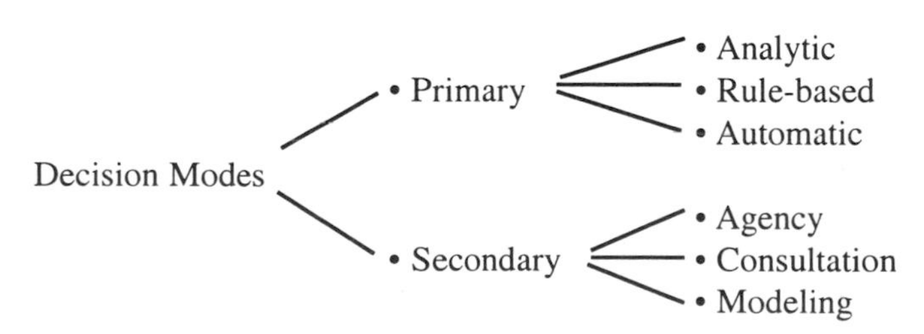

As suggested by the schematic, two broad classes of decision modes are recognized. *Primary modes* are ones in which the decision maker personally (possibly in a collective) derives the decision. In contrast, in *secondary modes*, someone else derives the (prospective) decision on the decision maker's behalf.

In the first of the primary decision modes, *analytic decision making*, the decision maker effortfully thinks things through and arrives at what seems to "make sense" to do in the given situation. Analytic decision making is the mode addressed by traditional decision scholarship as well as the one people tend to use in any situation they see as serious and unfamiliar, for example, deciding what house to buy. In *rule-based decision making*, the decision maker deliberately applies a rule of the form, "If Condition C holds, then take Action A." Klein's (1989) discussion of recognition-primed decision making provided excellent illustrations. But there are lots of others, including physicians applying standard protocols for treating patients whose signs and symptoms fit particular profiles and bankers awarding loans on the basis of qualification checklists. *Automatic decision making* is similar to rule-based decision making in that Action A results from a match between the current situation and Condition C of a C → A chain. The difference is that, in the case of automatic decision making, the process is effortless, uncontrollable, and often cognitively inaccessible. That is, if and when the decision maker "recognizes" (correctly or incorrectly, consciously or otherwise) that Condition C is present, Action A simply "pops out," and there is nothing the decision maker can do to prevent that from happening. Moreover, the decision maker might have great difficulty accurately describing how the decision was reached. Decision making in a fast-paced sport like tennis or basketball (in contrast to a slow, deliberate one like golf) offers ample examples.

Secondary decision modes include three major types, too. The first is *agency*, whereby the decision maker commissions someone else (or something else, like a computer) to make the given decision. Parents routinely serve as decision agents for their children. So do physicians for their patients, despite contemporary pressures from some quarters for "non-paternalistic" medicine (see, for example, Peters, 1994). Even further examples include lower ranked military personnel making tactical decisions within the framework of strategies set by their superiors, and trading programs automatically buying and selling financial securities when specified market conditions occur. Next among secondary decision modes is *consultation*, wherein the decision maker accepts or rejects recommendations developed by consultants. This is the standard mode by which corporate boards of directors tend to function (legally, at least), with professional managers taking on the consultant role. *Modeling* is the final secondary decision mode. There, the decision maker observes the decision of some other, respected decision maker and simply mimics it. The business world again provides plenty of examples, such as choosing the same vendor for a required service as the one chosen by an industry leader.

NDM provided a tremendous service to decision scholarship generally by crystallizing and bringing long overdue attention to the idea of decision modes, mainly through its discussion of Klein's RPD model. It is time now, however, for the field to move beyond that plateau. As suggested by the broader cardinal issue perspective view of modes, much more can and should be done with the concept. For instance, NDM researchers would do well to pursue the practical implications of modes, which they do not appear to have recognized fully. In my view, the main reason for acknowledging modes is that decisions made according to different modes demand different means for their improvement; approaches that work with one mode are often useless with others. This seems to have been well documented in the repeated experiences of NDM developers. For example, it is reasonable to assume that the kinds of flawed judgments that plague many decision makers' choices (e.g., ones that are overconfident) rest on automatic processes. If that is indeed true, then it is small wonder that merely explaining the biases to decision makers does virtually nothing to eliminate them (Fischhoff, 1982). Expecting otherwise would be like expecting an explanation of the physics of hitting tennis balls to be sufficient for correcting a player's flawed backhand.

The cardinal issue perspective on modes also brings to mind two important, related concepts that seem to have been neglected in NDM work—individual differences and the implications of basic learning principles. If there is any theme that characterizes NDM more than any other, it is "experience." Repeatedly, NDM authors emphasize that the key to understanding and improving decision making is the decision maker's experience level. Thus, the often poor performance of subjects in the experiments reported by traditional decision researchers is commonly attributed to those subjects (who are typically college students) being asked to perform tasks that are unfamiliar to them. And, the recommended way to train good decision makers is to accelerate the experience that they would normally acquire over many years (e.g., Cannon-Bowers & Bell, 1997). Furthermore, in the prototypical NDM study (e.g., most of those discussed by Klein, 1989), the focus is exclusively on experienced decision makers who are considered to be "proficient" or "expert" or on a comparison between experts, on the one hand, and novices, on the other. Other than in the case of chess players, there is no independent, rigorous documentation that the experienced decision makers being studied really are experts. (Of course, part of the difficulty is that, as discussed previously, NDM has given short shrift to the question of what decision-making expertise actually is.)

What is the problem here? Implicit in NDM analyses, there seems to be an assumption that repeated decision episodes are sufficient to yield the following progression:

$$\text{Incompetence} \rightarrow \text{Expertise}$$

Expressed in this stark way, almost no one would actually believe this to be true in any domain other than decision making. And it is doubtful that even the NDM investigators who have asserted it would accept the assumption outright. It amounts to saying that, if you perform a task often enough, you are guaranteed to become outstandingly good at it.

The first reason most of us reject this assumption is that we expect significant individual differences in the extent to which people are capable of reaching the highest levels of achievement in any domain. Thus, I know that, no matter how many rounds I played, I would never become a good enough golfer to make the professional tour. By the same token, it seems unrealistic to expect that everyone could become a stellar decision maker. In practical settings, those with ultimate decision authority often want (or need) to delegate decision-making responsibilities to others (i.e., pursue the secondary decision mode of agency). They would find it useful knowing how capably the agents they might appoint can decide right now or how easily they could be taught to decide proficiently if chosen for training. To date, the NDM community has expressed little interest in serving such needs. But it should.

The second reason for doubting the "mere repetition" assumption is that it amounts to saying that all decision episodes are equivalent in their ability to contribute to the development of good decision skills. This seems unlikely to be true. In the cardinal issue view, over repeated episodes, a class of decision that is initially made analytically is then made via a rule and eventually becomes automatic. That is, there is a mode progression of this form:

Analytic → Rule-Based → Automatic

(Incidentally, there is no claim that this is the only way that automatic decision making can emerge.) This expectation is based on the same kinds of studies that supported Fitts and Posner's (1967) well-known model of motor skill acquisition, which acknowledges this sequence of phases:

Cognitive → Associative → Autonomous

More recently, Anderson (1985) has argued that the same kind of sequence applies to the development of cognitive skills as well. What is easy to overlook is that the original empirical evidence for these sequential changes in the character of people's functioning says nothing about the proficiency of the behavior that is evolving. Thus, Fitts and Posner's model could just as well be used to describe the development of incompetence (or bad habits) as the development of expertise. Similarly, if a person again and again makes a decision in some particular dysfunctional way, we can expect that dysfunction to become automatized.

These observations have several implications. First, there is almost certainly such a thing as "experienced incompetence." Hence, it is risky to continue NDM research programs in which experience is, effectively, equated with expertise. Instead, whenever possible, expertise should be verified directly. A second implication is that NDM-inspired decision trainers need to be highly discriminative in what they include in their programs for accelerating decision-making experience. As far as possible, they should make certain that trainees are led to make repeated decisions using good strategies, not the bad ones that, ironically, arise naturally with some nontrivial frequency. For if they do not, they will simply accelerate the development of intractable incompetence. A third implication is that ingrained incompetence is likely to be an especially onerous practical problem for decision trainers. Suppose that (as seems guaranteed) a trainee already possesses some automatic but dysfunctional strategies for making the decisions the trainer seeks to have the trainer make well. Then, before the training effort can begin in earnest, those "bad habits" have to be eliminated. This will be a prodigious task because, by their nature, the already established automatized strategies will initiate themselves effortlessly, uncontrollably, and beyond the trainee's consciousness. Anyone who has tried to correct a firmly ingrained weak backhand used for 20 years of tennis matches has experienced the same thing. NDM specialists interested in decision training should anticipate it as well. The final implication is related. Because decision making is such a fundamental human activity, which each of us has practiced day-in and day-out all our lives, perhaps the majority of our decision strategies are automatized. This means that it is folly to think that it is even possible to fundamentally change a person's basic decision-making dispositions. Thus, consistent with a conclusion NDM trainers (and also decision support system developers) have reached on the basis of other considerations, one should not even try to do so. A more realistic goal is to help decision makers make judicious adjustments in the natural decision procedures already in place.

Cardinal Issues

Cardinal decision issues are questions that typically arise in some form or another in almost all decision problems, no matter how (e.g., via what mode) they are solved. The main reason the issues are important is that, when they are resolved inadequately, this translates directly into decision deficiencies, that is, poor decision quality:

Cardinal Issue Resolution → Decision Deficiencies → Decision Quality

The cardinal issue perspective makes no specific claims about how various issues are resolved. The list of issues is not a psychological theory per se. It is more accurately viewed as a theory about the nature of decision problems, their essential elements. What *essential* means is that, as I suggested, if any elements are defective, this bodes ill for the sufficiency of the resulting decision. This also implies that, at a particular high level of analysis, decision making *is* addressing (or failing to address, as the case may be) all the cardinal issues in a given instance. Thus, the descriptive value of the theory is that it specifies the classes of things that focused theories have to explain. Its prescriptive potential lies in the claim that decisions turn out badly because one or more cardinal issues are handled poorly, perhaps out of neglect or oversight rather than "active error." It therefore can be used as something of a checklist. For each issue, the decision maker can ask: "How well are we dealing with this question, or have we overlooked it altogether?" There are ten cardinal issues. There is insufficient space to consider them all. Instead, I will merely describe and comment on three that seem especially significant for NDM concerns.

The Options Issue. *"What* could *we do in this situation?"* This is a good, colloquial way of characterizing the options issue as it might come up in an actual decision episode. Put another way, the options issue is about how the decision maker comes to consider some options but not even think about others. Its significance is immediately apparent when we realize the obvious, that an option cannot be selected unless it is recognized. Everyone can recall instances in which they have said things like, "If I had only thought about doing that, I would have. And I'd be a whole lot better off than I am right now." Concern for such occurrences is what motivates the great effort decision makers in domains like marketing devote to the generation of lots of options, for example, via brainstorming (see, for example, Valacich, Dennis, & Connolly, 1994). But there are downsides to identifying many options. One is that the process of doing so is costly, in terms of time and other resources. Another is that appraising those alternatives is expensive. And then, ultimately, typically only one of those options will actually be pursued. Hence, from one vantage, all those resources will have been "wasted." Thus, in the ideal case, the decision maker should generate and consider only a single option—the best.

A major theme in NDM writings is that expert decision makers typically generate very few options, often just one. This is seen as problematic for traditional decision scholarship, which, in the view of NDM writers, argues for the inherent merits of generating large numbers of alternatives. The story is not that simple. It is unclear what should be meant by "traditional decision scholarship" (or "classical decision theory," as it is often termed). The expression is often used to refer to ideas within economic and statistical decision theory, such as expected utility theory and Bayesian analysis. If that is in fact the interpretation, then the attribution is unfair. Formal theories of that class are actually silent on option generation. They offer accounts and prescriptions for what is and should be done with the options that happen to be in hand. This is so even if there is just one alternative, as in the case of the accept/reject and evaluation decisions I mentioned before. Work on option generation of the brainstorming variety actually grew out of an applied area, marketing, more specifically, advertising. For the most part, applied cognitive and social psychologists, not decision theorists per se, have done the scholarly work on how well various techniques perform. And, occasionally, even those investigators have recognized the burdens of dealing with the vast numbers of useless options the techniques typically produce. Thus it seems that the original source of the claims NDM investigators have contested might be more narrow than suggested, perhaps not much beyond Janis and Mann (1977), who are the authors most often cited.

NDM writers' emphasis on experts generating single, good options carries with it a prescription: Ordinary decision makers should seek to emulate experts.

Do not waste time trying to generate a pool of options. Instead, try to come up with just one, one that satisfices if it does not maximize the decision maker's interests. By itself, advice like this is reminiscent of recommendations of the "Buy low, sell high" flavor. They are more like goals than advice. The devil is in the details. Here, there are two key questions of detail. The first is this: "When experts generate only a single, good option, how do they do it? That is, what operations do they execute?" The second question is just as important but apparently has not been articulated as such heretofore: "How, exactly, did the experts acquire their ability to perform those operations and their disposition to apply that ability as decision situations present themselves?" A corollary to this second question asks how that development process can be imitated and accelerated, for training purposes. Of course, there is another possibility, too. The pertinent abilities and dispositions could be inborn personal characteristics, in which case the practical issue would be how to identify easily people who have them.

NDM investigators have proposed that experts are often able to identify immediately a single sufficient alternative in two different ways. The first is via recognition (Klein, 1989). The basic idea is that the decision maker recognizes the given situation as essentially similar to ones seen and handled successfully before. Thus, the option that is pursued in the current situation is one that replicates those that worked for the earlier decision problems. In the terminology of the cardinal issue perspective, the decision maker adopts either a rule-based or an automatic decision mode. (Clearly, it would be impossible for novices in a given domain to apply such strategies because they have not had the opportunity to establish the pertinent rules.) NDM authors have tended to cite illustrations from domains such as fire fighting, where commanders formulate plans for handling current fires by drawing on their personal experiences with previous ones. In other domains, this kind of rule-based and automatic decision making is formally taught. Thus, in medical training, physicians are instructed that, when patients present with Sign/Symptom Configuration S, they should be given Treatment T. It seems plausible that the typical physician develops personalized rules continually over the course of his or her career as well. Individual developmental differences are a critical, neglected concern to which NDM investigators should direct their attention. There are undoubtedly substantial variations in how readily different experienced decision makers (be they fireground commanders or physicians) immediately generate adequate options for the decision problems they confront. How can we explain those differences? How did they come into being, from the earliest days in a field when, presumably, no novice could accomplish such feats? Are there constitutional differences (e.g., in personality characteristics and aptitudes) that contribute to the variations? What kinds of early decision experiences encourage the development of good rule-based decision making and what kinds do the opposite?

The second means by which, according to NDM researchers, experts can often go directly to a workable alternative in a decision situation is more interesting than the first. It applies particularly in circumstances that are relatively novel, where rule-based (and certainly automatic) decision modes cannot be invoked. NDM writers propose that the key is "situation assessment," which is, according to Klein (1989, p. 51), "the sense of understanding what is going on during an incident." Part of situation assessment entails pattern matching, determining whether the current decision situation matches the precondition of a C → A decision rule in the decision maker's repertoire. But there is more, which comes to the fore when the conclusion is that there is not a match. According to Gary Klein (personal communication), "a key process here is sizing up a situation, which triggers a set of plausible reactions ... time is better spent on trying to understand the situation than in generating large sets of options." Serfaty, MacMillan, Entin, and Entin (1997) proposed that the expert's experience leads to the construction of a rich mental model of the situation which, in turn, supports the development of an especially good initial prospective course of action. NDM research has yielded good evidence that decision makers often proceed in a serial manner, accepting the first alternative that appears to be good enough. If Serfaty et al.'s proposal is correct, this implies that experts are indeed quite likely to generate only a single, sufficient alternative for a given decision problem.

My interpretation of all this is that, when experts encounter decision problems in their domains of expertise that rule-based and automatic decision making cannot handle, they necessarily revert to analytic decision making. In principle, that decision making could entail choice, accept/reject, or construction decisions. NDM data suggest that, spontaneously, and at a rate greater than that for nonexperts, experts tend to pursue the construction decision route. That is, they do not search for and collect several alternatives from which they then pick one. Nor do they search for and serially deliberate the adequacy of candidates until one passes muster. Why? Novices to a domain by definition are incapable of intelligently generating plausible alternatives on their own—and they know that. Therefore, they have no choice but to adopt search strategies. (They would likely say, "Hey, I'm clueless here. Help me out with some suggestions, would you?") For this reason alone, by default, construction decision making would be more probable for experienced experts. But there are likely to be more positive reasons for this tendency, too. And my hunch is that these go beyond assessment of the given situation per se. Instead, they plausibly rest more heavily on experts' knowledge of the given domain as well as on cognitive economics (cf. Payne, 1982).

The argument is this: Suppose a person knows a domain well (or at least thinks that is so). Then that person has a detailed mental model of how that entire domain functions. For instance, physicians have highly articulated mental models of how various bodily systems work biologically. Now, suppose that a problematic situation arises, one that will eventually require a decision, for example, a patient presents with an illness. The decision maker can be expected to employ his or her model of how the domain works generally to assemble a model of how it is functioning in the given instance. Thus, physicians will construct an explanation for the biological basis of the patient's complaints, given their conception of how the pertinent systems function as well as the patient's history and test results. In at least some instances, the physician's broader mental model of the body and of medicine then makes it immediately obvious what treatment option makes sense: "That's simply the way things work," he or she would say. Now notice that, in this example, the patient's history and test results are what NDM normally refers to as the "situation assessment" and that it assumes a relatively small role in the enterprise. We also recognize how inefficient (indeed, unnecessary) it would be for the physician to search for reasonable treatment alternatives to consider. I would suggest that testing this domain knowledge/construction decision proposition is a worthy challenge for NDM researchers. And then, of course, there are the same kinds of questions repeatedly articulated before, for example, how some experts are more expert than others at using domain knowledge to construct good options.

The Realization Issue. *"What would happen if we did that?"* Suppose a decision maker recognizes that some significant occurrence is possible in the given situation. Then the next question is whether it actually would occur. Put differently, would that possibility be realized? The importance of the realization issue in virtually every decision problem is readily apparent. People obviously are inclined to pursue alternatives they expect to yield positive outcomes and eschew the ones they believe would produce negative ones (e.g., hire employees they think will do good things for their organizations and pass on everyone else). When such expectations are off base (e.g., when an employer consistently misjudges candidates' potential), this imposes a low ceiling on the quality of decisions predicated on those judgments. These facts are a major reason that studies of how and how well people make judgments continue to dominate the literature on decision behavior.

As I noted before, the NDM community was disappointed (and, I dare say, surprised) that the massive literature on judgment proved to have such little value to NDM concerns in the 1970s and 1980s. As I also noted, one plausible reason the attempt to apply insights from that literature has been so fruitless is that developers have failed to take into account the modes by which people arrive at their judgments. That is, they have assumed that those modes are analytic when they are more likely to be automatic. Work on what are sometimes called "low-cost," evolutionarily plausible judgmental heuristics (e.g., Gigerenzer & Goldstein, 1996) are likely to serve NDM needs better. Pursuing these possibilities would seem to make sense for NDM investigators. NDM research has offered its own special perspective on the realization issue. If NDM realization conclusions are correct, then other judgment findings suggest that there are additional fundamental and practical questions that need attention as well.

As best described by Klein and Crandall (1995), evidence for mental simulation is the most prominent contribution NDM research has offered that is pertinent to the realization issue. The idea is that, when a decision maker is deliberating the wisdom of pursuing a prospective course of action (e.g., attempting some particular maneuver for rescuing a person in danger), the decision maker mentally simulates carrying out that action. If the outcomes in the simulation are sufficiently appealing, the action is taken; otherwise, it is not. Mental simulation is a judgment process that is qualitatively different from at least the formal metaphors that are used to frame judgment questions in the traditional literature. In one version of the latter, the person is represented as examining discrete items of evidence, pondering their validity, and synthesizing their net implications for the events in question. Generally, this characterization is taken as nothing more than a metaphor for what people do rather than a description of what they do literally, what is commonly called a *paramorphic representation*. But, there is little doubt that there are occasions when people almost necessarily must make judgments according to such broad, "formalistic" schemes, for example, when presented with job candidates' application forms that simply list their credentials. Nevertheless, there has long been speculation that, whenever possible, people are more likely to rely on action-sequence procedures such as mental simulation (cf. Abelson, 1976; Yates, 1990, chap. 7).

Now, suppose that mental simulation is indeed the predominant means by which people address the realization issue. Then we could speculate that this would encourage judgments that are overly extreme. That is because the mental models that people "run" in their mental simulations seem likely to be deterministic ones that neglect statistical considerations or, more generally, uncertainty. Evidence for Pennington and Hastie's (1988) story model version of mental simulation illustrates the idea. The model (as well as the data) suggests that criminal trial jurors try to construct a story that accounts for all the available evidence. If such a story entails the defendant's innocence, then jurors are inclined to judge that the defendant is indeed innocent and support a "not guilty" verdict. But if the defendant is guilty in that story, they do the opposite. A salient feature of the kinds of stories people construct is that the "actors" in the stories literally and definitively do things, for example, stab other folks; they do not "probably" do them. Properly introducing uncertainty into such mental simulation requires a higher level of analysis that people seem unlikely to undertake spontaneously, as suggested by process studies on the planning fallacy, the tendency to be overly optimistic about project completion (e.g., Buehler, Griffin, & MacDonald, 1997). Evaluating this prediction of extreme judgment arising from mental simulation should be a priority in future NDM studies.

The Conflict Issue. *"Not all considerations favor the same action. So what should we do?"* At some point, in just about every practical decision problem imaginable, we get to the point where the conflict issue stares us in the face. Thus, when considering a new job, we realize that the work is terrific but the location is awful. And, when contemplating a new investment opportunity, we recognize that, although the potential returns are substantial, so is the risk. The conflict issue is the preoccupation of mainstream formal decision theory. For instance, expected utility maximization is essentially a rationalized scheme for resolving the conflict that might exist between our chances of achieving particular returns from a risky prospect and the value attached to those returns (e.g., the odds of losing a military engagement and the seriousness of the loss). Applied multiattribute utility theory is intended to help decision makers resolve the conflict that typically exists when the alternatives under consideration have multiple aspects, for example, when one prospective new business location is cheaper while another has better highway access.

The conflict resolution ideas of traditional decision scholarship have been singularly unpopular in the NDM community. Most objections have emphasized that the methods are hard to use, take too much time, and are seldom employed in practice, even when decision makers know about them. Implicit is a belief that they are not worth all the trouble they demand. What does this all mean? Have traditional researchers been wasting their time all these years? Or is the NDM community failing to recognize something that is in its own interests? There are many plausible answers to such questions. But one that is consistent with my original proposal about why NDM and traditional decision scholarship often do not see eye to eye is this. The conflict issue has unquestionable significance for the kinds of financial and one-off operational decision problems for which they were designed, for example, portfolio analysis and designing queuing systems for efficient service delivery. Yet, in the kinds of decision situations that have captured the attention of NDM researchers, other issues, such as the options and realization issues, objectively are probably far more important. In the larger picture painted by the cardinal issue perspective, we realize that each of these issues is but one of many that demand attention in typical real-life decisions.

A FINAL WORD

Perhaps the greatest contribution of NDM has been the attention it has brought to the inadequacies of prior paradigms of decision scholarship, particularly from the perspectives of decision research "consumers." NDM writers often have asserted that the key inadequacies were claims that were believed to be true but were actually false. In contrast, as suggested by my earlier comments, I believe that the inadequacies were more a matter of previous paradigms being incomplete rather than wrong. That is, they have been silent on significant elements of the decision-making enterprise as a whole as represented in, say, the cardinal issue perspective. It is now time for the NDM community as well as other decision communities to move beyond the "wake-up call" stage in a more positive direction. The new work that is required would seek to close the gaps in our understanding that are now more apparent. Throughout this commentary I have suggested explicitly and implicitly numerous priorities that might guide those efforts. In closing, I offer one final observation and its accompanying recommendation.

The focus of traditional scholarship has been on formal structures and procedures. In the practical arena, this is most evident in the nearly exclusive emphasis of decision analysis on providing clients with alternative operations for synthesizing new decision-relevant entities from prior ones. One example is recommendation of variants of Bayes's theorem for deriving posterior probabilities, another the suggestion of combining probabilities and utilities via expectation operators to yield choice prescriptions. The implicit assumption has been that, when clients' decisions go awry, shortcomings in the clients' normal (natural?) synthesis procedures are significant if not exclusive causes for these occurrences. NDM seems to have tapped into practical decision makers' beliefs that shortcomings in substantive knowledge are important, too, and probably more so. This belief is perhaps most strikingly apparent in the character of the decision aids (typically not known by that name) that predominate in increasingly broad swatches of contemporary life, from business (e.g., electronic commerce), to medicine (e.g., transplant allocation systems), to personal relationships (e.g., Internet introduction services). These tools are knowledge-focused rather than procedure-focused. The driving belief seems well captured thusly: "When people decide badly, it isn't because they manipulate their facts improperly. Instead, the bigger problem is that they know too few facts that really matter and too many about things that don't." A key challenge for NDM is to determine whether this belief is justified, and if it is, to pursue it to its logical conclusions.

ACKNOWLEDGMENTS

Work on this project was supported in part by an award from the Michigan Business School. I am grateful for the comments and suggestions of other members of the Judgment and Decision Laboratory, especially Winston Sieck, Jason Riis, and Richard Gonzalez. Discussions with Gary Klein and Eduardo Salas were especially helpful in sharpening the focus of the analysis.

REFERENCES

Abelson, R. P. (1976). Script processing in attitude formation and decision making. In J. S. Carroll & J. W. Payne (Eds.), *Cognition and social behavior* (pp. 33–52). Hillsdale, NJ: Lawrence Erlbaum Associates.

Anderson, J. R. (1985). *Cognitive psychology and its implications* (2nd ed.). New York: Freeman.

Buehler, R., Griffin, D., & MacDonald, H. (1997). The role of motivated reasoning in optimistic time predictions. *Personality and Social Psychology Bulletin, 23*, 238–247.

Cannon-Bowers, J. A., & Bell, H. H. (1997). Training decision makers for complex environments: Implications of the naturalistic decision making perspective. In C. E. Zsambok & G. Klein (Eds.), *Naturalistic decision making* (pp. 99–110). Mahwah, NJ: Lawrence Erlbaum Associates.

Edwards, W., Kiss, I., Majone, G., & Toda, M. (1984). What constitutes "a good decision?" *Acta Psychologica, 56*, 5–27.

Fischhoff, B. (1982). Debiasing. In D. Kahneman, P. Slovic, & A. Tversky (Eds.), *Judgment under uncertainty: Heuristics and biases* (pp. 422–444). New York: Cambridge University Press.

Fitts, P. M., & Posner, M. I. (1967). *Human performance*. Belmont, CA: Brooks/Cole.

Gigerenzer, G., & Goldstein, D. G. (1996). Reasoning the fast and frugal way: Models of bounded rationality. *Psychological Review, 103*, 650–669.

Janis, I., L., & Mann, L. (1977). *Decision making*. New York: Free Press.

Josephs, R. A., Larrick, R. P., Steele, C. M., & Nisbett, R. E. (1992). Protecting the self from the negative consequences of risky decisions. *Journal of Personality and Social Psychology, 62*, 26–37.

Klein, G. A. (1989). Recognition-primed decisions. In W. B. Rouse (Ed.), *Advances in man-machine system research* (Vol. 5, pp. 47–92). Greenwich, CT: JAI Press.

Klein, G., & Crandall, B. W. (1995). The role of mental simulation in problem solving and decision making. In P. Hancock, J. Flach, J. Caird, & K. Vicente, (Eds.), *Local applications of the ecological approach to human-machine systems* (pp. 324–358). Hillsdale, NJ: Lawrence Erlbaum Associates.

Klein, G. A., Orasanu, J., Calderwood, R., & Zsambok, C. E. (Eds.). (1993). *Decision making in action: Models and methods*. Norwood, NJ: Ablex.

Loomes, G., & Sugden, R. (1982). Regret theory: An alternative theory of rational choice under uncertainty. *The Economic Journal, 92*, 805–824.

Means, B., Salas, E., Crandall, B., & Jacobs, T. O. (1993). Training decision makers for the real world. In G. A. Klein, J. Orasanu, R. Calderwood, & C. E. Zsambok (Eds.), *Decision making in action: Models and methods* (pp. 306–326). Norwood, NJ: Ablex.

Payne, J. W. (1982). Contingent decision behavior. *Psychological Bulletin*, 92, 382–402.

Pennington, N., & Hastie, R. (1988). Explanation-based decision making: Effects of memory structure on judgment. *Journal of Experimental Psychology: Learning, Memory, & Cognition, 14*, 521–533.

Peters, R. M. (1994). Matching physician practice style to patient informational issues and decision-making preferences: An approach to patient autonomy and medical paternalism issues in clinical practice. *Archives of Family Medicine, 3*, 760–764.

Serfaty, D., MacMillan, J., Entin, E. E., & Entin, E. B. (1997). The decision-making expertise of battle commanders. In C. E. Zsambok & G. Klein (Eds.), *Naturalistic decision making* (pp. 233–246). Mahwah, NJ: Lawrence Erlbaum Associates.

Valacich, J. S., Dennis, A. R., & Connolly, T. (1994). Idea generation in computer-based groups: A new ending to an old story. *Organizational Behavior and Human Decision Processes, 57*, 448–467.

Yates, J. F. (1990). *Judgment and decision making*. Englewood Cliffs, NJ: Prentice-Hall.

Yates, J. F. (1994). Subjective probability accuracy analysis. In G. Wright & P. Ayton (Eds.), *Subjective probability* (pp. 381–410). New York: Wiley.

Yates, J. F., & Estin, P. A. (1998). Decision making. In W. Bechtel & G. Graham (Eds.), *A companion to cognitive science* (pp. 186–196). Malden, MA: Blackwell.

Yates, J. F., & Patalano, A. L. (1999). Decision making and aging. In D. C. Park, R. W. Morrell, & K. Shifren (Eds.), *Processing of medical information in aging patients: Cognitive and human factors perspectives* (pp. 31–54). Mahwah, NJ: Lawrence Erlbaum Associates.

Zsambok, C. E., & Klein, G. (Eds.). (1997). *Naturalistic decision making*. Mahwah, NJ: Lawrence Erlbaum Associates.

II

Tools: Training Methods and Systems Design

3

Decision Skills Training: Facilitating Learning From Experience

Rebecca M. Pliske
Dominican University
Michael J. McCloskey
Gary Klein
Klein Associates Inc.

The development of good judgment and decision-making skills is essential for individuals to function well in society. Although various approaches have been developed to train these skills, it has been difficult to empirically demonstrate the effectiveness of these approaches. In this chapter, we describe a decision skills training approach that attempts to facilitate the development of the decision maker's domain expertise. We briefly review the types of approaches that have been used previously to train decision-making skills before we present the rationale for the development of our approach. We also describe the training tools we have developed and some of the preliminary data we have obtained to evaluate the effectiveness of this approach. In addition, we discuss future directions for research on how to improve judgment and decision-making skills.

REVIEW OF PREVIOUS APPROACHES

Traditional approaches to training decision making have attempted to teach a set of generic decision-making strategies that could be used successfully in a wide variety of contexts. The theoretical model underlying the traditional approach assumes the decision maker is a rational/economic human (Simon, 1956) or vigilant decision maker (Janis & Mann, 1977) who systematically searches for relevant information in an unbiased manner and then carefully weighs the utility of each alternative before making a choice. Early efforts developed by traditional decision-making researchers focused on how to train people to use decision analytic techniques to select the best alternative course of action. As more and more research demonstrated that human decision makers often fail to follow the prescriptions of normative models (Kahneman, Slovic, & Tversky, 1982), training efforts became focused on how to "debias" decision makers. Examples of each of these approaches are described in turn.

Baron and Brown (1991) provided an excellent review of some of the traditional decision training programs that have been implemented with students. Although their book focuses on how to improve adolescents' decision-making ability, their approach, which they call Personalized Decision Analysis (PDA), has been used by adults in a variety of settings. The PDA approach has also been called "applied statistical decision theory," and the "decision analytic approach"; it is based on Expected Utility Theory (von Neumann & Morgenstern, 1947). The PDA approach assumes the decision maker is faced with a choice between options (i.e., courses of action). The decision maker must estimate the uncertainty of the possible outcomes associated with each option. Decision makers are also required to judge the desirability of each outcome in terms of the expected utility to be gained or lost should this outcome occur. The mathematical formulas of statistical decision theory (Savage, 1954) are then used to determine the option with the highest value based on the probability-weighted utility of its outcomes (Raiffa, 1968).

One of the most extensive decision training programs based on the decision analytic perspective is the GOFER course developed by Mann and his colleagues (Mann, Beswick, Alloache, & Ivey, 1989; Mann, Harmoni, & Power, 1991; Mann, Harmoni, Power, Beswick, & Ormond, 1988). The GOFER course is based on Janis and Mann's conflict theory of decision making (Janis & Mann, 1977), which claims that the ideal approach to decision making is that of the "vigilant" decision maker. Mann et al. (1991) explained that the acronym GOFER is formed from the criteria for vigilant decision making as follows:

G—goals (surveying values and objectives)
O—options (considering a wide range of alternative actions)
F—facts (searching for information)
E—effects (weighing the positive and negative consequences of the options)
R— review (planning how to implement the options)

Mann and his colleagues conducted several evaluation studies to test the effectiveness of the GOFER approach (Mann, Beswick, et al., 1989; Mann, Harmoni, Power, Beswick, et al., 1988). These studies used students' self-reported descriptions of their decision strategies and confidence in their decision-making abilities to measure training effectiveness. Mann and his colleagues noted that these measures are "soft" measures of the effectiveness of the GOFER course and that additional evaluation studies are needed to determine whether the decision skills taught in the GOFER course will transfer to real-life settings.

In contrast to the efforts of Mann and his colleagues, other researchers working from the traditional decision-making perspective focused their training efforts on attempts to "debias" the decision maker. These researchers were influenced by the work of Tversky and Kahneman (Kahneman & Tversky, 1972; 1973; Tversky & Kahneman, 1971; 1974) who identified cognitive heuristics (e.g., representativeness, availability), which decision makers use under conditions of uncertainty that can result in systematic biases in choice behavior (e.g., neglect of base rate, overconfidence, etc.). Recent efforts at improving decision-making skill have focused on reducing the overconfidence bias by requiring decision makers to think about the reasons underlying their judgments. For example, Yates and Estin (1996) described their efforts to improve the judgment skills of MBA students. They used a scenario in which the participant role-played the part of an attorney involved in the jury selection process. Among other things, the training encouraged the participants to describe explicitly their judgment strategies and to question the assumptions underlying their choices. Participants who received the training demonstrated less overconfidence bias than untrained controls.

Shanteau and his colleagues (Shanteau, Grier, Johnson, & Berner, 1991) developed a course for nursing students that included training on a variety of decision skills. They identified three deficiencies in the nursing students' decision-making skills: (a) The nurses were not making effective or efficient use of available information, (b) they were making errors in estimating risk and uncertainty, and (c) they were having difficulty selecting among alternative courses of action. Shanteau et al. taught decision skills concepts in a classroom setting. For example, they told students they should evaluate the relevancy and diagnosticity of each piece of information available for solving a particular decision problem and cautioned students against "information overload." In addition to the classroom instruction, decision training also took place in the hospital under the guidance of nursing instructors. Shanteau et al. evaluated the effectiveness of this training with a well-designed study that included 115 nursing students. Assessment materials included several different scenarios that were designed to test for improvement in the three deficiencies described earlier. Performance of students who received the training was compared to control groups who did not receive the training. In addition, Shanteau et al. had seven expert nurses complete the test scenarios. In general, the evaluation data supported the effectiveness of their decision training; students who received the training performed more similarly to the expert nurses as compared to students in the control groups. However, not all aspects of the training were equally successful; students did not show improved performance on the measures of accuracy of probability assessment.

Shanteau et al. stressed the importance of including both classroom and practical applications of the decision skills being trained.

Researchers associated with the Naturalistic Decision Making (NDM) perspective have also developed approaches to training decision-making skills. For example, Fallesen and his colleagues at the U.S. Army Research Institute for the Behavioral and Social Sciences (Fallesen, 1995; Fallesen, Michel, Lussier, & Pounds, 1996) have developed a comprehensive cognitive skill training program. They call their training "Practical Thinking" to contrast it with theoretical or formal methods. Fallesen and his colleagues were tasked to develop a Practical Thinking course to be implemented as part of the U.S. Army's Command and General Staff Officers Course. The goal of this course was to increase officers' skills for reasoning and deciding in battle command. Fallesen and his colleagues reviewed existing cognitive instruction programs to identify training approaches applicable to their goal. Based on this review, they developed course materials that included lessons on multiple perspectives (i.e., thinking outside the box), metacognitive skills that allow the individual to guide his or her thinking deliberately, techniques for identifying hidden assumptions, practical reasoning techniques (e.g., demonstrations of reasoning fallacies), and integrative thinking techniques to increase students' understanding of the relationships among events and concepts.

Cohen and his colleagues (Cohen, Freeman, & Wolf, 1996; Cohen, Freeman, & Thompson, 1998) have also developed an approach to training decision-making skills based on the NDM perspective, which they call Critical Thinking Training. Cohen et al. (1998) described the results of two studies they conducted with a total of 95 military officers (a mix of Navy, Marine, Army, and Air Force) to evaluate the effectiveness of this training. In both studies the officers were asked to play the role of the Tactical Action Officer in a Combat Information Center in a battleship. The training included several sessions on a low-fidelity, computer-based simulator of the Combat Information Center. During these sessions the participants learned how to use tools such as a "hostile intent story template" to facilitate the development of alternative explanations for potential threats (e.g., an enemy airplane or ship) that appeared on their computer screens. Cohen et al. assessed the effectiveness of their training by having their participants complete questionnaires in which they were required to assess the intent of a particular threat (e.g., an enemy airplane) and justify their assessments. Participants were also required to generate alternative possible assessments, identify conflicting evidence, and describe actions they would take at a specified point in the scenario. In general, the results of the Cohen et al. study support the effectiveness of the critical thinking skill training they have developed.

Previous efforts to improve individuals' decision-making performance (from both the traditional and NDM perspectives) have attempted to teach decision makers to use specific decision-making strategies (e.g., a hostile intent story template, the steps in the GOFER process). In contrast, we have developed an approach to decision-making training, which we call Decision Skills Training (DST), that does not attempt to teach decision-making strategies per se. Instead, we attempt to facilitate the development of the decision maker's experience base within a particular domain, which should, in turn, result in improved recognitional decision-making skills. We briefly describe the rationale for this type of training next.

RATIONALE FOR THE DST APPROACH

Traditional training approaches for improving decision-making skills have attempted to teach generic strategies, such as decision analysis, in which the decision maker analyzes the costs and benefits of alternative courses of action. One reason this type of decision training has not proven effective in improving performance in field settings is that many decision-making situations do not involve choices between alternative courses of action. Previous research has shown that skilled decision makers spend more time sizing up the situation than comparing alternative courses of action (Klein, 1998). These decision makers are able to quickly generate an acceptable course of action based on their previous experiences. For example, Klein (1989) presented a Recognition-Primed Decision model that described how people can use experience to make rapid decisions under conditions of time pressure and uncertainty and that preclude the use of analytical strategies.

The Recognition-Primed Decision model has been evaluated in a wide array of activities, including urban and rural firefighting, Army tank platoon operations, command and control in Navy AEGIS cruisers, hardware and software design, flight control in commercial airliners, chess tournament play, and nursing in Neonatal Intensive Care Units (Crandall & Getchell-Reiter, 1993; Flin, Slaven, & Stewart, 1996; Kaempf, Klein, Thordsen, & Wolf, 1996; Klein, Calderwood, & Clinton-Cirocco, 1986; Klinger, Andriole, Militello, Adelman, Klein, & Gomes, 1993; Randel, Pugh, & Reed, 1996). In all these domains, good decision-making skill depends on the domain expertise of the decision maker. Therefore, in order to improve decision-making performance we have developed an instructional approach for facilitating the development of the decision maker's domain expertise. The rationale for the development of this approach is described by Klein (1997) and is summarized below.

The DST program was developed based on a review of the literature on expertise (Chi, Glaser, & Farr, 1988; Ericsson, 1996; Ericsson & Charness, 1994; Glaser, 1996; Klein & Hoffman, 1993) to identify strategies that experts use in order to learn more rapidly and effectively. These strategies include: (a) engaging in deliberate practice, so that each opportunity for practice has a goal and evaluation criteria; (b) obtaining feedback that is accurate and diagnostic; (c) building mental models; (d) developing metacognitive skills; and (e) becoming more mindful of opportunities for learning. Based on this review, we developed a set of learning tools (described in the next section) that could be used in conjunction with low-fidelity simulation exercises, termed Decision Making Games, as well as with training exercises conducted in high fidelity simulators or in the field. The learning tools include the Decision Making Critique, the Decision Requirements Exercise, the PreMortem Exercise, the Commander's Intent Exercise, the Situation Awareness Calibration Exercise, and the Development of Decision Making Games. The mapping between these tools and the strategies experts use to learn more rapidly are shown in Table 3.1. The tools, which are briefly described next, are still under development. We are continually refining these tools, as well as developing new tools, based on feedback we receive from the participants in our training sessions.

DST TOOLS

Decision Making Games

Decision Making Games (DMGs) are low-fidelity, paper-and-pencil simulations of incidents that might occur in the field. They are intended to provide simulated, domain-relevant experiences and to allow participants to practice their recognitional decision-making skills. The DMGs provide the context for teaching and practicing the other tools described. Participants are presented with a dilemma in which a decision must be made; typically, some sort of action must be taken. This situation involves some degree of uncertainty, and the participants are given only a few minutes to determine their course of action. A participant is then chosen to brief his/her plan to other participants who role-play the positions of the subordinates for this exercise. For our work with the U.S. Marine Corps, the DMGs typically included a map and a brief textual description of the situation. These games were adapted from the Tactical Decision Making Games developed by Schmitt (1994). An example of a DMG, called Razorback Ridge, is shown in Fig. 3.1.

TABLE 3.1
Mapping of Experts' Learning Strategies to DST Learning Tools

Expert Learning Strategy	*DST Learning Tool*
Engaging in deliberate practice	Decision Making Games Commander's Intent Exercise
Obtaining accurate and diagnostic feedback	Decision Making Critique
Developing metacognitive skills	Decision Requirements Exercise Premortem Exercise
Building mental models	Situation Awareness Calibration Exercise
Becoming more mindful of practice opportunities	Development of Decision Making Games Decision Requirements Exercise

Situation

You are a squad leader in Lima Company. You are operating in a terrain characterized by narrow ridges separated by wide, flat valleys. The vegetation is low, sparse, scrub brush. Your platoon has been inserted after dark by helicopter and is moving to its squad release point where the squads will split off to conduct limited search and attack missions with supporting arms. An earlier mission showed the immediate area clear of enemy forces. The platoon is heading east, sideways along the slope of Razorback Ridge, a couple hundred meters from the crest. Your squad is bringing up the rear of the platoon column.

Suddenly, automatic weapons fire—you recognize it as 7.62 mm—erupts from the left, somewhere along the crest of the ridge. From your position, it's hard to see exactly where the fire is coming from. The other two squads, caught in the killing zone, try to find cover in the microterrain. You and your first fire team are just at the edge of the beaten zone, and you're able to dive for cover behind the crest of a low finger that runs diagonally down the slope. From the lack of return fire and the shouts of "Corpsman up!" you can tell that the rest of the platoon has been hit pretty hard. You hear calls for help, and somebody yells: "The L.T.'s been hit!" What do you do?

Requirement

In a time limit of three minutes, describe your actions in the form of any orders you will issue or reports/requests you will make. Draw a sketch of your plan.

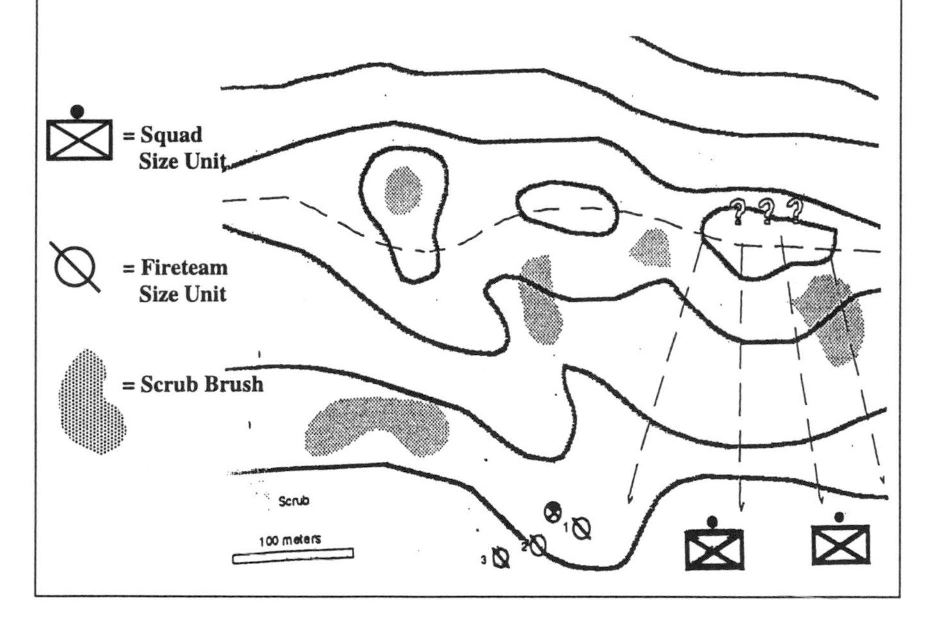

FIG. 3.1. Razorback Ridge: An example of a Decision Making Game (DMG).

Decision Making Critique

The Decision Making Critique facilitates thinking about what went well and not so well during an exercise. It can be used within the context of a DMG or an actual field training exercise. It consists of a series of questions designed to identify the difficult decisions made during the exercise. These questions explore important cues that might have been seen earlier, assessments that were mistaken, and the types of uncertainties encountered and how they were handled. The specific form of the questions depends on the particular group of individuals being trained. An example of the Decision Making Critique questions we used with U.S. Marine Corps squad leaders is shown in Table 3.2.

Decision Requirements Exercise

The Decision Requirements Exercise is intended to help the participants "unpack" the challenging decisions they faced during a DMG or field exercise in order to maximize the amount of learning that occurs. After using the Decision Making Critique to identify difficult decisions, participants are asked additional questions to determine what types of information they considered and why these particular decisions were so challenging. After performing the Decision Requirements Exercise, participants have a better sense of the judgments and decisions facing them, why they are difficult, and where people can go wrong in this type of situation. An example of a worksheet used for the Decision Requirements Exercise is shown in Fig. 3.2.

PreMortem Exercise

The PreMortem Exercise is used to identify key vulnerabilities in a plan. After someone has developed a plan, the group is told that by looking into a crystal ball it is determined that the plan failed. Individual group members then spend a few

TABLE 3.2
Decision Making Critique Questions

What were the tough decisions?
For each decision:
- Why was it difficult?
- Why did you choose that course of action?
- What one piece of missing information would have helped you the most?
- What other actions did you consider? Why didn't you choose them?

General discussion at the end:
- What would you do differently if you were in this situation again?
- What was our biggest weakness? What was our biggest strength?
- What are some important lessons learned from this exercise?

Key Decisions

List a Decision: Where to position my rifle teams after enemy fire erupted

Types of Information Used	Difficulties
Terrain *Vegetation available for cover* *Direction of enemy fire* *Report of wounded*	*Didn't know size of enemy unit*

FIG. 3.2. Example of a Decision Requirements worksheet from a training session with Marine squad leaders.

minutes independently writing down reasons why the plan failed. The facilitator then leads a discussion in which he or she elicits reasons from each group member until all the concerns have been identified. The intent of this process is to help planners and plan executors decenter from their current vision of the plan and view the plan from another perspective. In assuming the plan has failed and then attempting to identify the reasons for the failure, participants can uncover critical flaws that may have otherwise gone unnoticed. The PreMortem exercise also establishes a nonthreatening context in which these flaws can be uncovered. Discussion of these concerns typically results in an improved plan.

Commander's Intent Exercise

The Commander's Intent Exercise provides an opportunity for the participants to practice their skills for communicating Commander's Intent (which is a leader's rationale underlying a particular order or plan of action). This exercise is administered in conjunction with a DMG. Participants describe their solutions to the DMG in the form of a set of orders to their subordinates and also provide a description of their intent. The facilitator then identifies a plausible, but unexpected event that will interfere with that plan and assigns participants to role-play the leader and his or her subordinates. The participant role-playing the leader then writes down how he or she expects all of the subordinates to react. At the same time, the participants role-playing the subordinates write down how they would actually react based on the intent provided by the leader. Next, the two interpretations are compared. Typically, everyone is surprised by the different interpretations of intent. This exercise improves Commander's Intent statements, not by providing a checklist of what to say, but rather by providing direct feedback to enable the participants to find out how people are interpreting their orders and their intents.

Situation Awareness Calibration Exercise

Previous research demonstrated that it is important to gather and integrate multiple points of view when critiquing performance (Rouse, Cannon-Bowers, & Salas, 1991). The Situation Awareness (SA) Calibration Exercise is used to provide insight into how different team members perceive the same environment. This exercise is ideally administered in conjunction with an actual training exercise, but can be used with a low-fidelity simulation like a DMG. The training exercise is stopped at some point (preferably at a point in time when some of the team members appear confused or highly stressed), and each team member is required to independently answer a brief questionnaire that attempts to capture his or her current SA about the exercise. Examples of the types of questions that are typically included in this questionnaire are shown in Table 3.3. The questionnaires are collected and the training exercise resumes. The exercise facilitator tabulates the team members' answers to the questions and uses this information during the debrief after the training exercise is completed. The discussion during the debrief can focus on issues such as the roles and functions of the individual team members, prediction of future events, and indicators that a plan is either succeeding or failing.

Development of Decision Making Games

The use of DMGs is a critical part of our DST program because these low-fidelity simulations provide opportunities for trainees to practice their recognitional decision skills. We have also found that having participants develop their own DMGs is a useful learning tool when we conduct train-the-trainer programs for people who will be teaching others how to use our DST learning tools. To develop DMGs, we typically start by asking the trainers to identify difficult decision situations that their trainees may face in the future. We use these situations and the decisions within to help the trainers develop realistic scenarios, which typically include a map or other types of visual representations of the decision situation. The process of developing their own DMGs provides trainers with an opportunity to develop new insights into the complexity of the decisions involved in these situations, to appreciate the expertise they have in making these decisions because of their past experiences, and to recognize the opportunity DMGs offer for sharing their expertise with their trainees.

TABLE 3.3
Situation Awareness Calibration Questions

What is the immediate goal of your team?
What are you doing to support that goal?
What is our biggest worry?
What is the current threat's location, size, and intention?
What do I think this situation will look like in ____ minutes? Why?

PRELIMINARY EVALUATION OF APPROACH

We have conducted workshops using the training tools described earlier with firefighters, U.S. Marine Corps squad leaders, U.S. Marine Corps officers, U.S. Navy pilots, commercial pilots, and graduate students in a business management program. Most of these workshops were limited to one-day or two-day workshops; however, we used this approach to train U.S. Marine Corps squad leaders periodically over the course of several weeks, and in two cases, the training was conducted over a span of several months. We have conducted training with over 240 participants in these five different domains. These training opportunities have given us a chance to obtain some preliminary evaluation data regarding the effectiveness of this DST approach, and we describe these data briefly next.

Our DST approach was originally developed for U.S. Marine Corps squad leaders to prepare them to handle the variety of decisions they are faced with during field exercises. Our DST program was administered to the squad leaders over a $3^1/_2$ month period, from November 1996 to late February 1997. Only a few hours per week could be allocated to the DST program; the rest of the time the squad leaders were involved in other types of training activities.

The motivation for this training effort with U.S. Marine Corps squad leaders was the establishment of a new Special Purpose Marine Air Ground Task Force, which was designed to rely on a command post that might remain safely at sea during a military operation, thereby reducing the continual need to protect and move it. Small, squad-sized units, serving as the eyes of the command element, would covertly occupy the battlefield, so that fires could be directed on enemy targets from afar. However, a result of this new organization was that the squad leaders were going to have to function very independently.

Squad leaders are noncommissioned officers, who are high school graduates with limited (or no) college experience. They typically have 4 to 8 years of experience in the U.S. Marine Corps. Traditionally, the U.S. Marine Corps provides a great deal of command training to officers, but they have provided little if any such training to enlisted personnel. Accordingly, the request was to quickly improve the ability of the squad leaders to make judgments and decisions during an upcoming large-scale exercise (held in March, 1997). The three fielded companies each had 10 squad leaders. The squad leaders for each company were given the DST as a company, so three separate groups were trained in parallel. Platoon leaders attached to each company reinforced the tools we had introduced in the classroom when the squad leaders were in the field.

The subjective evaluations of the DST elicited from the squad leaders were very high. For the question "How useful is this overall project for your particular needs?" the mean response from the squad leaders was 2.68 on a 3-point scale (1, *not at all useful*, 2, *somewhat useful*, and 3, *very useful*). For the question "Are you likely to practice with the methods?" the mean response was 2.55 (1, *very unlikely*, 2, *may use them*, 3, *very likely*). The supervisors of the squad leaders had very positive impressions of that training as well. Their subjective reports indicated that the squad leaders' performance in the field had improved as a result of the training. The overall perception was that the training program had been very successful in preparing the squad leaders to make the types of decisions they are faced with during field exercises.

Based on the positive reactions to the initial DST program for the squad leaders, the U.S. Marine Corps asked us to work with them to develop additional training programs for squad leaders and also for junior officers. We also worked with the U.S. Marine Corps to develop course materials that will integrate aspects of DST into their on-going training of squad leaders.

We recently conducted train-the-trainer workshops with squad leader instructors. These workshops also generated positive feedback from the participants. The mean rating of usefulness of the training by the 20 instructors was a 4.59 on a 5-point scale (1, *not at all useful*, 3, *somewhat useful*, and 5, *very useful*). The instructors indicated they were very likely to use the DST tools with their squad leaders. Each of the tools received a rating between 2.43 and 2.98 on a 3-point scale with a mean rating of 2.69 (1, *low expectation of use*, 2, *medium expectation of use*, 3, *high expectation of use*). The instructors described a variety of ways in which they could use the materials with their squad leaders including holding additional classes focused on the material, supplementing topics already covered in training, using the material every chance they get including "down time," and teaching by example.

We have had similar positive reactions in the other domains in which we have conducted DST workshops. For example, the results from workshops we conducted with 21 Navy pilot instructors were as follows. The mean rating for usefulness of the training was 4.20 on a 5-point scale (1, *not at all useful*, 3, *somewhat useful*, and 5, *very useful*). The participants were asked to rate each DST tool on how likely they were to use it with their students. The majority of the ratings were between medium and high expectation on a 3-point scale (1, *low expectation*, 2, *medium expectation*, 3, *high expectation*). The Decision Making Critique was rated the highest with a mean of 2.65. The participants reported that they particularly valued the opportunity to practice making decisions and hearing how others reacted to DMGs. The instructors offered several examples of how they envisioned incorporating the tools in the fleet, such as using war-room training time to present DMGs and the Decision Making Critique.

Thus far, the only type of evaluation data we have collected are trainee reactions (Kirkpatrick, 1987), such as the data we have summarized. We have yet to collect data that demonstrate objectively measured changes in trainee behavior as a result of participation in our DST workshops. We are, however, strongly encouraged by the consistently positive reactions of our workshop participants.

FUTURE DIRECTIONS FOR RESEARCH

There are three primary areas in which future research needs to be conducted on our DST approach. First, we need to conduct a formal evaluation of the effectiveness of the DST approach. Second, we need to explore the relationship between the DST tools we have developed and training programs focused on improving trainers' "coaching skills." Third, we need to conduct additional research to determine the boundary conditions for implementing our DST approach—when is this type of decision-making training approach likely to be effective? Each of these topics is briefly addressed.

Based on the results of our preliminary evaluation efforts, it appears that our approach to training judgment and decision skills by facilitating the development of the decision maker's domain expertise is a promising approach. Clearly, additional evaluation data need to be collected in a more controlled environment to validate empirically the effectiveness of this approach. A field experiment, which includes random assignment of participants to experimental and control groups and objective measures of potential improvement in decision-making performance on the job, needs to be conducted. Unfortunately, these types of controls and measures were not available in the domains in which we have used this approach.

Our DST workshops typically include some exercises that facilitate the development of coaching skills. Coaching is an important adjunct to DST because our learning tools need to be implemented by a skilled facilitator who also has expertise in the domain of interest. Many of the participants in our workshops are team leaders in their organizations and want to improve not only their own decision-making skills, but also the decision skills of their team members. In order for team leaders to be effective in this developmental role, they need to be able to facilitate good team debriefs (Kozlowski, Gully, McHugh, Salas & Cannon-Bowers, 1996; Tannenbaum, Smith-Jentsch, & Behson, 1998). Our experience to date suggests that most of the subject matter experts who are currently providing training in military and paramilitary (e.g., firefighting) domains need additional training in order to improve their coaching skills. Our workshops include training on skills such as instructional strategies that allow the coach to share his or her expertise with the trainee, assessment strategies that allow the coach to determine the skill level of the trainee, and strategies for maintaining a positive learning climate. Research is needed to determine how best to integrate training on coaching skills and training on decision-making skills.

Research is also needed to identify the boundary conditions for implementing the DST program. We do not believe that DST is a universal solution for every performance problem in all types of organizations. There are places where it will work very well, and places where it is likely to have little value. The better we can understand these boundary conditions the less likely we are to waste effort in applying it in the wrong place.

We believe DST will work best in situations where people need to build their experience base in order to develop expertise for a specific domain. For example, U.S. Marine Corps squad leaders need to size up the combat situations they find themselves in and know how to react quickly. In these types of situations, we can define the types of judgments and decisions that have to be made and design DMGs that allow trainees to practice making these decisions.

Developing a DST program for an organization takes considerable time and effort. It is not a "one size fits all" type of training program. Prior to conducting DST workshops, it is necessary to conduct interviews with subject matter experts to identify the types of critical decisions that are relevant to that particular domain. It takes time to develop the decision-making scenarios that provide the context for teaching the DST tools. These scenarios (or DMGs) need to be developed in conjunction with subject matter experts so that they represent realistic and challenging situations.

The DST program is more likely to be successful when the sponsoring organization commits to maintaining the program as an on-going effort. Trainees who attend a two-day workshop to learn new strategies for developing their domain expertise must be encouraged to continue to practice these strategies after they return to their everyday job duties. Ideally, DST would be introduced into an organization from the "top down." Supervisors would learn how to be good coaches, how to use tools like the Decision Making Critique to share their domain-relevant expertise with their subordinates, and how to use the SA Calibration questions to improve team functioning. These supervisors would then be role models to less senior-level employees who would in turn learn how to use these types of tools to improve the decision-making skills of their subordinates.

It takes work to develop decision skills. Organizations that are looking for quick fixes to improve the decision-making skills of their employees have unrealistic expectations, whether they use the DST program we have developed thus far, or any other training program.

SUMMARY

We have developed an approach to train recognitional decision making skills and have used this approach to train decision makers in several different domains. Based on the evaluations made by the participants, this training appears to be an effective way to facilitate the development of recognitional decision-making skills. Unlike previous approaches that have attempted to teach generic decision-making strategies, such as decision analysis, our approach provides participants with simulation exercises and a set of learning tools to increase the amount learned from these simulations. The learning tools focus on the critical judgments and decisions made during the simulation and how uncertainties were managed. Future research needs to include a formal evaluation of the effectiveness of the

DST program, an exploration of the relationship between training decision-making skills and coaching skills, and a determination of the boundary conditions for the effectiveness of the DST program.

ACKNOWLEDGMENTS

This research was carried out under a subcontract from Synetics Corporation under NSWCDD Prime Contract #N00178-95-D-1008.

The authors would like to thank Patty McDermott, David Klinger, John Schmitt, and Doug Harrington for their assistance in the development of the Decision Skills Training program described in this chapter.

An abbreviated version of this chapter was published in the Proceedings of the Fourth Conference on Naturalistic Decision Making, Warrenton, Virginia, May, 1998.

REFERENCES

Baron, J., & Brown, R. V. (Eds.). (1991). *Teaching decision making to adolescents*. Mahwah, NJ: Lawrence Erlbaum Associates.

Chi, M. T. H., Glaser, R., & Farr, M. J. (1988). *The nature of expertise*. Mahwah, NJ: Lawrence Erlbaum Associates.

Cohen, M. S., Freeman, J. T., & Thompson, B. (1998). Critical thinking skills in tactical decision making: A model and a training method. In J. Cannon-Bowers and E. Salas (Eds.), *Making decisions under stress: Implications for individual and team training*, (pp. 155–159). Washington, DC: APA Press.

Cohen, M. S., Freeman, J. T., & Wolf, S. (1996). Meta-recognition in time-stressed decision making: Recognizing, critiquing, and correcting. *Proceedings of the 40th Human Factors and Ergonomics Society, 38*(2), 206–219. Santa Monica, CA: HFES.

Crandall, B., & Getchell-Reiter, K. (1993). Critical decision method: A technique for eliciting concrete assessment indicators from the "intuition" of NICU nurses. *Advances in Nursing Sciences, 16*(1), 42–51.

Ericsson, K. A. (Ed.). (1996). *The acquisition of expert performance: An introduction to some of the issues*. Mahwah, NJ: Lawrence Erlbaum Associates.

Ericsson, K. A., & Charness, N. (1994). Expert performance: Its structure and acquisition. *American Psychologist, 49*(8), 725–747.

Fallesen, J. J. (1995). *Overview of practical thinking instruction for battle command* (Res. Rep. No. 1685). Alexandria, VA: ARI. U.S. Army Research Institute for the Social and Behavioral Sciences.

Fallesen, J. J., Michel, R. R., Lussier, J. W., & Pounds, J. (1996). *Practical thinking innovation in battle command instruction* (Tech. Rep. No. 1037). Alexandria, VA: ARI. U.S. Army Research Institute for the Social and Behavioral Sciences.

Flin, R., Slaven, G., & Stewart, K. (1996). Emergency decision making in the offshore oil and gas industry. *Human Factors, 38*(2), 262–277.

Glaser, R. (1996). Changing the agency for learning: Acquiring expert performance. In K. A. Ericsson (Ed.), *The road to excellence* (pp. 303–311). Mahwah, NJ: Lawrence Erlbaum Associates.

Janis, I. L., & Mann, L. (1977). *Decision making: A psychological analysis of conflict, choice, and commitment.* New York: The Free Press.

Kaempf, G. L., Klein, G. A., Thordsen, M. L., & Wolf, S. (1996). Decision making in complex command-and-control environments. *Human Factors, 38,* (Special Issue), 220–231.

Kahneman, D., Slovic, P., & Tversky, A. (Eds.). (1982). *Judgment under uncertainty: Heuristics and biases.* Cambridge, MA: Cambridge University Press.

Kahneman, D., & Tversky, A. (1972). Subjective probability: A judgment of representativeness. *Cognitive Psychology, 3,* 430–454.

Kahneman, D., & Tversky, A. (1973). On the psychology of prediction. *Psychological Review, 80,* 237–251.

Kirkpatrick, D. L. (1987). Evaluation of training. In R. L. Craig (Ed.), *Training and development handbook: A guide to human resource development* (3rd ed., pp. 301–319). New York: McGraw-Hill.

Klein, G. A. (1989). Recognition-primed decisions. In W. B. Rouse (Ed.), *Advances in man-machine systems research* (pp. 47–92). Greenwich, CT: JAI.

Klein, G. (1997). Developing expertise in decision making. *Thinking and Reasoning, 3,* 337–352.

Klein, G. (1998). *Sources of power: How people make decisions.* Cambridge, MA: MIT Press.

Klein, G. A., Calderwood, R., & Clinton-Cirocco, A. (1986). Rapid decision making on the fire-ground. *Proceedings of the 30th Annual Human Factors Society, 1,* 576–580. Santa Monica, CA: HFES.

Klein, G. A., & Hoffman, R. (1993). Seeing the invisible: Perceptual/cognitive aspects of expertise. In M. Rabinowitz (Ed.), *Cognitive science foundations of instruction* (pp. 203–226). Hillsdale, NJ: Lawrence Erlbaum Associates.

Klinger, D. W., Andriole, S. J., Militello, L. G., Adelman, L., Klein, G., & Gomes, M. E. (1993). *Designing for performance: A cognitive systems engineering approach to modifying an AWACS human-computer interface* (Contract AL/CF-TR-1993-0093 for Department of the Air Force, Armstrong Laboratory, Air Force Materiel Command, Wright-Patterson AFB, OH). Fairborn, OH: Klein Associates, Inc.

Kozlowski, S. W. J., Gully, S. M., McHugh, P. P., Salas, E., & Cannon-Bowers, J. A. (1996). A dynamic theory of leadership and team effectiveness: Developmental and task contingent leader roles. In G. R. Ferris (Ed.), *Research in personnel and human resources management* (Vol. 14, pp. 253–305). Greenwich, CT: JAI.

Mann, L., Beswick, G., Alloache, P., & Ivey, M. (1989). Decision workshops for the improvement of decision-making skills and confidence. *Journal of Counseling and Development, 67,* 478–481.

Mann, L., Harmoni, R., & Power, C. (1991). The GOFER course in decision making. In J. Baron & R. V. Brown (Eds.), *Teaching decision making to adolescents* (pp. 185–206). Hillsdale, NJ: Lawrence Erlbaum Associates.

Mann, L., Harmoni, R., Power, C., Beswick, G., & Ormond, C. (1988). Effectiveness of the GOFER course in decision making for high school students. *Journal of Behavioral Decision Making, 1,* 159–168.

Raiffa, H. (1968). *Decision analysis: Introductory lectures on choices under uncertainty.* Reading, MA: Addison-Wesley.

Randel, J. M., Pugh, H. L., & Reed, S. K. (1996). Methods for analyzing cognitive skills for a technical task. *International Journal of Human-Computer Studies, 45,* 579–597.

Rouse, W. B., Cannon-Bowers, J. A., & Salas, E. (1991). *The role of mental models in team performance in complex systems.* Norcross, GA: Search Technology, Inc.

Savage, L. J. (1954). *The foundations of statistics.* New York: Wiley.

Schmitt, J. F. (1994). *Mastering tactics.* Quantico, VA: Marine Corps Association.

Shanteau, J., Grier, M., Johnson, J., & Berner, E. (1991). Teaching decision-making skills to student nurses. In J. Baron & R. V. Brown (Eds.), *Teaching decision making to adolescents* (pp. 185–206). Hillsdale, NJ: Lawrence Erlbaum Associates.

Simon, H. A. (1956). Rational choice and the structure of the environment. *Psychological Review, 63*, 129–138.

Tannenbaum, S. I., Smith-Jentsch, K. A., & Behson, S. J. (1998). Training team leaders to facilitate team learning and performance. In J. A. Cannon-Bowers & E. Salas (Eds.), *Making decisions under stress: Implications for individual and team training* (pp. 247–270). Washington, DC: American Psychological Association.

Tversky, A., & Kahneman, D. (1971). Belief in the law of small numbers. *Psychological Bulletin, 76*, 105–110.

Tversky, A., & Kahneman, D. (1974). Judgment under uncertainty: Heuristics and biases. *Science, 185*, 1124–1131.

von Neumann, J., & Morgenstern, O. (1947). *Theory of games and economic behavior*. Princeton, NJ: Princeton University Press.

Yates, J. F., & Estin, P. (1996, November). *Training good judgment.* Paper presented at the annual meeting of the Society for Judgment and Decision Making, Chicago, IL.

4

Identifying and Testing a Naturalistic Approach for Cognitive Skill Training

Jon J. Fallesen
U.S. Army Research Institute
Julia Pounds
Civil Aeromedical Institute

The emergence of the field of Naturalistic Decision Making (NDM) has raised issues that previously held little interest for the workplace. This is especially true for tradition-rich organizations such as the U.S. Army. Although doubt about prescriptive procedures for decision making and problem solving existed prior to the emergence of NDM, the field has directed more attention at the intricate and idiosyncratic nature of human thinking. While it has long been recognized that tactical decision making is more than a "black box" phenomenon, NDM has brought more attention to how specific thinking abilities are developed and sustained (Klein, 1989). The NDM movement has brought into question the applicability and sufficiency of rational and formal-based procedures taught and backed throughout Army doctrine. One implication that can be drawn from NDM research findings is that prescriptive, formal models provide limited guidance on how to think and how to prepare to think in complex, ill-defined situations. This has been corroborated through various studies on military decision making (e.g., Fallesen, 1993; Pascual & Henderson, 1997; Serfaty, MacMillan, Entin, & Entin, 1997). Alternatives to the formal, analytic-based approaches for training decision makers are feasible and would provide extended capabilities and increased versatility in thinking.

Training for more versatile thinking can support the requirement for a broadening of U.S. Army missions that are occurring due to striking world changes. For one, the decreased threat of superpower conflict has been replaced with an increase in the type and number of regional conflicts. While the force has been downsized, the variety of missions and roles that Army units serve has been increasing. These changes in the operational Army require that conceptual thinking capabilities be at peak levels. Thus, leaders need to have a wide range of knowledge and skills to respond to the wide range of missions. Furthermore, doctrine for operations other than war (OOTW) places unique demands on how leaders decide what courses of action are prudent. Also, the increase in weapon and information capabilities will necessarily change the procedures used by commanders and their staffs in the future. All of these shaping factors make it more important than ever to have a sound understanding of the actual ways military leaders think so that natural, individual strengths can be built on rather than ignored.

Better understanding of natural reasoning processes is not enough. Equally important is the question of how to reinforce and extend the natural thinking skills of military leaders. Consideration of how best to develop conceptual thinking abilities has identified two contrasting approaches. One is a formal, structured approach that prescribes formalized rules and procedures. Explicit teaching points are derived from the fields of logic, economics, and probability based on formal theories and models. In contrast, an informal approach has developed from behaviorally based NDM models. The typical NDM approach centers around the experience that encourages cue recognition and differentiation. Under the typical NDM approach, learning is primarily a tacit process with feedback (usually intrinsic) leading eventually to competent levels of performance. However, this approach offers few explicit teaching points.

An alternative to these approaches is a third approach, which is to focus on process but develop explicit training materials from the naturalistic study and assessment of people solving their problems. The naturalistic approach takes into account the decision task, the decision maker, and the environment. Tactical decision-making problems are characterized by the factors identified to be central to NDM situations, such as high uncertainty, dynamism, meaningful consequences, ill-structured goals, and complexity (Cannon-Bowers, Salas, & Pruitt, 1996). It is the unique war fighting and peacekeeping environments that compel the use of an NDM approach for tactical problem solving. Strategies used by individuals to solve actual or representative problems can be identified along with the strategies' strengths and weaknesses. By examining the variation of natural strategies, training programs can be developed to help individuals make their natural thinking processes initially more explicit. Then, individuals can be encouraged to learn how to intensify their thinking according to internal intellectual standards (Paul, 1993) rather than external procedures.

This chapter presents a test of this approach to cognitive skill training. In summary, the approach (a) surveyed strategies used by actual decision makers, (b) examined the variation of thinking strategies and correlated it to other performance indicators, (c) selected skills to focus on in training, and (d) developed the training to make the skills explicit. The purpose of the study was to understand better the actual problem-solving strategies used by U.S. Army officers and to perform a proof of concept test of this NDM approach to training. The study was conducted in two phases. The first phase collected strategy usage and importance data from 48 company and field grade officers. A focus for training was identified from this first phase. In the second phase, a training lesson was tested with 31 officers. The same data as in phase 1 were collected using a pretest-training-posttest design.

PHASE 1: IDENTIFYING A PROBLEM-SOLVING STRATEGY FOR SKILL TRAINING

Skilled problem solving is highly valued for command and tactical problem solving. Despite much historical speculation and research on what command consists of, there is no consensus on a single model to describe commanders, command skills, or the command process. With serious questioning about the applicability of the traditional, prescriptive models as a description of command processes came a need for some framework to identify cognitive behaviors important to military leaders. The NDM paradigm and its associated data collection methods provided the framework, but a specific level of cognition needed to be identified and stated in measurable terms if we were ultimately to improve cognitive behavior. Although this is not typically an issue with formal models, NDM models are more loosely structured. Their point of commonality seems mainly to be in their representation as flowcharts and diagrams (Shanteau, 1995); they do not consistently portray the same level of cognitive activity. For this study of military decision tasks, problem-solving strategies seemed to be the proper level of analysis.

We conceptualized *cognitive strategies* for problem solving as regularities in reasoning that guide thinking in order that mental resources can be allocated appropriately. Strategies consist of cognitive processes, which, in turn, are made up of relatively more distinct and stable subelements called cognitive operations. *Cognitive operations* do much the same thing in every context, such as comparing some object X to some object Y (Neisser, 1993). *Cognitive skills* occur as the activation of strategies and processes and reflect an individual's performance capability. Skills can be and are adapted to the conditions of specific problems. *Cognitive styles* represent longer term adaptation. Styles represent a link between cognition and attitudes.

A review of various literature, for example, psychology, judgment and decision making, artificial intelligence, operations research, economics, and education, identified more than 120 possible strategies (Pounds & Fallesen, 1994). These strategies were distilled into a more distinct set of 66 individual strategies, which were then described on five dimensions (e.g., definition, triggering conditions, example, strengths, and weaknesses) and grouped into three classes and nine categories. The classes included managing information, controlling progress, and making choices. Categories included considering hypotheses, beliefs, and uncertainty; combining information; managing the amount of information; ordering by hierarchical structure; sequencing; ordering by merit or payoff; managing the number of options; using compensatory choice; and using noncompensatory choice. With the assistance of an active duty U.S. Army officer as a subject matter expert, these strategies were transformed into 48 phrases free of decision-making jargon. Also, four higher order groupings called *approaches* were developed to reflect styles of problem solving. The groupings represented analysis, procedures, recognition, and dominance. The description for analysis involved structuring the problem in a constant way to allow linear comparisons (such as a matrix used in multiattribute utility analysis). The procedural approach emphasized following a sequence of steps or processes to reach a decision. Recognition (Klein, 1993) stressed recollection of familiar problems and familiar solutions. Dominance (Montgomery, 1993) conveyed beginning with a single option and modifying it until it was perceived better than any other that could be imagined.

Method

In a series of data collection sessions at three troop installations, 48 U.S. Army officers (lieutenants, captains, majors, and lieutenant colonels) were interviewed individually by one of two experimenters. First, officers were asked to describe the amount of time that they used each of the four approaches in job-related problem solving. Next, they were asked to recall and describe some critical decision situation from their past personal experiences.

After the situation was detailed sufficiently, the experimenter gave the participant a set of index cards, with a different strategy phrase printed on each. The participant sorted the cards into groups to indicate the usage and importance in the described situation. The sorting categories were later transformed into numeric ratings of 0 to 5 (*no use*, *uncertain*, *used but not important*, *important*, *very important*, and *most important*). After sorting, the participant also explained how the strategy applied or did not apply to the described situation. The explanation provided a way to verify the ratings and suggested to the participants that they would have to provide reasons for their ratings.

Participants completed two other questionnaires. The first asked them to estimate the quality of the outcome of the situation. Rated outcome was treated as an exploratory measure and was not relied on as a criterion measure; in effect, it provided an indication of the individual's confidence about his or her solution. The second questionnaire asked them to rate their level of familiarity, importance of the problem, time available, and effort required (relating to the NDM factors discussed by Cannon-Bowers, Salas, & Pruitt, 1997 and the FITE mnemonic proposed by Fallesen, Michel, Lussier, & Pounds, 1996).

Upon conclusion of the discussion about the past personal situation, participants were presented with two scenarios, one at a time, and were asked to role play the scenario's commander. The scenarios were Army modifications to Schmitt's Marine tactical decision games (1994), "enemy over the river" and "rescue the ambassador." The participant described aloud what he or she was thinking while considering the problem. Afterward, the participants again rated the usage and importance of strategies and gave a prediction of outcome for their solution and rated the situational characteristics. Audio recordings were made throughout the entire session and were later transcribed.

Findings

The results confirmed that informal approaches are often used in place of the doctrinal and instructional ones. For example, less than 5% of the participants used a concurrent option comparison strategy. Most of the time, they focused on only one option: defining it, evaluating it, and deciding whether to refine it or create another option. Only in this small number of cases did they follow doctrinal guidance and generate multiple options before evaluation and comparison.

Figure 4.1 shows a split in use between traditionally taught approaches (analytic and procedural) and naturalistic approaches (recognition and dominance). For familiar problems, a recognition approach was used most frequently. For unfamiliar problems, a dominance approach was used more frequently than recognition (Pounds & Fallesen, 1995). Five of the top seven strategies are components in each of the four approaches (identified goals and facts, visualized the problem, broke the problem into smaller problems, and employed criteria; see

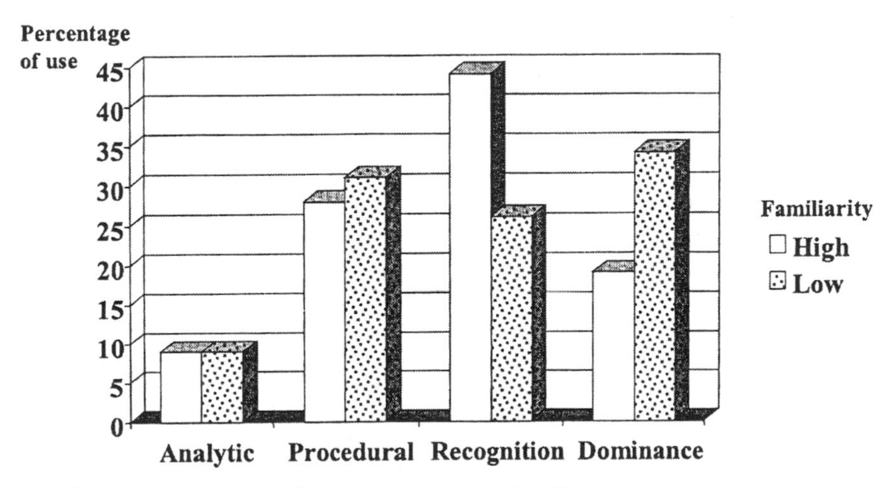

FIG. 4.1. Problem solving approaches by familiarity.

Fig. 4.2). Six of the seven lowest rated strategies were unique to the analytic, recognition, or dominance approaches. Multiple regression analysis showed unexpected relationships among strategies and approaches. The hypothesized model for the procedural approach accounted for significantly less variability than one determined by a best-fit model. Eleven strategies for the best-fit model predicted variance of the procedural approach with an R^2 of .61 compared to an R^2 of .36 for the hypothesized model with 20 strategies ($F_{(61,70)} = 1.864, p = .006$; Fallesen, 1996). Only three strategies from the hypothesized model showed up in the best-fit version. We interpreted this to mean that the strategies actually associated with a preference for a procedural approach are different from the strategies that are taught as part of a procedural approach. The individual strategy ratings and the multiple regression models showed that actual problem solvers do not perform the procedural approach in the same way as instructional and doctrinal materials convey it. Instead, participants used a mix of doctrinal and naturalistic strategies at the same time that they believed they were following a procedural approach.

In order to identify which strategies might be associated with an outcome difference, we looked for qualitative differences in solutions to the "enemy over the river" scenario. The situation starts with the battalion commander already immersed in action on his way to move into an assembly area (AA). After arriving at the AA, he is to prepare to lead the division's attack across a bridge the next morning. As the battalion is moving to the AA, the commander receives two

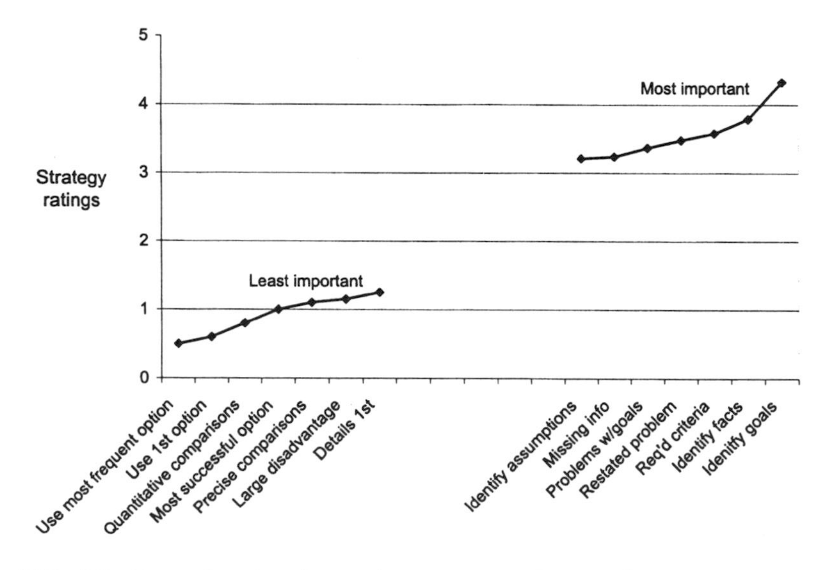

FIG. 4.2. Seven least and most important rated strategies.

reports that enemy infantry forces with light vehicles are already occupying his AA, enemy units are continuing across the bridge, and there are no friendly forces near the river. The officers offered a number of different types of solutions for this problem (four different ones are illustrated in Fig. 4.3). Two were judged to be clearly separable. A major difference is whether the "commander" focuses on the enemy in the AA or on the bridge and the enemy coming across there. The problem with going immediately to the bridge is that, because the last orders received were to go to the AA, going to the bridge ahead of schedule may conflict with the higher

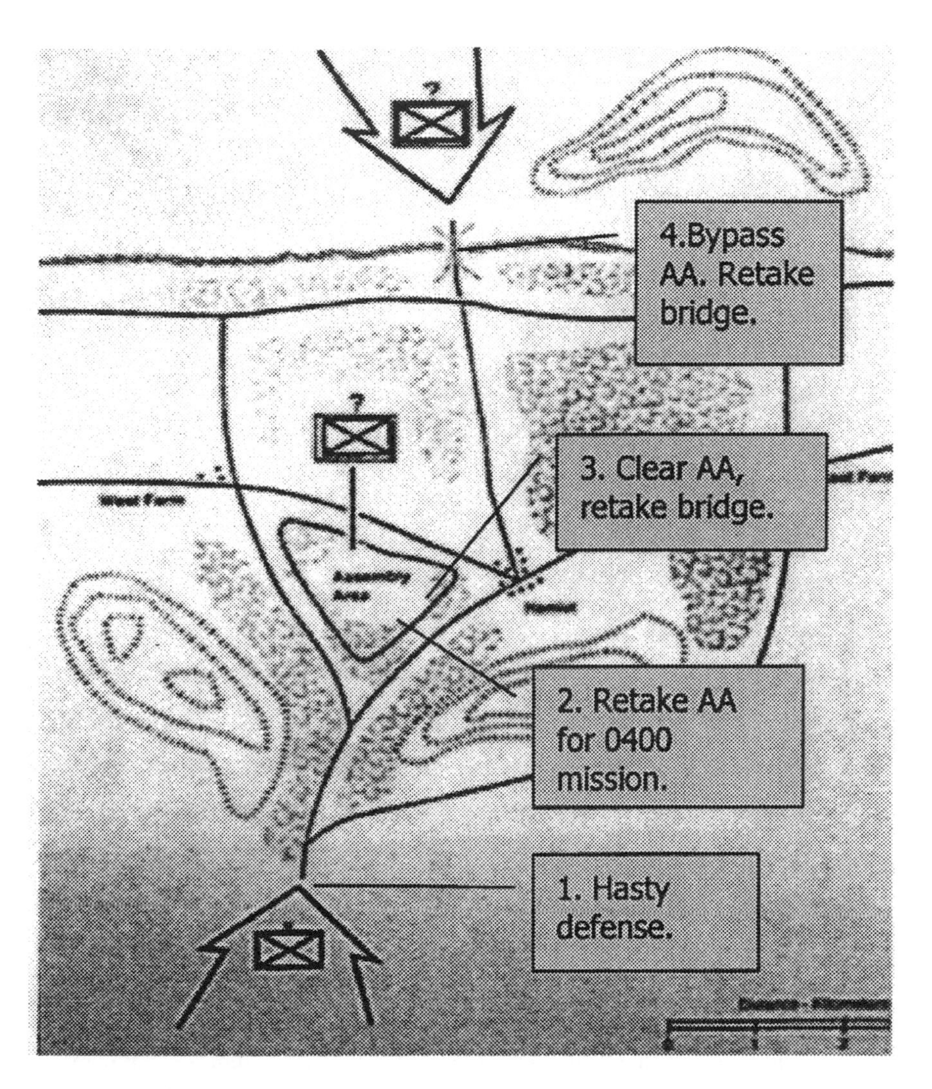

FIG. 4.3. Tactical sketch of *'enemy over the river'* with 4 solution types indicated.

echelons' plans. The problem with focus on the AA is that the location has little apparent future value, whereas the bridge is essential for the next day's mission. It is difficult to consider one choice over the other as ideal or correct. It brings into play several contrasts for tactical operations: mission accomplishment vice force protection, and short-term vice long-term goals. The scenario seems to be an excellent prototype for ill-defined problems with conflicting and uncertain goals.

Thus, to converge on a training theme we reasoned that the possibilities of both tactical options needed to be considered. The formal approach would suggest that the options be evaluated and compared to each other. But, using the formal approach did not guarantee that either of these two options were selected. Instead, we screened the 48 strategies to select one for training. We specifically examined the strategies of participants who considered going to the bridge. (All participants considered the AA, which was not too surprising because the scenario started with the units on their way to the AA). The only strategy of the 48 that had a significant relationship to considering the bridge was the strategy *considered the relevance of information* ($r = .53$, $p = .01$). Those participants rating *relevance of information* as an important strategy also tended to consider the consequences of going to the bridge, and those that rated that strategy lower tended not to consider the bridge at all.

Further Consideration for Training

Just identifying this strategy as important in the type of solution did not translate directly into how the strategy could be used in skill training or what size effect to expect from the training. Our derivation of training content followed the recommendations on cognitive task and cognitive performance analysis (Essens, Fallesen, McCann, Cannon-Bowers, & Doerfel, 1995) and was consistent with the principles used in an earlier experimental program of instruction on practical thinking (Fallesen et. al., 1996). The former guidance emphasized using actual knowledge of behaviors in designing system improvements instead of relying on theoretical models of desired performance. The latter reference implemented theory on adult education, constructivist learning, and reinforcement of critical thinking principles.

We should explain why the recommendation by Lipshitz (1993)—to study experts and use them as the model for training the ability to make high quality decisions—was not the specific direction that we took. Because the identification and measurement of expertise for unfamiliar, ill-defined, and adversarial problems is problematical regardless of the amount of domain expertise, we felt it could not direct us to an approach but could help confirm one determined through the measurement of strategies. The literature on expert problem solving revealed a number of principles that were consistent with our Phase 1 findings. Some models suggest that experts use more information than novices; however, studies of medical pathologists, nurses, auditors, and weather forecasters show that their judgments can be accounted for by relatively few cues (Shanteau, 1996). Experts seem better able to discriminate relevant from irrelevant cues, whereas subexperts lack the same diagnostic ability to determine relevancy. Experts are better at distinguishing what is important from what is not.

Training Program

A short training lesson was developed around relevancy checking. Relevancy checking was framed as a method of improving intellectual-thinking standards that influence everyday reasoning and as the premises of practical thinking (Fallesen et. al., 1996). The goal of relevancy checking was to improve the participant's situation understanding, identification of goals, planning actions, and assessment of reasons. The instruction included description of the meaning and importance of relevancy, ways to make relevancy checks, and short exercises to reinforce the concepts. The assumption was that it is desirable to think without much conscious intervention. However, during learning, conscious attention and reflection are useful to try new ways of thinking and to acquire additional capabilities. Conscious attention to a strategy is not the desired mode in real, "on-the-fly" problem solving. This rationale comes from understanding that flexible expertise will be required for novel problem solving, typical in tactical NDM, and that it cannot be prepared for simply with drill and practice. Flexible knowledge structures need to be formed from attentive learning, and leveraged during operational problem solving with a minimal strain on cognitive effort.

Relevancy is a characteristic of information and conclusions and to what extent they relate to the matter at hand. Usually, relevance checks probably occur naturally as a non-conscious comparison of the current situation against similar instances. In this training the relevancy checks were discussed as explicit possibility questions. Participants were asked to apply a set of relevancy checking questions in each of the three problem solving activities (see Fig. 4.4) primarily focusing on *what if* and *what else. What if* questions can stimulate both critical and creative thinking. In terms of relevancy, they prompt the consideration of other possibilities: what if the situation is not exactly what I think it is, what if goals or planned actions were different? *What else* questions can stimulate richer or alternate understandings: If something is not what I believe it is, what else could it be? *What else* questions can generate other reasons or explanations for conclusions, other understandings of the situation, and other possible goals and planned actions.

These questions are compatible with the recognition/metacognition framework of Cohen, Freeman, and Thompson (1997) and their training program for metarecognition skills. However, a difference is that ours posed the question of relevancy instead of questioning a belief or conclusion that participants were

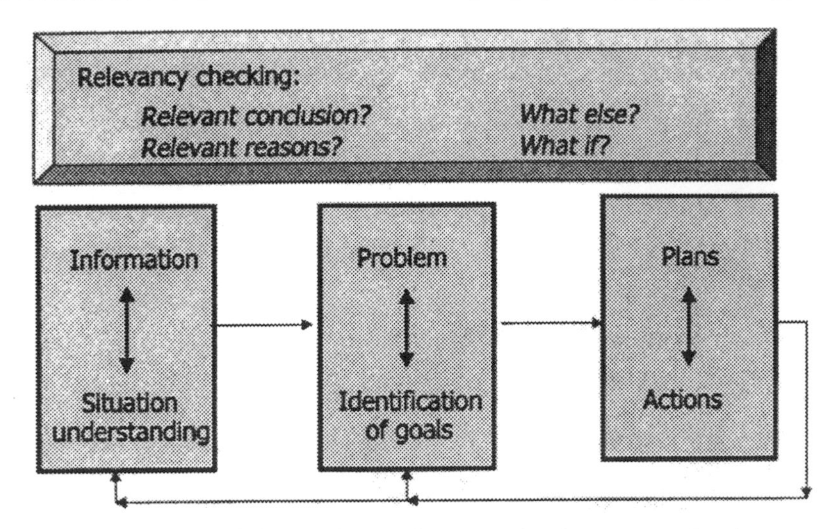

FIG. 4.4. Framework for applying relevancy checks to tactical tasks.

already certain about. In our NDM and cognitive skills approach, we established an empirical basis for the importance of the skill prior to the creation of the training. Both approaches use reflective questioning, but our training tries to push the questioning upstream during an individual's understanding processes, rather than during a critique or correction phase occurring later. Our training also tries to give the relevance check as a basis for why and when reflective questions would be used. The questioning technique of Cohen et al. is to be used in cases when the individual is most confident that they are correct.

Three exercise examples were used in the lesson to illustrate the teaching points and to capture the officer's attention, giving him or her a chance to try out the concepts, and to explore the presented ideas more deeply. The exercises included a problem on increasing the armor protection on aircraft, questioning the relevancy of reasons that an enemy might use chemical munitions, and a national security situation involving possible U.S. intervention in another country's domestic rebellion. This last exercise was based on an actual incident that occurred in 1989 (Woodward, 1991). The student is assigned to role-play a member of a special advisory staff. The staff provides advice to the National Security Council on quick reaction policy. With a brief background about the nature of the relationship between the notional country of Panglossia and the United States, students are given opportunities to assess the relevancy of goals and reasons given by others and to come up with relevant alternative goals and solutions on their own. Some of the concepts used in the lesson were translated into teaching points given in Table 4.1.

TABLE 4.1
Teaching Points Incorporated Into the Relevancy Checking Lesson

Use clear-headed thinking, so you don't just get caught up in the emotion of the moment.
Bring out considerations that wouldn't normally be thought of.
Develop critical thinking habits because habits are easy to rely on.
Check on relevancy early so you don't go off and solve the wrong problem.
Check on relevancy by considering how close or remote associations have to be to judge something as relevant.
Sometimes it is easier to jump ahead and consider a solution and see what some action would accomplish than trying to determine the best goal and start toward it.
Check the consequences of proposed plans and actions.
Use relevancy checks to help safeguard against poor solutions.
Use relevancy checks to help you to see other sides of an issue.
Use relevancy checks to consider what you already know, and to think more deeply where you might be uncertain.

PHASE 2: TESTING THE SKILL TRAINING OF THE RELEVANCY CHECKING STRATEGY

Method

Thirty-one officers from two troop installations participated individually in the learning and testing phase of the relevancy checking training. Rather than reporting on a critical incident as participants did in Phase 1, officers in the training phase were asked what approaches and strategies they used in general and how important they were. Then, they were pretested using one of the two scenarios, participated in the training, and completed the posttest using the second tactical scenario. (The order of scenarios was counterbalanced as it was in Phase 1.) There were no differences in characteristics of the trained and Phase 1 samples in terms of time in service, proportion of company and field grade officers, and U.S. Army schooling. We found no difference in baseline approaches between the relevancy-trained participants and the control group from Phase 1. There was some difference in the proportion of different branch specialties represented.

Administration of Training

Although we had found some reluctance on the part of midcareer officers in earlier classroom instruction to consider the merit of cognitive-based training (Fallesen et al., 1996), we found no such reluctance in this one-on-one mode. Only three students had questions indicating any reluctance toward the training. One mentioned that the questioning of relevancy works only to a point—that it cannot go on forever. Two others mentioned that consideration of relevancy needs to be considered from a group perspective rather than just from an individual one. The essence of their comments are included in Table 4.2.

Written Feedback

The 31 students gave the instruction an average rating of 85 using a rating scale from 0, *worst* to 100 *best*. Although the training period was no more than 60 minutes long, participants expressed a desire to receive more than four times that amount of training. They thought the training should be done regularly in officer professional development courses in their units and in officer advance and basic courses. The single concept that stood out most to them was the use of questions (*what else* and *what if*) to supplement their thinking. They told us that the part of the training that worked best to make the strategy real was the National Security

TABLE 4.2
Impromptu Feedback From the Students During Instruction

About the problem that the training addresses. . .
I see what you're talking about. I've jumped in and solved the wrong problem.
I like the way the 1st problem tricks you; we need that so we just don't go along.
We fall into these kinds of traps (treating assumptions as facts, assuming a cooperative enemy) in our orders process routinely.
The biggest learning point I see is that we don't address what our goals are.
I had a former S3 who also emphasized relevancy.
Yes I can see how to use this to check other people's reasoning or our own.
Sometime what you don't know is the most relevant thing.
Relevancy becomes even more critical as we get access to much more information than we've ever had to work with before.
About the focus on thinking standards. . .
Standards are important to thinking. I see how making them explicit can help.
You have to discipline yourself to think according to the standards.
Yes this needs to be "train yourself."
About the questions for relevancy checking. . .
These are pretty good questions for checking relevancy.
I really like the what if, what else questions.
I liked these questions. They become the methodology for solving problems.
These questions should also lead to creativity.
This can help so you don't make yourself predictable.
I like the way this encourages you to make deliberate choices.
General:
This idea allows you to stay focused.
It seems it can be a natural way of thinking.
I agree with the training whole-heartedly.
I try to make it a habit to think like this. It works pretty good, when it's a habit, you don't even need to think about it.
I could see using this in a CPX, but it also comes from studying history, asking questions about whether something someone else did would be relevant and how we can use it now.
There ought to be classes on this in officer basic and advanced courses (OBC & OAC).

Council decision. A close second was the use of tactical scenarios. We originally thought of the scenarios as only testing materials, but the card sorting of strategies done by the participants caused them to feel that the scenarios were useful to help them become more aware of how they solved problems. About one fifth (6 of 31) of the participants thought that more details in the exercises would be useful (e.g., more on commander's intent). But about one third felt that no improvements at all were needed in the training. Other suggestions included a better definition of relevancy, adding group discussion, receiving feedback on performance style, using different scenarios, imposing a time limit in the exercises, and giving more real life examples of success and failure of using the *what else* and *what if* questions.

Statistical Comparisons

A multivariate analysis of variance (MANOVA) was used to determine differences across the strategy ratings. An interaction between the training groups and the pretest–posttest factor would show a significant effect due to the instruction. Instead, we found an order (of scenario) by test interaction ($F_{(48,28)} = 1.77$, $p = .05$). The interaction of test and order suggests that the sequence of problems had a differential effect on the posttest. Also, the two scenarios could provide different opportunities and requirements for relevancy checking. Two separate MANOVAs, one for each scenario order, were performed on strategies that had been previously screened as most likely to show an effect from the relevancy training. For the battalion-company scenario ordering, there was a training main effect ($F_{(5,35)} = 3.11$, $p = .02$) and no training by test interaction. The training main effect without a test by training interaction suggests that the two training groups were different overall in the way they rated the strategies across the two scenarios. For the company-battalion ordering, there was no training main effect ($F_{(5,32)} = 2.44$, $p = .0533$) but a training by test interaction ($F_{(10,64)} = 2.07$, $p = .04$). This interaction can be explained by looking at differences in means for four strategies.

The strategy, *imagined the best outcomes*, showed a simple effects difference between the posttest control and trained conditions ($p = .04$) with the trained participants rating this strategy lower. A similar difference ($p = .002$) was found with the strategy, *identifying assumptions*, again with trained participants rating this strategy as less important. And, the same trend was found for the strategy, *choosing an option that had at least one important characteristic*. The strategy, *information relevancy*, showed a significant increase within just the trained group from pretest to posttest ($p = .04$) attributable to the training, while comparison of the posttest ratings between the control and trained conditions approached significance ($p = .07$). This corresponds to the relationship we saw in Phase 1 in which there was a significant relationship between the type of tactical solution and the importance of relevancy checking in the battalion problem.

We also considered how training on relevancy checking would be influenced by the individual's familiarity with the situation. We took the participants who had low familiarity ratings and looked for training by test interactions. We speculated that the relevancy checking strategy would be most applicable as explicit questions for unfamiliar problems. There were significant interactions for several strategies: *considered the relevance of information* ($p = .01$), *determined parts of the plan that would be prone to flaws* ($p = .05$), *chose the option that was most attractive on the most important characteristic* ($p = .04$), and *kept the set of things to think about as small as possible* ($p = .05$). Increases in strategy importance for the trained group occurred for the first three of these strategies, and a decrease in importance by the trained group occurred for the last. (The third of these strategies is related to a dominance approach.)

We looked for a relationship between relevancy checking along with other strategies and the types of solution given in each scenario. For example, we asked if relevancy checking or other strategies were associated with more Assembly Area alone solutions or more Assembly Area plus Bridge solutions, but found none using discriminant analysis. Earlier findings (Fallesen, 1996; Pounds & Fallesen, 1997) showed complex interactions among sets of strategies, and we were not surprised by the lack of relationships here. The findings underline the idiosyncratic nature of individual solutions and the absence of best or ideal solutions for complex problems.

Discussion

The research has led to the conclusion that cognitive skill training has merit for advanced problem solving by midcareer Army leaders. This corresponds to earlier findings reported by Cohen et al. (1997) and Halpern (1996). Gains in quality of thinking can be achieved by directing the focus on natural ways of thinking and ways that might not be a part of an individual thinking style but compatible with natural ways nonetheless. The findings suggest that intellectual thinking standards can be tuned by reminding officers to consider their beliefs and standards of thinking and to try to use more critical thinking.

The effect these behaviors have on the quality of decisions is difficult to assess for complex situations. There was no learning retention phase in this effort, and, unfortunately, we do not know whether performance changes will persist. Practical utility is likely to occur only after time to allow individuals to integrate the new skills into their habitual conceptual abilities. However, this issue is not unique. It reappears whenever the issue is situated performance; that is, to what extent does training transfer, generalize, and last? It was anticipated that any change would require at least initial recognition of the direction of improvement and initial understanding of the concepts. Thus, one important indicator of success in future studies will be whether explicit cognitive skill training can prepare adults to learn how to learn to improve their thinking.

Our results showed that NDM methods—those aimed at establishing a baseline of natural decision behaviors—are useful for providing a basis for cognitive skill training. Previous NDM results implied training improvements by calling for opportunities for more or richer experience. This cognitive skill approach demonstrated the feasibility of building training from the deliberate study of cognitive strategies. New studies are underway to further our understanding of the role of intellectual thinking standards, attitudes, and beliefs on conceptual strategy usage.

REFERENCES

Cannon-Bowers, J. A., Salas, E., & Pruitt, J. S. (1996). Establishing the boundaries of a paradigm for decision-making research. *Human Factors, 38*, 193-205.

Cohen, M. S., Freeman, J. T., & Thompson, B. B. (1997). Training the naturalistic decision maker. In C. E. Zsambok & G. Klein (Eds.), *Naturalistic decision makin.* (pp. 257–268). Mahwah, NJ: Lawrence Erlbaum Associates.

Essens, P. J. M. D., Fallesen, J. J., McCann, C. A., Cannon-Bowers, J., & Doerfel, G. (1995). *COADE: A framework for cognitive analysis, design, and evaluation* (Tech. Rep. No. AC/243 (Panel 8) TR/17). Brussels, Belgium: NATO Defence Research Group.

Fallesen, J. J. (1993). *Overview of Army tactical planning performance research* (Tech. Rep. No. 984). Alexandria, VA: U.S. Army Research Institute for the Behavioral and Social Sciences.

Fallesen, J. J. (1996). Understanding and improving tactical problem solving. In *Principal Scientist Colloquium Special Rep. 26*. Alexandria, VA: U.S. Army Research Institute for the Behavioral and Social Sciences.

Fallesen, J. J., Michel, R. R., Lussier, J. W., & Pounds, J. (1996). *Practical thinking: Innovation in battle command instruction* (Tech. Rep. No. 1037). Alexandria, VA: U.S. Army Research Institute for the Behavioral and Social Sciences.

Halpern D. F. (1996). *Thought & knowledge: An introduction to critical thinking.* Mahwah, NJ: Lawrence Erlbaum Associates.

Klein, G. A. (1989). Strategies of decision making. *Military Review*, 69(5), 56–64.

Klein, G. A. (1993). A recognition-primed decision (RPD) of rapid decision making. In G. A. Klein, J. Orasanu, R. Calderwood, & C. E. Zsambok (Eds.), *Decision making in action: Models and methods* (pp. 138–147). Norwood, NJ: Ablex.

Lipshitz, R. (1993). Converging themes in the study of decision making in realistic settings. In G. A. Klein, J. Orasanu, R. Calderwood, & C. E. Zsambok (Eds.), *Decision making in action: Models and methods* (pp. 103–137). Norwood, NJ: Ablex.

Montgomery, H. (1993). The search for a dominance structure in decision making: Examining the evidence. In G. A. Klein, J. Orasanu, R. Calderwood, & C. E. Zsambok (Eds.) *Decision making in action: Models and methods* (pp. 182–187). Norwood, NJ: Ablex.

Neisser, U. (1983). Components of intelligence or steps in routine procedures. *Cognition, 15,* 189–197.

Pascual, R., & Henderson, S. (1997). Evidence of naturalistic decision making in military command and control. In C. E. Zsambok & G. Klein (Eds.), *Naturalistic decision making* (pp. 217–226). Mahwah, NJ: Erlbaum.

Paul, R. (1993). *Critical thinking: How to prepare students for a rapidly changing world.* Santa Rosa, CA: Foundations for Critical Thinking.

Pounds, J., & Fallesen, J. J. (1994). *Understanding problem solving strategies* (Tech. Rep. No. 1020). Alexandria, VA: U.S. Army Research Institute for the Behavioral and Social Sciences.

Pounds, J., & Fallesen, J. J. (1995). *Familiarity effects on strategy use in tactical problem solving*. Poster presented at Society for Judgment and Decision Making Conference. November 11–13, 1995. Los Angeles, CA.

Pounds, J., & Fallesen, J. J. (1997). *Problem solving of mid-career Army officers: Identification of general and specific strategies* (Research Note 97–21). Alexandria, VA: U.S. Army Research Institute for the Behavioral and Social Sciences.

Schmitt, J. F. (1994). *Mastering tactics: A tactical decision game workbook*. Quantico, VA: Marine Corps Association.

Serfaty, D., MacMillan, J., Entin, E. E., & Entin, E. B. (1997). The decision-making expertise of battle commanders. In C. E. Zsambok & G. Klein (Eds.), *Naturalistic decision making* (pp. 233–246). Mahwah, NJ: Lawrence Erlbaum Associates.

Shanteau, J. (1995, August). *Why do experts disagree?* Paper presented at the 15th Bi-Annual Conference on Subjective Probability, Utility, and Decision Making, Jerusalem, Israel.

Shanteau, J. (1996, April). *The upper limits of expert performance and the search for world records: What is the question?* Invited paper presented at the meeting of the Oklahoma/Kansas Judgment and Decision Making Society, Norman, Oklahoma.

Woodward, B. (1991). *The commanders*. New York: Schuster.

5

Solving the Problem of Employee Resistance to Technology by Reframing the Problem as One of Experts and Their Tools

Lia DiBello
Workplace Technology Research Group
CUNY Graduate School

This chapter addresses the relationship between human cognition and tools as it applies to the problem of rapidly changing information technology, an issue I have been studying for a number of years (e.g., DiBello, 1996a, 1996b, 1997; DiBello & Kindred, 1992; DiBello & Spender, 1996; Scribner, Sacks, DiBello, & Kindred, 1991; Scribner, DiBello, Kindred, & Zazanis, 1992). Although we focus on relatively complex information technologies, we continually find that the fundamental mechanisms of tool mastery and appropriation remain deceptively the same, regardless of tool complexity. We believe that most of what human beings do with complex tools could be predicted by close observation of how people use and modify even the simplest tools.

Since the 1970s large information technologies have been fundamentally changing many industries. Of particular interest to us are those that are changing work by making the manipulation and analyses of information easier and more widespread. These systems are usually large, highly integrated information systems that capture relatively "live" data for purposes of analysis and make possible rapid changes in business strategy. Two examples we have studied in depth are Materials Requirements Planning (MRP) and MRPII and Computerized Maintenance Management Systems (CMMS). A brief description of these two types of systems will help clarify some of our later points.

THE IMPACT OF COMPLEX TOOLS

MRPII

MRP has been characterized as a theory of inventory and material management. It instantiates certain key economic concepts such as *zero inventory* and *just-in-time* production and is based on principles of manufacturing (for example, formulas regulating how future orders are forecast) developed over the last several decades (Harrington, 1974; Timms & Pohlen, 1970). In many ways it is considered a somewhat counterintuitive approach to material planning in that it "begins" in the future and moves backward in time. However, when properly executed, on-hand inventory can be reduced by as much as 70%, freeing a large part of a company's capital.

In general, MRP approaches are in contrast to traditional "aggregate" planning. Aggregate planning methods evolved after WWII and were influenced by material scarcity concerns. The goal of most aggregate planning methods was to accumulate as much raw material as possible to cope with growing demand. With improved distribution and material availability, inventory surplus soon became a significant and costly business problem. So called just-in-time approaches evolved as a result of a better understanding of fixed or known demand.

However, in actual practice MRP systems are often used to automate aggregate planning practices. In these cases, the implementation is not considered successful by industry standards. In fact, it is actually easier for employees to make things worse when they misuse the system.

CMMS

CMMS are a rapidly growing set of systems for managing activity. Industry leaders estimate that there are now about 80 different products on the market. In some ways, they begin where MRP systems leave off. MRP systems are concerned with planning "things" (raw material, components, assemblies) to meet anticipated and very specific demand. The activities associated with the material are implied by MRP in the form of material routings. CMMS, on the other hand, is somewhat of a mirror image of this process: these systems plan activity, and material, components, and assemblies are only locations for activity.

CMMS systems are appropriate when activity and its details are the focus. Just as MRP and MRPII have become important in manufacturing, CMMS has become important in industries where maintenance or monitoring activity with an enduring asset ensures smooth delivery of a product or service. Some examples are railroads, bridges, or power generation plants.

CMMS systems assume that assets and their internal components and assemblies have fixed life cycles based on chronological time, operational time (such as service hours) or mileage. The underlying assumption is that life cycles are a function of equipment-environment interactions and that these interactions must be tracked in order to empirically derive and predict the length and nature of life cycles. Therefore, most of these systems are built to collect data from work orders (to do something to equipment) usually triggered by a "symptom." Over time, these systems link symptom types with defect types, and ultimately link both to a piece of equipment or one of its components. The systems then use this information to predict the time span of "natural" life cycles and plan when equipment needs to be replaced or serviced due to their completion. The ideal assumption underlying these systems is that the entire out of service time of a piece of equipment due to component failure can be predicted and prevented by these systems. Failure trends are then used to plan pre-failure "change outs," ensuring virtually seamless operation.

The approach represented in CMMS is very much opposed to traditional methods for asset maintenance, which are—at least formally—highly reactive. As with aggregate material planning methods, reactive methods of maintenance are also an outgrowth of post-WWII scarcity. They also assume that life cycles of equipment are unpredictable and that the most cost-effective approach to maintenance is to milk an asset for all it is worth by running it to failure. Now that components can be procured easily, however, reactive methods of maintenance are considered to be unnecessarily costly. In fact, recent data provided by the Society of Automotive Engineers indicate that in the 1960s the parts/labor ratio was 2/1. In the 1990s the ratio completed a full reversal (parts/labor = 1/2). Maintenance practices that emphasize reactive repairs also require redundant systems, large spare factors, and significant loss of revenue opportunity when equipment is down for repair. Further, when large fleets of buses or trains support the economic functioning of large metropolitan areas, and reliable service is expected, running to failure introduces unacceptable uncertainty.

Important as they are, technologies such as those described above have enjoyed mixed success in workplaces. Current literature on their failure (e.g., Boldt, 1994, 2000) indicates that large information technologies are hard to implement. Typical implementation times for the introduction of complex information technologies such as MRPII and CMMS are on the order of 12 to 18 months (per site) and success rates have been low, especially for systems that offer the opportunity for company-wide resource management through planning. For example, studies of technologies of this kind by the Gartner Group and others (for the transportation industry) have shown that as many as 50% of new systems are abandoned in the first year and possibly 90% never reach their full potential. In the manufacturing arena, success rates are as low as 20%. I have been told by senior managers for vendors of MRP systems that these methods have a return rate as high as 76% for the software.

Most businesses now recognize that the failure of these kinds of technologies is essentially a "user" problem rather than a technology problem. However, we think that businesses, especially, grossly misunderstand the nature of this user problem. The most striking misunderstanding involves the idea of "resistance."

For example, one commonly given reason for the failure of these technologies is workers who fear and are resistant to change. In general, most firms now acknowledge that technologies such as CMMS and MRP require 99% data accuracy and are sensitive to level-of-detail issues. Therefore, many attempts have been made to deploy technologies among frontline workers (such as mechanics and assemblers) who are in contact with the details of day-to-day operations or have detailed knowledge of equipment. However, this group has not responded well. Often, they are seen as not having adequate computer skills. When training has been attempted, frontline workers have not learned much from the (usually vendor provided) classroom instruction. They may not learn what is needed, do not transfer what they learn to practice, or resist the training experience itself.

I propose that a cognitive analysis of behavior that looks like resistance yields a deeper understanding of how people change and learn and offers greater opportunities for productive technology deployment. For example, frontline workers' resistance to technology instruction is actually partly rooted in their success as craftspersons, which both selects and develops a learning style that is based on problem solving, experimentation, and hands-on contact. As will become clear below, when pedagogy is modified to fit their learning style, frontline workers show themselves to be superior learners with resistance acting as a catalyst.

THE THEORETICAL PERSPECTIVE

The focus of my work concerns the cognitive impact of the introduction of technology into the workplace. Specifically, I am interested in exploring how workers' ways of thinking and understanding are affected by changes in the nature of work and workplace organization. Many of my questions have been addressed using a number of different models, such as "novice-expert shift" (e.g., Chi, Glaser, & Farr, 1988), "situated cognition" (e.g., Rogoff & Lave, 1984), or "naturalistic decision making" (e.g., Orasanu & Connolly, 1993) and my work has been influenced by the methods and theoretical models from all of these various approaches. However, since the focus of my inquiry concerns the *development* of different ways of thinking in different domains, my research has been most influenced by the theories and methods of developmental psychology and particularly by the developmental theories of Vygotsky (1987) and Scribner's application of them to workplaces and workplace cognition (e.g., as summarized in Scribner et al., 1996).

Cognition and skills develop in the service and support of activities at work (DiBello, 1996a, 1996b). This is the principle difference between school learning and ongoing learning at work. As one participates in a particular industry or occupation, specific strategies and ways of understanding the business at hand are selected and reinforced as they prove over time to have a direct bearing or accomplishing important goals (DiBello & Kindred, 1992; Scribner et al., 1992). Workers typically understand they are learning the right things when they are more fully able to participate in meaningful problem solving and are recognized for their value by their coworkers (Scribner et al., 1991; Scribner et al., 1996). This set of skills and these ways of understanding work comprise the culture of any workplace. Over time, the culture of practice takes on a life of its own, being passed on to new workers as they "learn the ropes."

When looking at workplace culture as really being about skills developed collectively (and over time) in service of accomplishing goals, it becomes clear why the culture of a workplace becomes the main impediment when a widespread rapid change in business practice is being introduced. This is even more the case when the sudden changes are represented in an information technology that affects every job. A long-standing culture of practice can become suddenly obsolete, at least in part.

When a change is being introduced, change agents (i.e., new management, consultants or a process improvement team) will often disregard any usefulness that previous strategies may add in the process of change. They are often unaware of the important role that prior knowledge can play in the "new" vision (Chamberlain & DiBello, 1997). Their strategy is often to replace all legacy practices, by either "selling" the change or eliminating key resisters. This does not acknowledge the importance of content knowledge employees accumulate over the years. The process of integrating useful aspects of legacy skills with practices that support new and changing business goals is required for any positive change (DiBello, 1997a, 1997b).

The Role of Constructive Activity in Learning

In my first studies of workplaces undergoing technology changes my colleagues and I made a small discovery at a plant north of New York City that influenced a great deal of my subsequent work. In a study of workers using MRP (Scribner, Di Bello, Kindred, & Zazanis, 1992) in two different factories—one with a successful implementation and one with an unsuccessful implementation—classroom instruction was shown to be a poor way of preparing workers to use MRP effectively at either plant (DiBello & Glick, 1993; Scribner et al., 1991; Scribner et al., 1992). Despite this, at one plant many individuals managed to master MRP and reduce their inventory by 72%. It turned out that on-the-job activity proved to be critical to developing the necessary skills. An analysis of day-to-day job activity by people in three comparable titles and levels of responsibility revealed two distinct patterns of activity: constructive and procedural. Briefly, *constructive* activities are those that have clearly defined goals and poorly defined means. The employee is compelled to "construct" a procedure, form, tool, or artifact that accomplishes some meaningful goal in an iterative fashion. In contrast, *procedural* activities are those that have clearly specified means and order of execution but goals that are either clearly conveyed or not. Important to note, constructive activities were associated with an in-depth understanding of MRP's underlying logic while procedural activities were not. In fact, when several variables—job title, years of experience, level of formal education, and number of opportunities (weekly) for constructive activities—were correlated with measures of in-depth grasp of MRP principles, only number of opportunities for constructive engagement was found to be significantly associated with mastery ($r = .69$, $p = <.01$; see Di Bello and Glick, 1993, for discussion). However, this study also showed that opportunities for constructive activities are usually fortuitous and ill structured. For example, they often occur because the person who knows what to do has left the job without documenting procedures for others to use.

After this study ended, my aim was to better understand the role of constructive activity when it comes to technology and to find the means to increase opportunities for it in the workplace. In the effort described below my colleagues and I tried to systematize opportunities for constructive technological activity through specially designed exercises. These exercises were developed to help employees better understand that they were participating in a particular set of practices that may have become obsolete, and to help them construct a new set of practices more relevant to their company's goals. As becomes clear, my colleagues and I found we had to develop an in-depth understanding of the company's legacy domain of practice in order to design these exercises.

MRPII and Transit Workers

The initial attempt to bring the benefits of accidental on-the-job constructive activity into an intentional intervention involved transit mechanics learning MRP. As I have already detailed in other publications (e.g., DiBello 1996, 1996b, 1997a), we provided real workers with an opportunity for accelerated constructive activity in manipulative simulations in which they gained an understanding of MRPII by having to construct a manual version, act as MRPII using manual means, and then implement whatever plan they had made. One small study was sponsored by the Spencer Foundation and conducted in 1993 among mechanics in the compressor shop at the New York City Transit Authority subway department. The outcome was that the typical long learning curve for MRP systems was greatly shortened among personnel who normally would not be targeted as users. This occurred because designing a very simple exercise simulating the mechanics' workplace and inventory concerns worked well and seemed to bypass the need for prerequisite knowledge of computers. In fact, the exercises did not focus on the "computerness" of MRPII, but rather on the conceptual differences between aggregate planning and MRP methods of planning. In previously published articles about the experiment I have explained how this training activity is different from other simulation training in two specific ways: 1) the hands-on nature of the simulations provides more avenues of engagement; rather than being a virtual workplace on a computer screen, physical production and manufacturing are required and 2) participants first go through the simulation with very little guidance, in other words, although constraints and goals were made clear, procedures for accomplishing them are not. Participants in the exercise were constrained to select the development of MRP methods according to the goals and with the resources (or tools) provided; all manner of methods were available (including MRP-like and traditional planning sheets and the information needed to use them) but the miniature business could only operate within budget by using MRP methods of material planning.

As elaborated elsewhere, in-depth knowledge of conceptual domains is constructed. However, part of this construction is a kind of "deconstruction" of existing expertise. This leads to the third difference with the hands-on simulations: in order for construction to happen properly, it seems vitally important that the simulation engage the participants' implicit ways of thinking (by introducing time pressure) and allow them to systematically fail.

The idea here is that learning via construction is actually about reorganizing existing knowledge, and existing intuitive expertise could not be reorganized for a new purpose without significant engagement and, ultimately, an activity-based challenge. In a sense, the challenge weakens the a priori nature of expert knowledge (as the learner notices the failure and begins to reassess the situation), and, therefore, fundamental reorganization in one's ways of thinking may have to involve failure.

THE MAINTENANCE INFORMATION DIAGNOSTIC ANALYSIS SYSTEM EXPERIMENT

For many years, NYCTA management wanted to implement a centralized life cycle-based maintenance system, or CMMS, as described in this chapter. Transit professionals had long known that CMMS could help them reduce costs and increase service, but few successful applications existed, and none of them were in the public sector. NYCTA made a number of heroic attempts to bring this approach to its maintenance divisions. Manual systems proved unwieldy, however, given the size of the fleet (over 8,000 buses and subway cars), and it was widely acknowledged that early information technologies failed for many of the same reasons cited for MRPII failure. The information needed to make the system work had to be extremely accurate at just the right level of detail. Ideally, the information needed to be inputted by the mechanic him- or herself. However, efforts to train mechanics on computers were not successful. At NYCTA, many of the workers neither spoke English as a first language (about 80%) nor knew how to use a computer keyboard. Also, frontline workers, in general, usually threatened by information systems on the shop floor, can cause widespread system sabotage or damage to expensive computer equipment.

Eventually, with the availability of powerful relational databases, interest in using new technology for maintenance purposes was renewed. In the early 1990s, NYCTA began planning its version of CMMS and called it a Maintenance Information Diagnostic Analysis System (MIDAS). At the onset of the design process, senior management in the Department of Buses decided to have the hourly Bus Maintainers, Class B (BMBs), enter their own repair data into the system without clerical assistance. There were two reasons for this decision. First, budget cuts forced a reexamination of redundant work; asking mechanics to record information in longhand and then have clerks type the same information into a computer represented a particularly costly practice. Second, considerable evidence showed that original handwritten records were much more accurate than what clerks (or supervisors) eventually entered. Therefore, senior management moved all data entry responsibility directly to the shop floor. The decision was occasion for considerable nervousness in middle-management. In general, this approach had never succeeded anywhere except in a few private transportation companies (such as UPS), where workers are carefully screened before hiring. Since making floor workers responsible for a management technology had never been tried before on such a large scale, the MIDAS team believed that any traditional education approach based on this very different kind of user would be inappropriate at best.

Our relatively minor success with MRPII and the "compressor gang" in 1993 were the impetus for making large-scale, frontline computer systems a reality in the NYCTA Bus Division. Specifically, senior management saw our project as successful in achieving mechanic acceptance and for mitigating system sabotage. During our first conversations, management did not recognize that the mechanics accepted the system because they had learned its business purpose and were using it as a tool for their work. Nor did they agree that user knowledge of the buses might be critical to the successful use of the system from a management perspective. That is, they did not recognize that understanding the reason for the system might affect the quality and nature of the data entered by mechanics, and that this level of quality would, in turn, affect the analytic results of the system's pattern analysis capability. In short, the frontline mechanic was not seen as a person with knowledge of the buses that could be critical to cycle identification.

Rather than attempt to convince management that these factors were important, my colleagues and I proposed a "training pilot" at one location, ostensibly to increase user acceptance and prevent system sabotage. The project we designed and eventually rolled out to 19 locations actually addressed user understanding of the reasons for the system. In fact, our exercises engendered the theory behind cycle-based preventive replacement and techniques for trend analysis and planning. That is, rather than emphasize procedures for using the system, we emphasized conceptual context. We believed the mechanics' mental model of maintenance was the root cause of their resistance, or poor learning, in the first place; we thought that once they understood life cycle-based maintenance theory and the system's assumptions, the computer would seem like any other tool in the shop. Further, to add to our own research, we designed measurements to examine the relationship between user knowledge, data quality, and financial impact.

Learning About the Culture of Fixing Buses

As indicated earlier, the entire study rests on the assumption that efforts to change a workplace culture most often fail because there is an already-functioning, cohesive culture that is actively competing with the change. In order to effect change, one must know as much as possible about the competing culture of practice.

On the surface, it seemed, in the case of the NYCTA study, that our "competition" was reactive maintenance and the attendant belief that parts do not have natural life cycles. However, this still did not tell me what actual practices instantiate these beliefs. From my experience, I knew that on the frontline of the business (usually the shop floor level) the picture is more complicated. Often there is some tension or inefficiency at the front line of the business that has led decision makers to consider alternatives. In these situations, legacy methods are already failing to meet challenges and new things are being tried. This informal domain of practice is usually the real source of competition and the real source of culture change failure.

Many ask how to "get at" the legacy domain of practice. Very few people in a given workplace are explicitly aware of the dominant domain of practice, but most are aware of when they operate effectively within its parameters. They know who is effective, who knows what is going on, and they are able to assess the meaning and significance of situations that are baffling to outsiders. The trick is to tap into the ways that these workers understand their workplace and its business. There is ample reason to believe that people who have implicit expertise in a given area are not the best at narrating their processes of working and making decisions (Dreyfus, 1997), especially in dynamic settings (Klein, 1999) such as vehicle maintenance.

Therefore, in order to understand more about how mechanics actually think about the business of fleet maintenance I felt we needed to begin by observing them on the job, but in such a way that led to understand from their point of view what it is like to do the job. In order to make this a natural and comfortable observation while still allowing us to ask questions as they worked, my colleagues and I each did our fieldwork as a "quasi-apprentice." In this role, it is normal to ask questions, want explanations for decisions, and be curious about the underlying reasons for doing things. Also, it put the experienced worker in the position of "master" or "teacher," which are roles they have to assume many times when breaking in new workers.

Following is one sample of the recorded dialogue taken during an observation of a periodic inspection.

Ed: Well, there is not much to this. We just go down the checklist. Nothing to it really.

Lia: So we start at the top and just go down . . .

Ed: No, I don't do that. I mean, I skip around the list.

Lia: Why is that?

Ed: Well, the order doesn't make sense. See that guy back there (points to rear of bus), I'll be in his way if I start back there. And if I follow the list exactly, I'll be running around the bus all day, literally. So I begin with the things in front. And since I have it up on the lift, I begin with the things underneath first.

Lia: Okay.

Ed: (Looking at steering arm bushing under bus.) Here, hold this flashlight for me. (Picks at dirt and rust around bushing.)

Lia: What's that?

Ed: that's the bushing. What's bothering me here is that it looks like some rust here. That's not good. Shows me there's a problem. Let's look and see when this is due back in. (Looks at schedule of inspections and picks more at the dirt and rust around bushing.)

Lia: What's up?

Ed: Well, see this bushing over here. Shine the light right here. This is good. See, no rust mixed in with the dirt. Now look at this one. There is some rust in here. But not too much. Not very red. See that? (Researcher sees no difference.) That bushing really needs to be changed. But given that this is coming in in 3000 miles for an A inspection, we can take care of it then. It's got at least that much time on it left. And they need this bus this afternoon. It's gotta wait. So we will make a note of it.

Lia: How do you know it has another 3000 miles left on it.

Ed: Well, it's obvious. By the color of the dirt. The amount of rust in there.

As can be seen from this transcript, even though the mechanic reported on an earlier occasion that he doesn't "think" but rather does what he is told to do, there are a significant amount of situation assessment, analysis, and information coordination (and life cycle-based maintenance) being done here. What this and other observations reveal is that experienced mechanics have an intuitive understanding of the life cycles and the coordination of life cycles among components within one piece of equipment. In other words, there is already an informal culture of preventive, coordinated maintenance operating when the formal practices of reactive maintenance threaten the depot's ability to make service requirements. However, it not yet explicit or consistent and has a "plan B" status as a practice.

I also was able to observe how the mechanics learn during the course of doing their work. None of the workers we observed considered themselves to be strong classroom learners or "read and write" types. Most contributed to their own ongoing learning by "puzzle solving," and when stumped, drew on the opinions and observations of peers to help them understand the equipment through systematic group experimentation.

These two observations—the existence of an implicit scheduled maintenance domain of practice and the mechanics' evolved method of learning—greatly influenced the next design decisions of the research project, that is, the cognitive probes designed to tap into individuals' ways of thinking about maintenance and the training exercise to move people into a new way of thinking.

The Construction of the Cognitive Probes

The next task was to understand the mental models being used in the daily business of doing work. I had found that the best way to do this is at the level of the individual using cognitive probes. These probes are very similar to those originally used by Klein et al., 1989, and resemble in spirit his critical decision making interview method. However, there are some important differences. Klein's method is an attempt to get at an expert's implicit knowledge and situation assessment skills by asking him or her to tell a work history story and explore the methods by which he or she reasons it through. Our approach was to constrain the problem-solving context and see how our interviewees view and handle the constraints we have defined. This involved setting up the problem and the tools available for solving it in a uniform way, while at the same time having a situation that invited the interviewee's implicit skills and situation assessment biases. The method involved the following steps:

1. Identify the strategies and practices associated with each domain that make sense only within the worldview of that domain. For example, in vehicle maintenance catalogue all the strategies associated with proactive life cycle maintenance (as one domain) and all the strategies associated with reactive run-to-failure practices (as a contrasting domain).
2. Identify behaviors associated with the strategies in the workplace in which I am doing the research.
3. Design a meaningful problem situation that can be solved using the strategies and behaviors from either domain, or from a mix of both.
4. Design one or more additional problem situations that are similar to those in step 3 but which are more abstract and generic than the site-specific versions.

5. Develop a scoring form that permits a coder to easily check off the strategies/behaviors and calculate the proportion of the strategies used from each domain to produce a solution.

For NYCTA, my colleagues and I constructed two basic tasks, each of which had three variations. The first task was an "active" task: given a pile of work orders, we asked the interviewee to look them over and then make five piles for each day of the work week. In other words, schedule the work. Below is a small sample of the strategy/behavior pairs for solving this task arranged according to domain of practice.

The same task was also given in two other forms: using another piece of equipment that is commonly known (bicycles) and using "Machines A-N," which were purely made up items with meaningless codes as defect or component indicators (such as defect Mu8).

TABLE 5.1
Strategies for Scheduling Work

Cycle-Based Scheduled Maintenance	*Traditional Reactive Maintenance*
Strategy: Coordinating work within asset **Behavior:** Interviewee sorts work-order cards first by equipment ID number. Or asks: can the same asset be taken out of service only once to satisfy multiple problems?	**Strategy:** Coordinating work by type or craft **Behavior:** Interview does initial sort of work-order cards by type of job or type of trade needed to do job, regardless of equipment ID number.
Strategy: Maximizes inservice time **Behavior:** Interviewee compares number of assets coming into shop with number needed for service. Assigns work accordingly. Brings in asset twice on two different days only as necessary to make service.	**Strategy:** Coordinating work within shop capacity **Behavior:** Interviewee distributes work-orders evenly among the days, regardless of the type of work needed to be done
Strategy: Identifying component life cycles based on empirical data **Behavior:** Asks if there are any historic records available that might help differentiate "normal" wear from "abnormal" failure.	**Strategy:** Attempts to reduce maintenance costs by identifying defective components or warranties **Behavior:** Asks if there are historic records in order to determine if a component was recently replaced and is therefore defective or under warranty.
Strategy: Coordinating cycles with each other into clusters **Behavior:** Looks at "what else has been failing" on work histories and speculates about clustering preventive replacement for components with similar life cycles.	**Strategy:** Assumes no life cycles but recognizes "infant mortality" **Behavior:** Looks at the symptom remarks on the work-orders to ascertain if problem is repeater and if part will need replacing.

A second group of tasks also took three forms that were more passive and required the interviewee to interpret information. The objects varied in the same way as the first set: interviewees interpreted bus repair histories, bicycle repair histories, and those for "machines A-N." A similar set of strategies (for interpreting data) to those shown above was used to code the protocol. Photographs were taken of the interviewee's piles and of any drawings or writing and all talking and "thinking aloud" were audiotaped.

Within each task, about 70% of each interviewee's strategies were reactive and about 30% were proactive, indicating that some proactive planning skills had developed in the workplace. There was also striking homogeneity among interviewees in the pilot depot, suggesting a strong workplace cultural effect.

An Education Process That Would Enable the BMBs to Do Data Entry

Based on the fieldwork and the cognitive battery results, my colleagues and I decided to construct a three-part manipulative simulation of a miniature depot, constraining the goals and resources in such a way that, in order to "win" (i.e., make service requirements and stay within budget), the participants had to use proactive strategies that were logically consistent with CMMS (and MIDAS in particular). Our previous work suggested that constructive activities in real workplaces lead to learning because they elicit the implicit knowledge that the worker has to bring to the problem and at the same time select against nonworkable strategies (through experiences of failure). Therefore, the first part of the exercise was designed to "engage the default" within the context of new business goals.

Specifically, teams of 8 participants were asked to run a depot of 40 plastic buses with relatively complex interior components. The goals were to maintain 32 buses in service at all times (limiting the number out of service to 8), order all the materials (within a budget) needed for doing so, and evaluate daily operator reports (each "day" being 20 minutes) that might indicate potential problems (e.g., noisy engine). The activity was rigged so that the only way to meet these goals was to predict what was due to break next. The breakdown patterns of all components followed time/mileage cycle rules and were precalculated using a computer. The toys were then actually "broken" according this pattern. The participants were given adequate tools to predict and calculate this breakdown (including printouts of every bus's repair history, among other items), but were given other tools as well, including those similar to those used to do "reactive" maintenance.

The trainers also played a role. One acted as dispatcher, regularly demanding buses to satisfy routes, while the other acted as a parts vendor and a Federal Transit Administration inspector, looking for safety violations or abuses of public funding, such as overspending or cannibalizing.

Despite loud disclaimers, people tended to construct solutions to even novel problems that fit with their experience, even when explicitly instructed to avoid doing so. In fact, the participants were rarely aware they were replicating their normal methods.

Rather than interfere with this tendency, the trainers allowed the participants to wing it, while carefully documenting the cash flow, labor flow, inventory acquisitions, and the number and type of on-the-road failures that resulted from failing to predict problems. Meanwhile, heavy fines were levied for expensive reactive problem-solving strategies, such as cannibalizing an entire bus for a few cheap parts in order to get other buses back on the road. As the activity progressed, participants were continually shown the financial consequences of their decision making patterns and asked what they were thinking by the vendors/inspectors and dispatchers. In general, by the end of the first day of these sessions, the depot was in crisis and the participants realized their budget was being expended to react to mounting problems. At that point, the activities were stopped and the team was sent back to work or to lunch.

On the second day of the exercise, participants reflected on what they did, as recorded by the trainers. The participants were asked to discuss among themselves what thinking led to various decisions and to begin to identify practices that lead to bad outcomes versus practices that are preventative. Only at this point were participants truly open to new ideas about how to solve the problems of vehicle maintenance. They also began to understand in detail the ways that their "gut feel" decisions revealed how they have actually misunderstood preventative maintenance.

In the last hour of the second day, the trainers facilitated the participants in building a manual scheduled maintenance system. The participants identified cyclical patterns from histories (which were available from the first but which now took on new meaning) and set up predictive data structures, identifying true cycles and—most importantly—coordinating cycles, so that their system brings in a bus only once to satisfy several cycles at the same time. For example, the participants quickly realized that a 15,000-mile cycle and a 30,000-mile cycle can be coordinated so that at least half the time the 15,000-mile cycle co-occurs with a 30,000-mile component cycle. The participants were then given materials to construct a maintenance allocation chart for the whole fleet over a number of months and evaluate the stress this would put on the shop. After doing this, they entered these data on an actual test region in MIDAS and created and assigned the preventive work orders according to this schedule.

During the second day, the participants completed their data entry and printed out their work assignment sheets and work orders. They ran their miniature depot again using MIDAS and saw the difference in profits and ease of workflow. Usually after only 5 "days," the team could afford to buy an additional bus to add to the fleet and thereby increase their fare-box revenue.

The last activity of the workshop involved entering the data on work orders (paying attention to detailing the components, defects, and symptoms involved) and closing out both work orders and work assignment sheets. At this point, participants also learned how to get various reports they now realized they wanted, such as a 30-day history on a bus. Most participants were no longer thinking of MIDAS as a computer per se, but rather a tool for doing what they had been developing manually over a number of days. After operating as MIDAS and then with MIDAS, participants navigated through the actual system more easily, knew what to look for, and asked informed questions. Even computer illiterate individuals showed little hesitation when exploring the system on the third day.

As indicated above, about 80% of the mechanics were not native English speakers and fewer than 20% were computer literate. Many midcareer individuals had not completed high school. None wanted to attend the training and most were resistant to the idea of having to do their own data entry on the first day of training. Despite these features, the trainees mastered the system at record speed. Rather than requiring the expected 12 months for implementation, the hourly staff reached independence with the system in 2 weeks and line supervisors (who do more) took 6 weeks. The one exception occurred at a location that received classroom training but no simulation exercises (this last site was not included in our original contract scope of work). After 8 months, the implementation was declared a failure.

After examining the success of the project and comparing it to the one failure, my colleagues and I believe classroom approaches (which involve explanation, simplification, and instruction) have not worked because individuals have different prior perspectives that must each be taken into account. It seems that learning through "constructive activity" actually involves building on, or reorganizing, the way that one already understands something. Therefore, it is critical to engage prior knowledge, if only to make sure it is eventually changed or reorganized. In this context, resistance to learning may actually be best understood as the assertion of existing expertise and may actually be necessary to learning. Whenever a teacher simplifies material for his or her students, he or she is really anticipating the entry point of the learners. This method often fails with experienced workers because the entry point is not always predictable or universal (for example, *simplified* is often not helpful for those experienced in thinking through vast amounts of detail). The learners—when allowed—actually do better at breaking it down for themselves in a way that is useful to them. What looks like resistance is actually an attempt to construct an entry into a new way of understanding something by beginning with what one has. In a sense, a "wrong" idea is used as raw material for a new idea, with the challenges of the exercise acting to remold the operating knowledge of the learner.

EVALUATION

Because the participants in the exercises were depending on us to orient them to the system, we could not arrange for a true control group. Rather, those who were out of work on the first day of their scheduled exercise (due to illness, personal days, or other reasons) were rescheduled to go through the exercise after everyone else and were measured as a kind of control group until they were trained. When the system went into use, they did receive classroom-based vendor training that differed from the others' only in that they did not through our exercises. There were 12 of these individuals out of about 150 total participants at the pilot site (which is where the cognitive probes were conducted).

After six months, we conducted cognitive postprobes on samples of the participants from both the pilot site (the same people who were preprobed) and the controls. Figure 5.1 shows the cognitive battery results of the pilot trainees. While the controls produced the same profile as they (and all mechanics) had before the training, those who went through our exercises produced a mirror image result compared with their former approach. Rather than solving the scheduling and data interpretation tasks with a reactive dominant approach, they exhibited about 70% of their strategies in the proactive domain. Further, when asked about how this compared to their prior performance, most did not remember doing it another way, and several could not replicate their former solution to the problem.

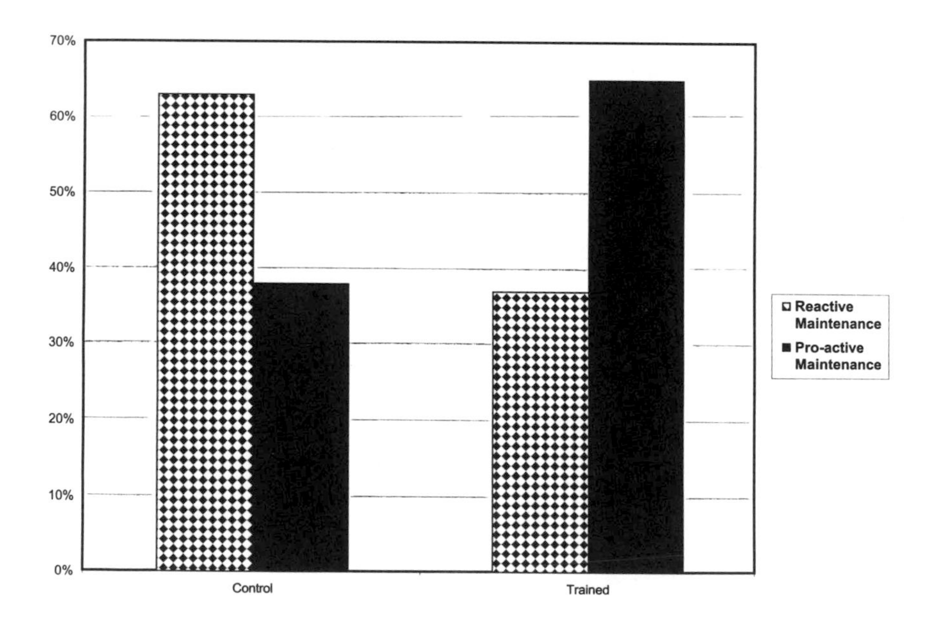

FIG. 5.1. Cognitive battery after 6 months.

Data Entry Patterns

Two measures used to determine the quality of the system use concerned the data inputted by the mechanics. I know from industry standards that coding the location of an equipment defect at a subsystem level is required for trend analysis. When users do not understand the level of detail required, they code at too general a level for the data to be useful. Figure 5.2 shows that our trainees were coding at the fourth level of the equipment template most of the time, which indicates that they are identifying a root cause component rather than a more general assembly. This is an unprecedented result in the transit industry. It indicates that the users were knowledgeable of the uses the system makes of the data and were coding appropriately. As can be seen, the control group's codes reflect a flatter, more general pattern that does not support root cause failure analysis. This indicates that the control group users do not understand the purpose of the data entry (although it was explained to them during classroom kinds of training).

Code Variation Measures

Downloads of workers' navigation through the system and data entry practices were also analyzed for component code variation and homogeneity. For the first, I measured the frequency with which any component code was chosen from a finite universe of about 2,000. In general, the data from systems such as MIDAS have been considered poor in quality, or inaccurate when the same symptom, defect, and component codes are chosen over and over because they are both general and

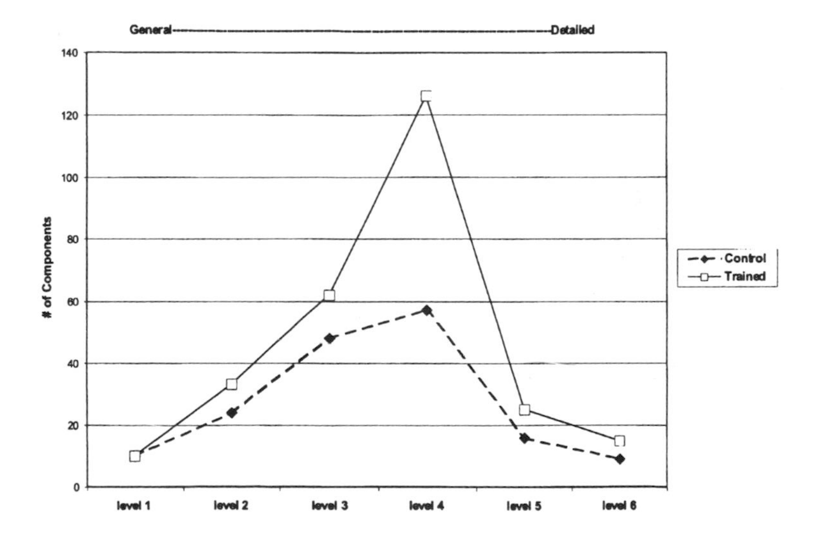

FIG. 5.2. Level of component detail.

easy to remember. An example of inputting a code would be "malfunctioning" (symptom), "broken" (defect), and "fuel system" (component). When analyzed by the system for patterns that indicate life cycles, these kinds of data are basically useless. When users understand this, they tend to code with more specific kinds information. A more detailed example of coding would be: "sporadic power surges" (symptom), "cracked" (defect), and "Injector valve-aft" (component). Simply put, a low average component code frequency indicates greater variability. Therefore, my colleagues and I looked for low hit rates per code, per user.

Figure 5.3 below shows the average frequencies per code, per person in two groups: the trainees and the 12 controls. The trainee group maintained the lowest frequency while the control group scored higher in both frequency and variance.

The low standard deviation among the trainees suggested a homogeneity effect. As a test of homogeneity, we conducted a Scheffé test of the standard deviation. The significant Scheffé indicates there was considerable within-group consistency in the type of detail entered even though the input was much more complex.

These data were initially collected on the 200 maintenance personnel at the pilot site. The striking success of the exercise was used to make a compelling case for deploying MIDAS among all 3400 frontline maintenance workers. For the full rollout, the training exercise was conducted with over 3600 people in a period of about a year. The evaluation process continued as well, with monthly downloads of data entry patterns being analyzed for each mechanic at each site as each location began using the system. The data were analyzed for two years after the rollout began and no degradation in quality was seen. Further, once the full scale implementation was underway, we also evaluated MIDAS success by measuring Mean Distance Between Failure (MDBF).

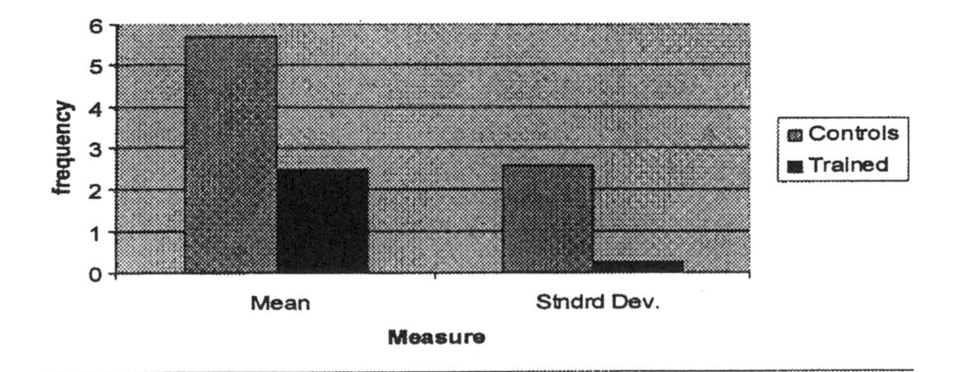

FIG. 5.3. Component frequency.

MDBF

MDBF is obtained by taking the number of in-service vehicle failures divided by the in-service distance traveled.

Rises in MDBF mean that the revenue earning asset is out earning money and is not incurring maintenance cost from repair labor. Simply put, therefore, the higher, the better. As can be seen from Figure 5.4, the MDBF rose system wide at the same rate that the MIDAS mechanics participated in the exercise. The savings from the increased MDBF are estimated to be about $40 Million. The savings in field supervisor time (handling the return of broken-down buses) is estimated to be 208,000 hours times a fully loaded hourly rate of $70, or $14,560,000. These numbers represent the financial benefits that occurred in the first year, before there were enough data collected to support the trend analysis needed for true preventive replacement based on life cycles. That analysis is just beginning.

DISCUSSION

The main point is simple: the way that workers understand their work and their role in the workplace acts as a kind of operating theory that affects how they do their job and what actions they choose at various decision points. Workplace

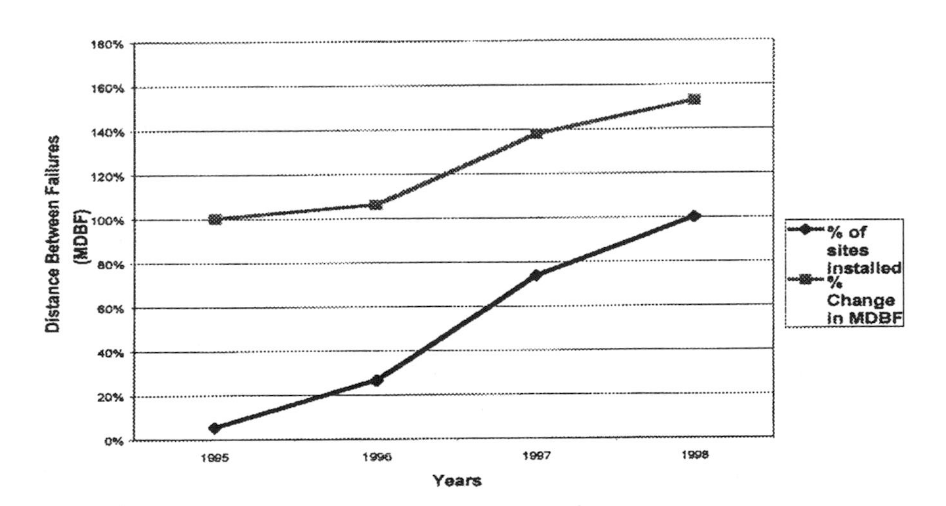

FIG. 5.4. Performance Increase with computerized maintenance management.

culture manifests as a set of activities, practices, and procedures that have evolved historically in response to having to accomplish important goals with specific resources. Ways of thinking also develop in response to goals and existing practices but in different ways depending upon experience. For new workers, the goal is to become a participant in a set of practices already in place for others; in a sense, their struggle is to understand the workplace in the same way as everyone else. For experienced workers, the goal is to get the work done and to contribute to maintaining, evolving, and refining the practices that accomplish this.

Only experienced workers can recognize the new opportunities that new resources provide, because they already intuitively contribute to practice and see the relationship between specific practices and specific outcomes. The barrier, of course, is that they have a goal-practice framework already in place, and struggle with the initial "gestalt switch" required to use their rich experiential knowledge in a new way made possible by a complex technology. Normally, their contribution is more gradual than complex technologies will tolerate. However, as the MIDAS project has shown, there may be more than one way to enlist the considerable knowledge capital of an organization for an aggressive change in goals.

As already indicated, after the MIDAS system was put into place, MDBF rose dramatically and, in truth, I do not really know why. Furthermore, system notes appeared that were increasingly in the private language of mechanics. In a sense, it might be said that the language of mechanics itself developed somewhat as a result of a wider reading and writing audience sharing the same conversation. In any case, as the mechanics grew more comfortable with the system, it became harder for us to know what they were doing with it. In other words, they grew beyond us in their understanding of what the data were saying.

There is an excerpt from the "notes" section of a work order at the New York City Transit Authority:

> Worked on 7016, which came from ENY minus the following items: one entrance door partition, one station upwright and grabrail, one dome light partition cover and front dest sign lock. Remove dest compart locks from bus 7033—which is waiting for other parts—to meet req. <u>All other items listed were obtained from spare buses at yard.</u> Tap-out damage Riv-nuts installed new ones on same. Interior close tbc.

There are two striking features in this passage. The first is the admission of cannibalism, stealing parts from one bus to get another into service, a practice that could have led to dismissal before MIDAS was implemented. Using MIDAS, mechanics soon realized that indicating parts shortages in the components fields helped MIDAS correct parts ordering forecasts, making cannibalizing unnecessary. Telling other mechanics where the stolen parts came from helped them address missing parts problems in the cannibalized buses later. Other notes helped the mechanic on the next shift begin where the previous one had left off. The second striking feature is that I cannot decipher very clearly what is going on. In other words, the notes are not useful to nonmechanics. This trend became stronger as the system produced increasing financial benefits. What happened here is that the system has become a tool for the mechanics, and perhaps the absence of this practice had been the problem all along with failed technologies.

Possbily, the focus of design should be making new goals more visible to workers, with technologies acting as tools for examining how performance compares with or impacts on target goals. I suspect that MDBF is increasing at NYCTA because the mechanics have found a way to use repair histories (which can be accessed instantly by any user on any bus or system of a bus from any terminal) as feedback on their own diagnostic and repair decisions. Mechanics can now address a symptom a particular way and then follow the performance of a vehicle afterward to evaluate the result. The most important thing about this shift may be that it is entirely user controlled and initiated.

As with most works in progress, this issue will involve further study. Just as training turned out not be a straightforward issue, the feedback that supports learning has, and will continue to have, unforeseen subtleties as well.

REFERENCES

Boldt, R. (2000). *TCRP Synthesis of Transit Practice #35, Information Technology Update.*

Boldt, R. (1994). *TCRP Synthesis #5, Management Information Systems, A Synthesis of Transit Practice.*

Capelli, P., Bassie, L., Katz, H., Knoke, D., Osterman, P., & Useem, M. (1997). *Change at work: How American industry and workers are coping with corporate restructuring and what workers must do to take charge of their own careers.* New York: Oxford University Press.

Chamberlain, E. S. III, & DiBello, L. (1997). Iterative design and implementation: A model of successful scheduled maintenance technology deployment. *Transportation Research Record.* August 1997, no. 1571. pp. 42-49.

Chi, M., Glaser, R., & Farr, M. (1988). *The nature of expertise.* Hillsdale, NJ: Lawrence Erlbaum Associates.

Cole, M., Engestrom, Y., & Vasquez, O. (1977). *Mind, culture and activity: Seminal papers from the laboratory of comparative human cognition.* New York: Cambridge University Press.

Davydov, V. V. (1988). Problems of developmental teaching: The experience of theoretical and experimental psychological research. *Soviet Education 30.*

DiBello, L. (1997a). Measuring success in non-trivial ways: How we can know that a DSS implementation has *really* worked. In *Proceedings of the 1997 IEEE International Conference on Systems, Man and Cybernetics* 3(5) pp. 2204-2209.

DiBello, L. (1997b). Exploring the relationship between activity and the development of expertise: Paradigm shifts and decision defaults. In C. Zsambok & G. Klein (Eds.), *Naturalistic decision making* (pp. 163-174). Mahwah, NJ: Lawrence Erlbaum Associates.

DiBello, L., & Spender J-C. (1996). Constructive learning: A new approach to deploying technological systems into the workplace. *International Journal of Technology Management 11*(7/8), pp. 747-758.

DiBello, L. (1996). Providing multiple "ways in" to expertise for learners with different backgrounds: When it works and what it suggests about adult cognitive development. Special Issue of the *Journal of Experimental and Theoretical Artificial Intelligence* 8, pp. 229-257.

DiBello, L., & Glick, J. (1993). *Technology and minds in an uncertain world.* Paper presented at the Annual Conference of the National Society for Performance and Instruction, April 12-16, 1993, Chicago, Il.

DiBello, L., & Kindred, J. (1992). *Understanding MRPII systems: A comparison between two plants* (Tech. Rep. for Laboratory for Cognitive Studies of Work). New York: City University Graduate School.

DiBello, L., Kindred, J., & Zazanis, E. (1992). Third Year Annual Report prepared for the Spencer Foundation by Laboratory for Cognitive Studies of Work. New York: City University Graduate School.

Dreyfus, H. (1997). Intuitive, deliberative, and calculative models of expert performance. In C. Zsambok & G. Klein (Eds.), *Naturalistic decision making.* Mahwah, NJ: Lawrence Erlbaum Associates.

Dreyfus, H. L., & Dreyfus, S. E. (1986). *Mind over machine: The power of human intuitive expertise in the era of the computer.* New York: The Free Press.

Harrington, J. (1974). *Computer integrated manufacturing.* New York: Industrial Press.

Harvey, D. (1990). *The condition of postmodernity: An inquiry into the origins of cultural change.* Oxford, England: Blackwell.

Hedegaard, M. (1988). *The zone of proximal development as a basis for instruction.* Aarhus, Denmark: Institute for Psychology.

Heidegger, M. (1977). *The question concerning technology.* New York: Harper & Row.

Hoffman, R. R., Crandall, B. E., & Shadbolt, N. R. (in press). A case study in cognitive task analysis methodology: The critical decision method for the elicitation of expert knowledge. *Human factors.*

Klein, G. (1999). *Sources of power: How people make decisions.* Cambridge, MA: MIT Press.

Klein, G., Calderwood, R., & MacGregor, D. (1989). Critical decision method for eliciting knowledge. *IEEE Transactions on Systems, Man, and Cybernetics 19*(3).

Martin, L. M. W. (forthcoming). Introduction. In V. V. Rubtsov (Ed.), *Learning in children: The organization and development of cooperative actions.* Commack, NY: Nova Science Publishers.

Martin, L. M. W., & Scribner, S. (1991). Laboratory for cognitive studies of work: A case study of the intellectual implications of a new technology. In *Teachers College Record 92.*

Moll, L. C. (Ed.). 1990. *Vygotsky and education: Instructional implications and applications of sociohistorical psychology.* New York: Cambridge University Press.

Murnane, R. J., & Levy, F. (1996). *Teach the new basic skills: Princples for educating children to thrive in a changing economy.* New York: The Free Press.

Nardi, B., & O'Day, V. L. (1999). *Information technologies: Using technology with heart.* Cambridge, MA: MIT Press.

Norman, D. A. (1999). *The invisible computer: Why good products can fail, the personal computer is so complex and information appliances are the solution.* Cambridge, MA: MIT Press.

Orasanu, J., & Connolly T. (1993). The reinvention of decision making. In G. A. Klein, J. Orasanu, R. Caderood, & C. Zsambok (Eds.), *Decision making in action: Models and methods* (pp. 3-20). Norwood, NJ: Alex.

Rogoff, B., & Lave, J. (Eds.). 1984. *Everyday cognition: Its development in social context.* Cambridge, MA: Harvard University Press.

Postman, N. (1993). *Technopoly: The surrender of culture to technology.* New York: Vintage Books.

Scribner, S., DiBello, L., Kindred, J., & Zazanis, E. (1992). *Coordinating knowledge systems: A case study* (Res. Rep. for the Spencer Foundation). New York: Laboratory for Cognitive Studies of Work, City University of New York Graduate School.

Scribner, S., Sachs, P., DiBello, L., & Kindred, J. (1991). *Knowledge Acquisition at Work* (Tech. Rep. No. 22). New York: National Center on Education and Employment, Teacher's College, Columbia University.

Scribner, S., Tobach, E., & Falmagne, R. (Eds.). 1996. *Mind and social practice: Selected writings by Sylvia Scribner (Learning in Doing).* Cambridge, England: Cambridge University Press.

Sternberg, R., & Wagner, R. (1986). *Practical intelligence: The nature and origins of competence in the everyday world.* Cambridge, England: Cambridge University Press.

Timms, H. L., & Pohlen, M. F. (1970). *The production function in business: Decision systems for production and operations management (3rd ed.).* Homewood, IL: Irwin.

Winograd, T., & Flores, F. (1987). *Understanding computers and cognition.* New York: Addison Wesley.

Zsambok, C. E., & Klein, G. (1997). *Naturalistic decision making.* Mahwah, NJ: Lawrence Erlbaum Associates.

Vygotsky, L. S. (1987). Thinking and speech. In R. Rieber & A. S. Carton (Eds.), *The collected works of L. S. Vygotsky: Vol. 1. Problems of general psychology.* New York: Plenum.

Zuboff, S. (1988). *In the age of the smart machine: The future of work and power.* New York: Basic Books.

6

A Cognitive Approach to Developing Tools to Support Planning

Thomas E. Miller
Klein Associates, Inc.

Computer-based technology can rapidly generate detailed plans for complex situations. However, these plans may be rejected by planning experts, who judge dimensions such as the "robustness" of the plan using operational rather than computational criteria. Klein Associates' goal in this research was to first capture the tactical and strategic concerns of air campaign planners, then incorporate this understanding into planning technology to assist with filtering out the unacceptable options as a plan is being developed. Specifically, we focused on identifying characteristics of quality plans and how these characteristics are judged in operational settings. We relied on Cognitive Task Analysis (CTA) knowledge elicitation techniques to identify the process of plan evaluation and the factors underlying judgments of plan robustness. The research team drew on observations and interviews in a variety of settings. The primary data sources were from joint military exercises, in-depth interviews, and from a simulation exercise with Pentagon planning staff. The insights from the CTA formed the foundation of a software tool, the Bed-Down Critic, which highlights potential problem areas, vulnerable assumptions, and summarizes aspects of quality as the plan is being developed.

Supporting the tactical and strategic concerns as plans are developed and evaluated was central to this project. Technological advances in available computing power and advances in artificial intelligence have produced very impressive automated and semiautomated planning systems, yet a large gap remains between determining what is technically feasible and building systems that truly support the needs of the human planner. Many advanced planning systems focus more on the information architecture as opposed to the functional architecture of the human decision maker.

The air campaign planning domain presents an opportunity to move the Naturalistic Decision Making (NDM) field into the area of planning. Since inefficiencies in planning during the Gulf War have surfaced, the planning community has been expanding and evolving both conceptually and operationally to improve the process. Also, the sheer number of new technological developments has increased the potential for considerable improvements to the planning process. However, the complexities and intricacies of this domain require more than a technological solution. For example, the current time lag for developing a plan is too long. Shortening this period with current technologies and procedures will significantly increase the time pressure on human planners. Furthermore, there are many moving parts (e.g., aircraft preparation), many distributed planners, and constantly changing constraints on the plan (Klein & Miller, 1999).

One objective of this project was to understand and document the cognitive aspects of human planning at the strategic level. We specifically examined the planning done by strategic planners at the Pentagon and by Joint Force Air Component Commander (JFACC)-level planners. This understanding included the types of decision making that were necessary in order to develop sound air campaign plans. A second objective was to use this understanding as a basis for developing planning and evaluation tools that would support the human planner in creating more robust plans and that would take advantage of artificial intelligence technologies where appropriate.

METHODOLOGY

A core belief in the NDM community is that a user's cognitive experience of a task must be thoroughly understood before either systems or training approaches are developed. Rasmussen (1997) described three types of task models that can be used in designing systems. *Normative* models prescribe how a system should behave; *descriptive* models describe how systems behave in practice; and *predictive* models specify requirements for systems that behave in a desired way. Vicente (1999) summarized that researchers from the anthropological (Hutchins, 1995; Suchman, 1987), activity theory (Bodker, 1991; Nardi, 1996),and naturalistic decision-making research communities (Klein, Orasanu, Calderwood, & Zsambok, 1993; Zsambok & Klein, 1997) have all pointed out that users do not, and should not, always follow normative prescriptions. These researchers rely on descriptive approaches to current tasks to identify opportunities for improved performance through better system design or through better approaches to training.

In this work, we used the descriptive approach to modeling air campaign planning. We employed CTA to allow us to understand the cognitive aspects of those who plan the air portion of large-scale military campaigns. In particular, these aspects included the judgments and decision-making and problem-solving skills that are critical in the time-pressured, uncertain, and ever-changing air campaign planning domain. CTA provides the methods for eliciting general domain knowledge and specific knowledge pertaining to the cognitive requirements for the critical decisions and judgments made in the air campaign planning environment. As illustrated in Fig. 6.1, specific CTA methods, coupled with the NDM theoretical orientation, lead to the identification of cognitive requirements for planning and, ultimately in this project, to the design of a planning tool.

To allow us to understand both the high-level conceptualizations and the low-level information gathering and interpretation that takes place in this domain, the team used several specific knowledge elicitation tools, which are reviewed next:

1. *observations* at joint military exercises,
2. *interviews* using the Knowledge Audit and Critical Decision Method (CDM), and
3. a *simulation exercise* titled "Air Counter '97."

Exercise Observations

We attended three live exercises where Air Force and Navy planning staffs developed air plans that were actually flown during the exercise. The first exercise we attended was "Roving Sands," where we had one observer stationed with the planning staff at Roswell, NM. Because this was the first exercise we observed, it was used primarily to make contacts within the planning community and to become familiar with the planning process.

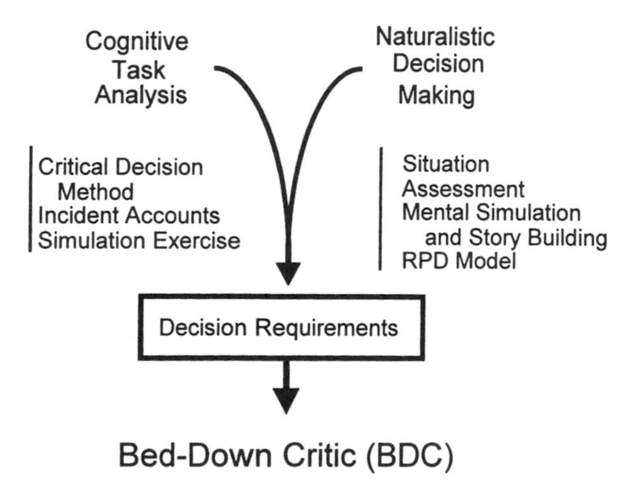

FIG. 6.1. NDM orientation for development of Bed-Down Critic.

The second exercise we attended was a joint task force exercise where the planning staffs were stationed aboard the USS KITTYHAWK, which was docked at San Diego for the exercise. Two researchers observed this exercise and conducted interviews with the planning staff as opportunities arose. This second exercise allowed us to explore issues identified from the first exercise and to deepen our understanding of the existing process.

The third exercise was conducted by the USS GEORGE WASHINGTON Carrier Battle Group at sea in the Atlantic Ocean. One researcher spent 7 days with the planning staff aboard the USS MOUNT WHITNEY, which is the command and control ship for the carrier battle group. This was the highest fidelity exercise observed, because being at sea adds to the stress, fatigue, and reality of 24-hour operations. Again, this exercise was used to further explore issues identified in earlier exercise observations, and also to specifically focus on the characteristics of quality plans, how the planning staff knows when they have a good plan, and where in the process there are opportunities to evaluate the quality of the plan. During this exercise, we also made contacts with the planning staff at Barksdale Air Force Base, who was responsible for developing the air plan. We conducted follow-up interviews with six members of the planning staff at Barksdale to explore specific plan evaluation issues.

In all of these exercises, the planning staff was separated into several teams, each performing one or more tasks within the overall goal to develop air plans. Such an arrangement made it difficult for individual members of the planning staff to maintain a broad understanding of the intent behind the plans. The observers of these exercises were in the fortunate position of not being assigned to any particular planning team, but rather were able to make observations across the various teams and at different times in the daily cycle.

Individual Interviews

To help us understand both the high-level conceptualizations and the low-level information gathering and interpretation that takes place in planning, the research team used two specific knowledge elicitation tools in the individual interviews: the Knowledge Audit and the Critical Decision Method (CDM).

The Knowledge Audit is organized around knowledge categories that have been found to characterize expertise (Militello, Hutton, Pliske, Knight, & Klein, 1997). These include: diagnosis and prediction, situation awareness, perceptual skills, development and knowledge of when to apply tricks of the trade, improvisation, metacognition, recognition of anomalies, and compensation for equipment limitations.

The Knowledge Audit employs a set of probes designed to describe types of domain knowledge or skill and elicit appropriate examples. The goal is not simply to find whether each component is present in the task, but to find the nature of these skills, specific events where they were required, strategies that have been used, and so forth. The list of probes is the starting point for conducting this interview. Then, the interviewer asks for specifics about the example in terms of critical cues and strategies of decision making. This is followed by a discussion of potential errors that a novice, less-experienced person might have made in this situation.

The examples elicited with the Knowledge Audit do not contain the extensive detail and sense of dynamics that more labor-intensive methods, such as the Critical Decision Method incident accounts, often do. However, they do provide enough detail to retain the appropriate context of the incident.

CDM interviews rely on eliciting memories of particularly challenging situations, especially where one's expertise has been challenged. Although memories cannot be assumed to be perfectly reliable, the CDM approach has been highly successful in eliciting perceptual cues and details of judgment and decision strategies that are generally not captured with traditional reporting methods (Crandall, 1989). The CDM provided this information from the perspective of the person performing the task and was particularly useful in identifying cognitive elements that were central to the person's proficient performance. Detailed descriptions of CDM and the work surrounding it can be found in Klein, Calderwood, and MacGregor (1989) and Hoffman, Crandall, and Shadbolt (1998).

My colleagues and I conducted interviews with three groups of experienced military personnel. The first were opportunistic interviews during the exercise observations. We call them "opportunistic" because these interviews were conducted with the exercise planning staff when they were not performing other duties.

The second set of interviews were with planning staff at Barksdale AFB. These were follow-up interviews with planning staff who had put the initial plans together for one of the exercises we observed and who modified and developed the plans as the exercise was conducted. The interviews were semistructured, CDM sessions lasting approximately 2 hours. The sessions began with a discussion of the interviewee's background in the Air Force and path by which he or she became an 8th Air Force planner. Our goal was to understand the knowledge and experience base possessed by the typical planner. We then continued questioning by eliciting specific incidents experienced by the interviewee either during the course of an exercise or in a real-world operation. The goal here was to probe into events in which the planning process had fallen apart in order to gain an understanding of the barriers to effective planning. Finally, we initiated a discussion of the characteristics of quality plans. We asked the interviewee about incidents in which they felt their plans were strong and incidents in which their plans had to be altered significantly following its briefing to higher level personnel. The purpose of this line of questioning was to identify aspects of plans that make them robust and, thus, acceptable for use in an operation.

The third set of interviews was conducted with high-ranking military officers, some of whom had intricate involvement in developing plans for the Gulf War in 1991. For example, we interviewed General Charles Horner, who was in a key position during the Gulf War to evaluate air campaign plans as they were developed. In his role during the war, no plans were implemented without his approval. For our purposes, we interviewed him to discuss specific incidents where plans were not approved in order to gain an understanding of what criteria were not met. We were interested in what was missing or erroneous about these plans and how these plans could have been better.

Simulation Exercise

The Air Counter '97 (AC '97) exercise was the last and most extensive of the knowledge elicitation work. In knowledge elicitation work prior to AC '97, my colleagues and I found that there is no formal (explicit) aspect of the planning process where evaluation takes place. In order to study plan evaluation, we designed AC '97 to explicitly engage experienced planners in evaluating plans. We did this by having two teams of human planners separately develop plans against the same challenging scenario. The situation in the scenario demanded rapid Air Force response to the massing of troops from a non-Allied country along the border of an Allied country. After the teams developed their respective plans, the two teams commented on and evaluated each other's plans.

During the exercise, research team members focused their observations on the incremental development of the plans and on any indication that the developing plans were being evaluated. The planning process was observed to document informal evaluation processes and criteria and to identify areas where human planners had difficulties within the planning process. The exercise also helped to identify a specific problem area (i.e., initial "bed-down" of resources) for technology development to support one aspect of plan development and evaluation.

AC '97 was a 3-day planning exercise hosted by a planning office in the Pentagon. The exercise was designed to simulate many attributes of actual planning events. Data produced from the planning teams in the exercise included their maps of the "Initial Preparation of the Battlefield," the bed-down plans, and briefing vugraphs containing the plans from the two teams.

During the first part of Day 1 of the exercise, the planners were all present and briefed on the situation. Planners jointly performed the Initial Preparation of the Battlefield and an analysis of possible enemy courses of action. After the planners had a shared understanding of the situation, the planners were split into two teams of three to four people each. During the second part of Day 1, the planning teams worked on the concept of operations and a strategic approach to the scenario. Each team worked independently in developing a plan. The planning continued through the morning of Day 2.

On the afternoon of Day 2, each team briefed their plans to the other team for critique. The briefings were open to questions and discussion by both the planners and the observers in order to provide insights into specific issues primarily involving evaluation.

On the morning of Day 3, the research team conducted individual interviews with all of the planners who participated in the exercise. The purpose of the interviews was to (a) clarify issues raised during the observation portion of the exercise, (b) elaborate on the evaluation criteria used both when developing a plan and when a plan is briefed formally as well, and (c) elicit opinions and issues not expressed in the open forum the day before.

The first stage of analyzing the data from the exercise was to do a preliminary sweep through the entire data set. The purpose was to identify a comprehensive set of variables and content categories for inclusing in a systematic examination and coding. The team also developed thematic categories in order to examine higher level processes and decision events that are not captured by more discrete measures (Hoffman et al., 1998). For example, the results of this thematic analysis resulted in the following data categories:

- The planning process
- Enemy course of action analysis
- Use of the map
- Timing and phasing
- Resource utilization and availability
- Tasks
- Flexibility
- Measurability
- Marketing and briefing the plan

FINDINGS

This section briefly summarizes our CTA findings, which are organized around three aspects of planning: aspects of quality plans, plan evaluation, and the planning process. For detailed results, see Miller, Copeland, Heaton, and McCloskey (1999).

Aspects of Quality Plans

In order to be robust, a quality air plan must strike a balance between various characteristics of the plan. Failure to explicitly consider these aspects of plans can yield a plan with vulnerable approaches to achieving the objectives stated. A key component to evaluating a plan during its development includes an understanding of the inherent trade-offs between certain characteristics of a plan. Specifically, the CTA data suggest that robustness of air campaign plans needs to consider at least the following plan characteristics:

- Risk to forces
- Coordination between plan components
- Flexibility in applying the plan
- Sensitivity to geo-political issues
- Resource utilization
- Communicating intent of the plan
- Measurability of plan progress
- Ownership of initiative.

Considerations of risk include understanding risks to friendly forces (i.e., location and proximity, and configuration of friendly command and control structures), risks to bases (i.e., location and proximity to Special Forces, missile, and other air attacks), risks to aircraft and mission success (i.e., mobile air defenses), etcetera. Planners are faced with the task of knowing where the sources of risk are and then balancing this risk to friendly forces with the potential consequences of not taking certain actions.

Coordination issues are difficult for any large-scale operation, but they are greatly complicated by joint or coalition military forces. Coordinating activities in the planning stages can help alleviate confusion and miscommunication that can occur during the execution of a plan.

Having adequate flexibility is the next aspect of quality plans. The ability to change with dynamic conditions and not be locked to certain actions, regardless of inopportune timing or conditions, is imperative to successful planning. One general we interviewed described this as being a prisoner to the plan. Using plan components in multiple ways is one way to build flexibility into a plan. In the air campaign planning domain, for example, one way to build in flexibility is to use aircraft that can perform multiple missions. That is, specific aircraft are tasked with certain missions, but if the need arises, these same aircraft can be redirected to another, more time critical task such as providing combat air support to ground troops. If the need does not arise, the aircraft performs its primary mission.

Understanding geopolitical, cultural and political motivations of the opponent (i.e., "being able to plan through the enemies' eyes") enables the planner to predict the opponent's reaction to actions taken. Without this understanding, actions may not have the intended effect.

A critical aspect of any quality plan is the efficient use of resources. Knowing which resources have limited availability and are, therefore, constraints to the set of feasible plans is critical to building sustainable plans. For example, during the Gulf War, cruise missiles were a limited resource, and planners needed to use them sparingly and efficiently.

Throughout the planning process it is also important that the planners consider how to communicate the intent behind the development of the plan. Planners may develop a series of tasks that they have clearly linked to stated objectives, but they may fail to document or otherwise communicate these links. Later in the planning and execution process, those who implement tasks without an understanding of the intent behind them will be unable to improvise if the situation requires adaptation of the plan.

The ability to measure the progress of plan implementation should be built into the plan. Tasks should be stated in such a way that realistic observations can be made that are relevant to assessing progress. Measuring plan progress is an area that is evolving. For example, measures of effectiveness of the plan have been tied traditionally to the reduction of enemy forces or their resources. However, current thinking links implementation of tasks with an effect on opponent functionality or capability. For example, it may not be necessary to reduce an opponent's aircraft supply by 50% if a few well-placed precision weapons can disable command and control of those aircraft. This way of thinking about measures of effectiveness has increased the need for clearly specified measures that are built into the plan.

Several participants interviewed during the CTA discussed the importance of having and maintaining the initiative in a plan. While very difficult to measure, *having the initiative* refers to the ability to force the opponent to react to your actions and not being forced to react to your opponent's actions.

Plan Evaluation

We found that plan evaluation takes place both informally and in more formal settings. The most obvious place where plan evaluation occurs is in formal settings, such as when a plan is briefed to the commander. Briefing the plan gives superior officers the opportunity to weigh whether the given plan meets the required military, political, and national objectives. Thus, during formal evaluations, it is imperative not only that the plan address the necessary objectives, but also that the plan is presented such that it is understandable to those who are doing the critique. Thus, marketing the plan is as much a part of the evaluation process as is satisfying the evaluation criteria.

We also found that evaluation can be informal in the sense that it is iterative, is tightly coupled to plan development, and is continuous throughout the planning cycle (Klein & Miller, 1999). Evaluation occurs within the development of the plan, and we found that members of the planning staff take the responsibility for questioning the plan or pointing out discrepancies within it. Therefore, computer-based planning systems that generate plans automatically cut out this vital linkage with the human planners.

This incremental evaluation is concurrent with plan development. We observed team members noticing details in the plan that could become problems downstream in the process, such as having the wrong kind of support equipment for F-15s at a certain air base. We also observed instances of backing up in the process to reconsider aspects of the plan or of reviewing the entire plan to maintain situation awareness. Another form of incremental evaluation is the use of specialized software tools. For example, there is software available to help planners "de-conflict" flight paths of aircraft in a crowded sky.

Bed-Down Planning Process

One of the goals of this work was to use results from the CTA to identify where artificial intelligence techniques could be used to support the human planners with respect to plan evaluation. We identified a slice of the planning problem, called "bed-down," on which to base a prototype support tool that would assist planners in critiquing the quality of their bed-down plan. *Bed-down* refers to the initial placement of resources in theater. For purposes of this work, this refers primarily to the placement of aircraft and logistic support at friendly air bases.

AC '97 produced a wealth of information on the planning process, sources of data used to develop plans, and the evaluation of plans. This exercise was instrumental in shaping the development of a computer-based planning system. One of our observations from this exercise was that the placement of initial resources in the theater of operation occurs very early in the planning process and that errors made here will propagate throughout the rest of the planning process until detected and corrected. If detected far downstream in the planning process, these errors can significantly delay implementation.

During Air Counter '97, it became clear that the initial bed-down of resources is important to the development and refinement of an air plan. Situational, political, and logistical factors all contribute to the unique problem of bedding-down resources in a theater of battle. The bed-down problem has become more important since the end of the Cold War as a result of the increase of rapid response missions that the United States now participates in. During the Cold War, U.S. assets were already forward positioned. However, since the Cold War, there has been a dramatic reduction in U.S. forces and in the number of overseas bases that are available to U.S. forces. These current circumstances make the initial bed-down plan more difficult to develop and more critical to the success of the mission.

The bed-down plan is not simply placing assets at various bases. The bed-down involves having an understanding and strategy for how limited assets can be used most effectively and where those assets should be located to accomplish tasks in the most economical and compatible fashion, while maintaining low levels of risk to the mission success (e.g., loss of life, loss of assets, loss of time, etc.). Just as the development of strategy will drive the initial bed-down, refinements or changes to the bed-down can also facilitate changes in the strategy. In this sense, the bed-down plan is instrumental to the transition between the strategy and the implementation of the plan.

During the development of the bed-down plans in AC '97, the planners relied on maps on the walls to help answer their questions about the developing plans. From observational data, and from the interview data, we found that these questions fall into the following four categories:

- Risk to air bases and to the mission that can be flown from the bases
- Time to complete objectives in the plan
- Logistics constraints
- Compatibility of aircraft and aircraft support to airbase capabilities.

We developed a prototype planning system to explicitly assist the human planners to evaluate these characteristics of the initial plan. This Bed-Down Critic is described next.

THE BED-DOWN CRITIC

The prototype Bed-Down Critic (BDC) is a planning system that assists planners in allocating resources and provides high-level plan evaluation functions. The concept emerged from the Air Counter '97 simulation exercise. The BDC can either be used to modify an existing bed-down or support the development of an initial bed-down plan. The BDC system also supports a strategy-to-task approach. It is a tool that allows the human planner to evaluate the planning process as the bed-down is being developed.

The plan evaluation functions are driven by manual (interactive) inputs on a map-based interface. The BDC provides estimates of feasibility, effectiveness, and threat in order to guide force allocation. The tool provides high-level evaluation estimates, as well as the intermediate information on which these estimates are based. The user can see more of the process and understand the trade-off issues and other factors that were created within the development of the bed-down plan.

The Bed-Down Critic supports the development and evaluation of the bed-down process. It provides feedback for (a) total aggregate risk, (b) time-to-completion, (c) logistics constraints, and (d) compatibility issues. An influence diagram of the system architecture is shown in Fig. 6.2.

The system uses a map-based interface where objects such as squadrons, bases, target sets, and threats are displayed as icons that can be directly manipulated by the user. The user can also access information about objects and modify object information (e.g., Desired Mean Points of Impact, DMPI; make up of target sets base preparation times; etc.).

The user can physically move squadrons to different locations, and create one-to-one squadron to target pairings (assign squadrons to targets; see Fig. 6.3). The user can also and label squadron, base, target and threat names on the map for rapid identification.

The user has the ability to set and modify theater-level variables, such as the likelihood of attacks from special forces, supply reserves, and resupply rates, and the required level of destruction required when attacking targets. Most of the theater-level variables are qualitative judgments that should (and would normally) be provided by the human planner.

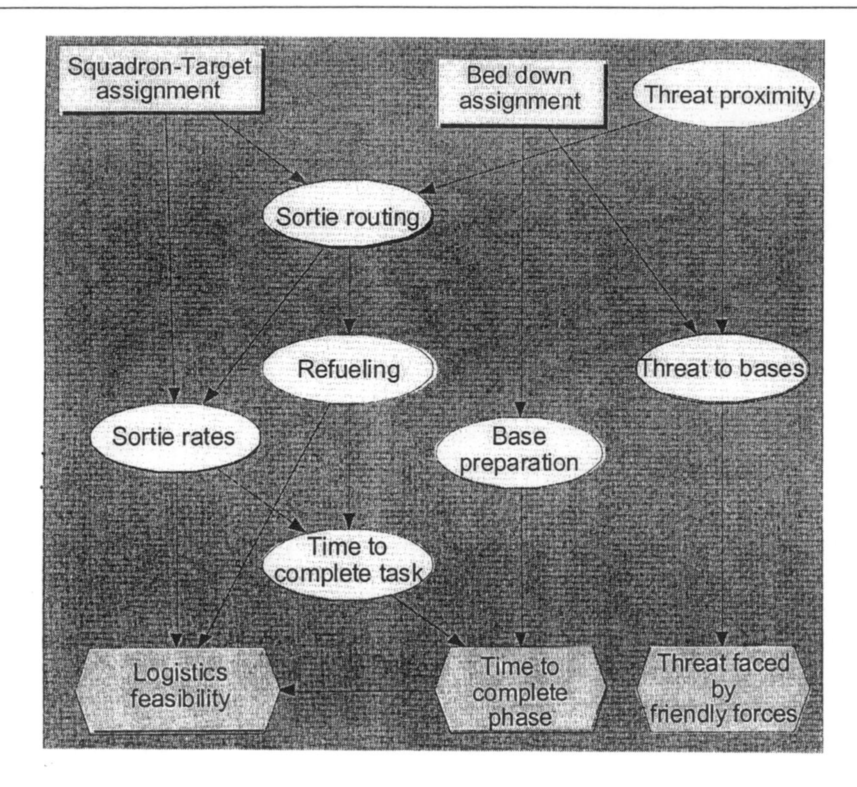

FIG. 6.2. System architecture.

The BDC supports rapid assessment and evaluation of the bed-down via graphical representations of evaluation criteria. An overall evaluation window is shown in Fig. 6.4. Within this window, the user can see aggregated evaluation of the logistics availability, time-to-completion, and aggregate risk for the currently planned bed-down. This window displays text-based feedback in a browser-like environment that allows the user to delve into the details of higher level, aggregated assessments.

For example, in Fig. 6.4, the planner sees that the completion time for Visalia Soc (an enemy target) is 5 days. However, in the test scenario from which this screen was taken, the commander specified that this target should be nonoperational within 2 days. The user goes to the Visalia Soc by clicking on it to reveal more detail. The user can see that the Visalia Soc is targeted using F-15e-1, that it will take 36 sorties, and that there will be nine sorties per day. The human planners know that nine sorties per day is a very low rate and that this is the source of the long completion time. The planner now has several options to increase the sortie rate, including actions such as locating the fighter aircraft closer to the target or allocating more fighter aircraft to the target.

Other forms of rapid evaluation feedback are implemented in the form of tripwires and agendas. Besides the aggregated evaluation window, the system also

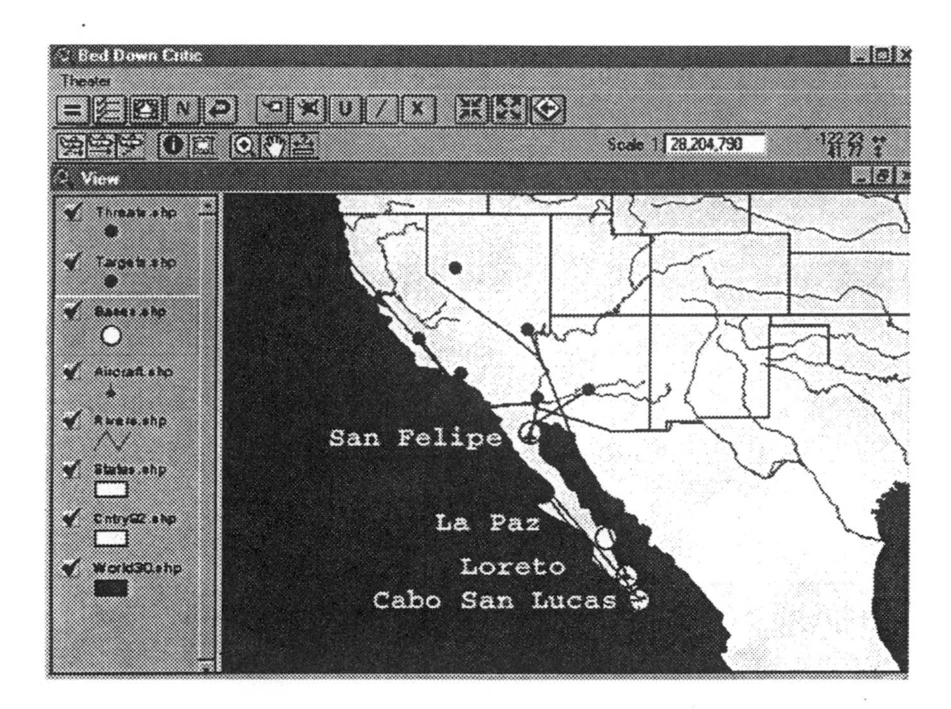

FIG. 6.3. User interface.

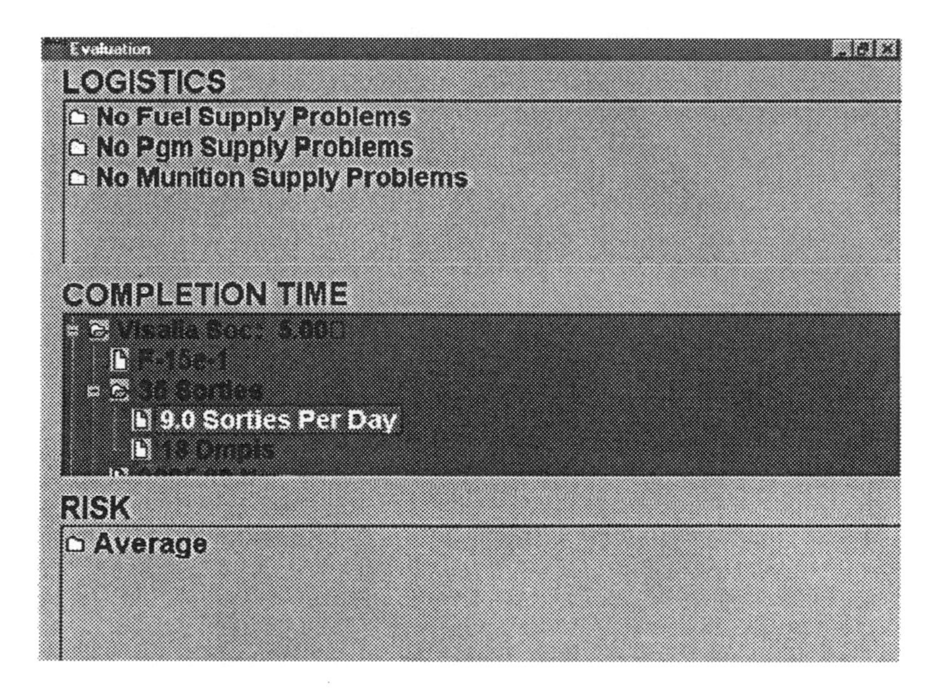

FIG. 6.4. Evaluation window for BDC.

alerts the user when specific constraints or trade-offs have been identified. These warnings are displayed when specific actions are taken (i.e., the user moves a squadron to another base) and a constraint is violated (i.e., there is not enough ramp space to accommodate a squadron).

These warnings do not require the user to correct the problem immediately. The BDC provides an "agenda" feedback. All warnings or alerts that represent a constraint violation and are yet to be resolved are recorded in the agenda. As a warning is corrected or addressed, that issue is removed from the agenda list. The user can refer to the agenda periodically to determine which issues have yet to be resolved.

The use of the agenda allows the user to keep track of factors that violate some specific constraint. The implementation of this feature does not force the user to solve each problem as it arises. Users can review the agenda when they are ready and address each item in an order they choose. The BDC system does not force the user to address each and every item (unaddressed items will remain on the agenda, but this will not prevent the user from continuing to develop the bed-down). This is especially important when the user wishes to continue with a bed-down for which the BDC detects potential problems, but which may be suitable for the user's needs.

The tripwire and agenda features are implemented using intelligent agents, where each agent is responsible for specific constraints. They monitor user actions and, when one of these actions violates a constraint, the agent immediately informs the user (tripwire). The agenda queries every agent simultaneously and presents these warnings to the user.

PLANNED FUTURE FUNCTIONALITY

There are several enhancements that should be considered for the next version of the Bed-Down Critic. One such enhancement involves changing the map-based architecture. The current mapping software implemented is Arcview Geographic Information System. This software enabled rapid prototyping of the BDC system, but it constrained the amount and type of functionality that would have been included otherwise.

The map images used were from ESRI's Digitized Map of the World and were limited in the detail that could be displayed. The Defense Mapping Agency (DMA) has developed more versatile mapping software that would be compatible with most digitized mapping formats and would enable more dynamic functionality and integration.

The first step in adding future functionality would be to integrate and refine the current functionality with the DMA's Mapping, charting, and geodesy Utility Software Environment (MUSE). In addition to a new map-based software, other functionality should include:

- Squadron toolbar
- Dynamic loading/integration
- Object modification
- Extending and refining evaluation functionality.

A squadron toolbar would enable the user to start with an initial basing, as it is implemented currently, or draw from assets located outside the theater. These assets could be located either in the United States or at any other staging area outside the theater of interest.

Dynamic loading and integration would involve linking the BDC system to a dynamic database. As the BDC is currently implemented, setup information (i.e., base location, locations of squadrons, targets, threats, etc.) must be loaded by the developer. In subsequent versions of this prototype, this capability would be available to the user and linked to the area of interest.

In later versions, more detail and object modification, such as threat severity, threat ranges, and squadron size, would be available to the user. Much of this could be accomplished through dynamic loading and integration, but refinements could be made by the user as well.

Extending and refining the evaluation functionality would include factors such as assessing collateral damage; including enemy aircraft in threat considerations (and adding Combat Air Patrols as defense); providing graphical supply rates over time; adding offensive capabilities, completion time, and logistics refinements, and more specific considerations for uncertainties; expanding considerations for dependencies among targets and target sets; and exploiting the currently implemented evaluation functions and integrating them with other automated planning systems.

SUMMARY

The Cognitive Task Analysis in this work identified a need to support time-pressured replanning, and it identified factors that planners use to evaluate certain types of plans. For example, the initial bed-down plan is evaluated in terms of (a) risk to air bases and to the missions that can be flown from the bases, (b) the time to complete objectives in the plan, (c) logistics constraints, and (d) compatibility of aircraft and aircraft support to airbase capabilities.

We further discovered that the most common form of plan evaluation occurs during plan generation, not after the plan is developed. This finding is problematic for automated plan generation systems because the automated systems short-circuit the most common form of evaluation that currently takes place in this domain. Automated plan generation not only leaves the human planner out of the loop as plans are being developed, but also denies the planner opportunities to critique the plan as it is developing.

The CTA on the bed-down problem allowed us to build a prototype plan evaluation system within the context of a critical, yet constrained, portion of the overall planning process. The Bed-Down Critic (BDC) is a planning system that assists planners in allocating resources and provides high-level plan evaluation functions. The BDC is work in progress. Future directions of this work will move beyond the bed-down problem and address plan evaluation issues in the context of the overall plan.

ACKNOWLEDGMENTS

This research was supported by Rome Laboratory, contract# F30602-95-C-0216, and by the Naval Research Laboratory (NRL) and the Office of Naval Research (ONR), contract# N00014-95-C-2016.

The author would like to thank Jennifer Phillips, Michael McCloskey, and Lewis Drew for their assistance throughout this work.

An abbreviated version of this chapter was published in the Proceedings of the Fourth Conference on Naturalistic Decision Making, Warrenton, Virginia, May, 1998.

REFERENCES

Bodker, S. (1991). *Through the interface: A human activity approach to user interface design.* Mahwah, NJ: Lawrence Erlbaum Associates.

Crandall, B. (1989, June). A comparative study of think-aloud and critical decision knowledge elicitation methods. *ACM SIGART, 108*, 144–146.

Hoffman, R. R., Crandall, B. W., & Shadbolt, N. R. (1998). Use of the Critical Decision Method to elicit expert knowledge: A case study in cognitive task analysis methodology. *Human Factors, 40*(2), 254–276.

Hutchins, E. (1995). *Cognition in the wild.* Cambridge, MA: MIT Press.

Klein, G. A., Calderwood, R., & MacGregor, D. (1989). Critical decision method for eliciting knowledge. *IEEE Transactions on Systems, Man, and Cybernetics, 19*(3), 462–472.

Klein, G., & Miller, T. E. (1999). Distributed planning teams. *International Journal of Cognitive Ergonomics, 3*(3), 203–222.

Klein, G. A., Orasanu, J., Calderwood, R., & Zsambok, C. E. (1993). *Decision making in action: Models and methods*. Norwood, NJ: Ablex.

Militello, L. G., Hutton, R. J. B., Pliske, R. M., Knight, B. J., & Klein, G. (1997). *Applied Cognitive Task Analysis (ACTA) Methodology* (Final Report) (Contract No. N66001-94-C-7034 prepared for Navy Personnel Research and Development Center). Fairborn, OH: Klein Associates Inc.

Miller, T.E., Copeland, R., Heaton, J.K., & McCloskey, M.J. (1999). *A cognitive approach to developing decision support tools for air campaign planners* (Tech. Rep. under contract F30602-95-C-0216 for Rome Laboratory). Fairborn, OH: Klein Associates Inc.

Nardi, B. A. (1996). *Context and consciousness: Activity theory and human-computer interaction.* Cambridge, MA: MIT Press.

Rasmussen, J. (1997). Merging paradigms: Decision making, management, and cognitive control. In R. Flin, E. Salas, M.E. Strub, & L. Martin (Eds.), *Decision making under stress: Emerging paradigms and applications* (pp. 67–85). Aldershot, England: Ashgate.

Suchman, L. A. (1987). *Plans and situated actions: The problem of human-machine communication.* Cambridge, England: Cambridge University Press.

Vicente, K. J. (1999). *Cognitive work analysis: Towards safe, productive, and healthy computer-based work.* Mahwah, NJ: Lawrence Erlbaum Associates.

Zsambok, C. E., & Klein, G. (Eds.). (1997). *Naturalistic decision making.* Mahwah, NJ: Lawrence Erlbaum Associates.

7

Designing a First-of-a-Kind Group View Display for Team Decision Making: A Case Study

Emilie M. Roth
Roth Cognitive Engineering

Laura Lin
Logicon, Inc.

Steven Kerch
Westinghouse Electric Company

Stephen J. Kenney
ESK Instrumentation Associates

Nubuo Sugibayashi
Mitsubishi Electric Corporation

One of the main tenets of naturalistic decision-making research is the importance of studying how actual decision makers perform in realistic situations as a basis for understanding the characteristics of expert decision making and the requirements for effective support. Studying the performance of actual decision-makers in realistic contexts can be an important tool for uncovering the demands of the domain and the knowledge and skills that underlie the performance of expert practitioners (Potter, Roth, Woods, & Elm, 2000; Mumaw, Roth, Vicente, & Burns, 2000). However, there are some situations in which empirical investigation of practitioner performance in the existing environment is not sufficient in itself to characterize the requirements for effective support. A case in point is the design of first-of-a-kind systems that, when implemented, are intended to change dramatically the cognitive and collaborative activities entailed by the work environment.

This chapter describes a case study that falls in this category—the design of a large wall-mounted group view display intended to support broad situation awareness of individuals and teams in a compact computerized control room for power plants (Rusnica, Kerch, Thomas, Kenney, Brockhoff, Morris, Roth, & Sugibayashi, 1999).

There have been numerous empirical studies examining operator decision making given existing control room technology (e.g., Mumaw, Roth, Vicente, & Burns, 2000; Roth, 1997; Roth & Woods, 1988). These studies provide insight into some of the fundamental operational goals to be achieved, inherent sources of performance challenge (e.g., complex process dynamics, multiple interacting systems) and strategies that operators have developed to cope with task demands. However, the goal of new control room designs is to increase the ability of operators to get an accurate assessment of plant state and to formulate effective response strategies. These new designs depend on and are intended to reinforce more sophisticated mental models of the plant and more adaptive response strategies than are observed in older control rooms. The mental models and response strategies of operators in existing control rooms are constrained by the limitations of current technology and, as such, cannot provide the foundation for radical new designs intended to create step changes in performance. Other sources of insight are needed.

We confronted this problem in the design of the wall-mounted group view display. The approach we took was to perform a function-based cognitive task analysis to ground the design. A function-based cognitive task analysis begins with analysis of the demands inherent in the domain (a work domain analysis). It uses a functional goal-means representation of the domain to guide the identification of human decision-making requirements and supporting information needs (Roth & Mumaw, 1995). Function-based cognitive task analysis falls in the general class of cognitive work analysis methods (Vicente, 1999).

This chapter traces the design process that was used to establish the content and organization of the group view displays and to evaluate empirically the resulting design. The design effort serves as a case study in design of first-of-a-kind systems. It illustrates the use of a cognitive work analysis to define the cognitive and collaborative activities to be supported by the new system and to establish the display content and organization that is required to support these activities. The results also contribute concepts for the design of group view displays intended to support situation awareness of individuals and teams.

THE DESIGN CHALLENGE

One of the trends in the process control industry is to replace traditional hardwired control rooms with compact computerized control rooms. Although computerized control rooms have distinct advantages, they also introduce new design challenges. One of these design challenges is how to maintain broad situation awareness of individual operators and the team as a whole (Roth & O'Hara, 1999).

Traditional hardwired control rooms present information in the form of individual indicators (e.g., dials, meters, alarm tiles) laid out in parallel on large display and control boards. Operators monitor plant processes and take control actions from the control board, and supervisors can maintain an overview of plant status by looking over the display and control boards from their desks. In a computerized control room, operators as well as the shift supervisor are each at their own workstation with multiple video display units (VDU) used to access information displays.

One clear advantage of computerized control room is the ability to create displays that integrate multiple pieces of datum into visualizations that are tailored to the particular operational context and task to be performed. Each operator can bring up displays, procedures, and controls tailored to his or her individual task.

Although computerized control rooms offer some clear advantages, there are features of traditional hardwired control rooms that naturally support maintaining broad situation awareness. These features are not inherently preserved in a computerized control room. In a traditional control room, displays and controls are available in parallel, dedicated positions. This enables operators to notice changes and rapidly shift their attention to areas of interest. Additionally, a conventional control board provides an "open" environment where everyone can see the same information and everyone can see what each other is doing. Operators can get some idea of what each other is doing by noting what displays and controls they are close to. This allows operators to maintain awareness of each other's activities and (because of their domain knowledge) to anticipate what influence those actions will have on plant state. It also allows new people coming into the room to quickly assess plant conditions and understand what the crew is doing.

There is a risk that these positive features of traditional control rooms will be lost when moving to computerized control rooms. Because procedures, indications, and controls can all be accessed from a single position, operators are more likely to remain focused at the workstation. They view isolated images of plant state data through a small set of "windows" (the video displays) and, therefore, must integrate information mentally unless a broader view of plant state is made available. Second, plant state data are no longer "spatially dedicated," instead, data exists in software space. As a result, operators' ability to maintain a continuously updated overview of plant state may be affected. Additionally, with each operator working at his or her own workstation, the ability to observe what each other is doing and share a common view of plant state is reduced.

Large wall-mounted group view displays are displays that are intended to be viewed simultaneously by multiple people. Group view displays provide a potential mechanism to reproduce the "big picture" view and shared situation awareness that are inherent features of traditional control rooms. As part of a program to design an advanced compact control room, we were tasked to develop and test a conceptual design for a large wall-mounted display referred to as a wall panel information system (WPIS). The WPIS is intended to support individual operators in maintaining the big picture with respect to plant state. It is intended also to provide multiple operators with a common frame of reference with respect to plant state to facilitate crew communication, cooperation, and coordination and to enhance team performance (Endsley, 1995; Salas, Prince, Baker, & Shrestha, 1995; Stubler & O'Hara, 1996).

The design challenge was to establish, at a general level, the kinds of information that need to be presented on the wall panel to support broad situation awareness of individuals and teams, and then to indicate more specifically the plant state information that needs to be included on the WPIS. The kinds of questions confronted included: What information should go on the wall panel? How should it be organized? How will operators interact with it? The next sections trace the design process that was used to answer these questions and the test program that was performed to assess the adequacy of the answers the design team came up with.

DEFINING REQUIREMENTS FOR SUPPORTING SITUATION AWARENESS OF INDIVIDUALS AND TEAMS

A theory-driven approach was used to specify the design basis for the WPIS. We began with an analysis of the requirements for situation awareness of individuals and multiperson teams. The analysis drew on the existing research base to identify issues and lessons learned across domains about the factors that contribute to situation awareness and support cognitive and collaborative performance (e.g., Cannon-Bowers & Salas, 1998; Endsley, 1995; Hutchins, 1995; Patterson & Woods, 1997; Patterson, Watts-Perotti, & Woods, 1999). These provided the basis for defining human performance support requirements for a group view display. In turn, the human performance support requirements served to establish the design basis for the group view display and to define the major elements that need to be included in the WPIS displays.

Endsley defined situation awareness as "the perception of the elements in the environment within a volume of time and space, the comprehension of the meaning, and the projection of their status in the near future" (1995, p. 36). The concept was first introduced in the context of aircraft pilot performance and has since been extended to cover other domains, including nuclear power plants (Hogg, Folleso, Strand-Volden, & Torralba, 1995)

The first step was to define the key elements of situation awareness as applied to the context of nuclear power plant operations. We defined situation awareness to include:

- Awareness of current plant state
- Awareness of changes in plant state (and projection of their status in the near future)
- Awareness of task state
- Awareness of situation-relevant workstation displays and access to them (i.e., navigational links from WPIS displays to seated workstation displays).

A design requirement of the WPIS is that it supports all these aspects of situation awareness.

Awareness of plant state and changes in plant state map closely with the concepts of perception of elements in the environment, their meaning, and projection of their status in the near future. Operators need to know not only the values of individual plant parameters, but also the implications of those values for plant state (e.g., Is the plant stable? Are we achieving electric production goals? Are there any malfunctions? What are the consequences of those plant malfunctions? What are the safety implications? What will happen next if no actions are taken? What will happen if mitigating actions are taken?).

Awareness of task state was added because in a power plant, as in many other domains, there is always on-going goal-directed activity. Operators need to understand the goals to be achieved, the steps needed to achieve those goals, which steps have already been completed and which remain, and whether the steps being taken are appropriate to the situation and progressing toward goal achievement. They also need to be aware of the goals and actions of the other crew members. In many cases, goal achievement depends on the coordinated action of multiple individuals. Effective coordination depends on a common understanding of task state (Serfaty, Entin, & Johnston, 1998). Furthermore, because of the potential for system interaction, it is important to be aware of the goals and actions of others in order to avoid taking an action that is counterproductive or has unintended negative consequences (Roth, Mumaw, & Lewis, 1994). Awareness of task state and the goals and actions of the other crew members also supports detection and correction of each other's errors (Hutchins, 1995).

Awareness of situation-relevant workstation displays was added to counter the potential "key-hole" effect associated with computer display of information—where there may be hundreds of potentially relevant displays but knowing when and how to access them may be difficult. The ability to recognize the availability of relevant information and to know how to rapidly access that information is a necessary element of effective situation awareness.

In addition to supporting broad situation awareness of individuals, the wall panel is intended to support multiperson team performance. There are typically at least three individuals in a control room: two operators and a supervisor. During particularly difficult maneuvers (e.g., startups) or in emergencies, the number of individuals in the control room can increase substantially. Furthermore, in emergencies, expert personnel from outside the control room are called in. These personnel arrive some time after the event has started (e.g., a technical advisor will arrive within 10 minutes of a plant shutdown) and have to be rapidly brought up to speed. As a consequence a key design requirement of the group view display is that it should facilitate communication, coordination, and cooperative problem solving of multiperson teams (Cannon-Bowers & Salas, 1998; Patterson, Watts-Perotti, & Woods, 1999; Stubler & O'Hara, 1996). This includes supporting the ability of new people on the scene to rapidly gain an understanding of plant state and task state (Patterson & Woods, 1997).

Operationally, we defined requirements to support situation awareness of multiperson teams to include:

- Providing a common frame of reference by having a display that everyone in the control room can see at the same time, so that operators (as well as new people entering the control room) have a shared understanding of plant state and task state
- Allowing operators to see how their control actions affect plant parameters that other operators are trying to control
- Allowing operators to see how the control actions of others are affecting what they are trying to do
- Allowing supervisors to monitor individual operator's actions by determining if expected changes occurred.

ESTABLISHING THE CORE ELEMENTS OF THE WPIS

After the design basis requirements associated with situation awareness of individuals and teams were specified, the next step was to establish how the WPIS would meet these requirements. In order to support the various elements of situation awareness, the WPIS displays were designed to include three core elements:

- A plant overview and status area intended to support situation awareness of plant state and changes in plant state
- An area devoted to information on task state
- An operator configurable area that allows operators to put up situation specific displays of their choice.

The plant overview area is intended to provide multiperson teams with a common frame of reference with respect to plant state and changes in plant state. The goal for the plant overview portion of the WPIS was to provide operators with sufficient information to assess overall plant status without requiring navigation to other more specific displays.

The area devoted to task state information is intended to provide the control room staff with a common frame of reference with respect to the state of currently ongoing tasks. It is designed to support operators in maintaining awareness of what tasks are to be accomplished, what has already been completed, what is currently going on, and what objectives remain to be accomplished. During normal operations, this area of the WPIS is used to display the current plant schedule of activities that are under the supervision of the control room staff (e.g., planned and ongoing system tests, maintenance activities, power change maneuvers). Keeping track of currently ongoing activities is critical to interpreting changes in plant state (e.g., an alarm may come on because of a currently ongoing test) as well as avoiding the potential for conducting tasks that can interact with negative consequences (Vicente, Mumaw, & Roth, 1997). In emergencies, all normal operations are suspended, and control room activities are governed by emergency operating procedures. In those cases, the task state portion of the WPIS is used to display a high-level overview of the current procedure being followed. It is intended to support the control room crew in maintaining a common understanding of the procedure currently being executed, the goals to be achieved, and the progress being made in working through the procedure. Presenting an overview of the procedure in the context of an overview of plant state is intended to support the operators in assessing whether the procedure in effect is appropriate to the situation and is achieving the intended goals and catching and correcting cases where the procedure is going down the wrong track (Roth, Mumaw, & Lewis, 1994; Roth & O'Hara, 1999).

The operator configurable area is a portion of the WPIS where operators can project any display from their individual workstations on the wall panel. This is intended to increase the context sensitivity of the WPIS by allowing operators to tailor the information available for shared viewing in response to the specific requirements of a situation. The inclusion of an operator configurable area reflects the explicit acknowledgment that it is not possible to anticipate and provide for all potential situations that the person on the scene is likely to confront. It is important to include facilities that enable operators to tailor the visualization to the specific situation (Mumaw, Roth, Vicente, & Burns, 2000; Vicente, 1999). Although the plant overview portion of the WPIS is intended to provide a principle-driven representation that should provide a solid foundation for monitoring the plant under most conditions, the operator configurable area is intended to enable operators to tailor the set of information that is available for shared viewing to the specifics of the situation.

The WPIS also included navigation mechanisms to allow operators to access more detailed display and control displays. By moving a cursor to areas of the wall panel, operators are able to bring up more specific displays or controls on their own local workstation display units.[1]

USING A FUNCTION-BASED COGNITIVE TASK ANALYSIS TO DEFINE DETAILED DISPLAY CONTENT

After the core elements of the WPIS were defined, the next step was to specify the particular plant status information to be included on the plant overview portion of the WPIS. Several key design decisions were made with respect to the content and organization of the plant overview.

One of the major design decisions was to organize the content of the plant overview portion of the WPIS around a goal-means functional representation of the plant (Roth & Mumaw, 1995). Theoretical foundations for function-based displays and their role in supporting supervisory control of complex systems and decision making can be found in Rasmussen (1986), Vicente and Rasmussen (1992), and Vicente (1999).

A functional representation allows one to rapidly assess whether the major plant functions are being achieved and the state of active plant processes that are supporting those plant functions. In cases of plant disturbances, where one or more of the plant goals are violated, a functional representation allows one to assess what alternative means are available for achieving the plant goals. For example, one of the primary plant goals is to maintain Reactor Coolant System (RCS) pressure. There are a number of engineered systems available for controlling Reactor Coolant System pressure. There are heaters that can be turned on and off to control pressure. There is a spray system that can be turned on to decrease pressure by injecting cold water. If those malfunction or fail to be sufficient, there are also emergency systems available. There are relief valves that can be used to reduce pressure and an emergency safety injection system that increases pressure by injecting lots of water rapidly. The idea of functionally organized displays is to provide the operator with a visualization that can enable him or her to rapidly assess:

1. The plant goals that need to be achieved and whether they are currently being achieved (e.g., Is RCS pressure being maintained at the desired value?);
2. The plant processes that are currently active and affecting goal achievement (e.g., Are the heaters on?);

[1]Each operator had his/her own cursor projected on the wall panel that was controlled via a mouse at his/her local workstation. More detailed displays could be brought up on the local workstation by pointing to an area on the wall panel and depressing a key on the mouse.

3. Alternative means that are available for achieving the goal? (e.g., What are my options for increasing pressurizer pressure? What is the status of the charging system? The pressurizer relief valves? The safety injection system?).

One of the primary benefits of functionally organized displays is that they allow an individual to take action to restore plant functions without needing to fully diagnose the malfunction. For example, an operator does not need to fully diagnose what is causing RCS pressure to decrease to take action to restore RCS pressure. Similarly, these displays allow an individual to take action to restore a plant function in situations that system designers had not foreseen or planned for (Vicente, 1999). In most cases, operators have procedures available to support diagnosis of malfunction and restoration of plant functions. However, should situations arise for which preplanned procedures are not available, functionally organized displays would enable operators to recognize that an important plant function is not being satisfied and to assess the engineered means available to him or her for restoring that plant function.

Function-based displays are intended to foster and reinforce new ways for operators to reason about the plant. They are intended to foster "goal-means" mental models of plant functions that support different lines of reasoning than mental models of the physical configuration of systems in the plant.

The content and organization of the plant overview portion of the WPIS was specified based on a function-based cognitive task analysis (Potter, Roth, Woods, & Elm, 1998; Roth & Mumaw, 1995). A function-based cognitive task analysis falls in the class of Cognitive Work Analysis methods (Vicente, 1999). The analysis begins with a work domain analysis that is intended to capture the goals to be achieved in the domain and the means available for achieving those goals. This provides the framework for identifying the tasks to be performed by humans and the cognitive activities those entail. Displays can then be created to support those cognitive activities.

A goal-means decomposition of the plant was developed that specified the primary goals of the plant, major plant functions in support of those goals, and plant processes available for performing the plant functions (Roth, 1996). Figure 7.1 provides a representation of selected parts of the highest four levels of this goal-means decomposition.

The two top-level goals of the plant representation are maintaining plant safety functions and generating electrical power. Beneath the two high-level goals are the plant functions designed to achieve these high-level goals. The different levels in the representation describe the plant functions at different levels of abstraction. The fourth level specifies the major engineered control functions available for achieving plant goals. This is the level at which manual and automatic control actions can be specified to affect goal achievement.

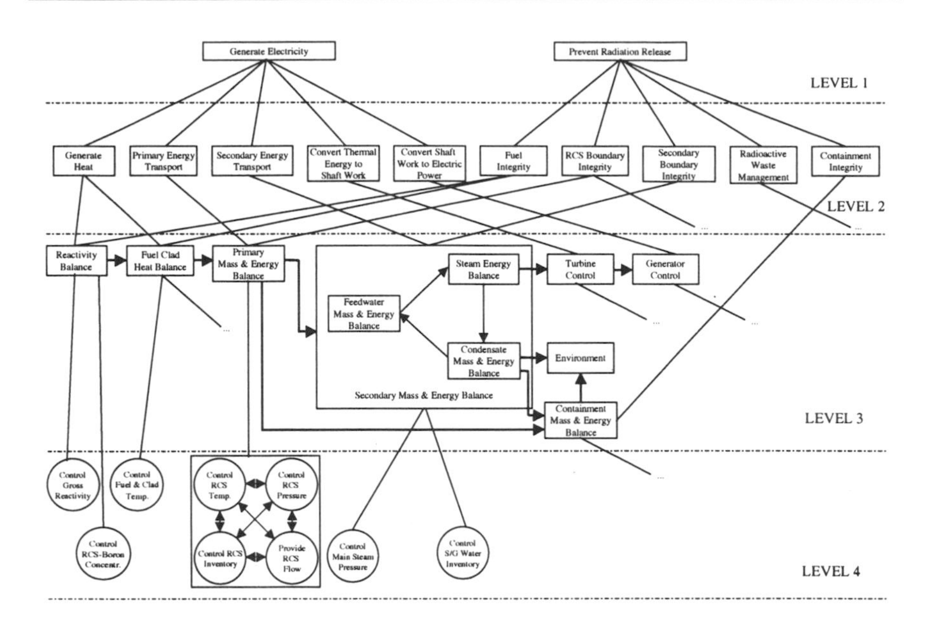

FIG. 7.1 Representation of selected portions of the first four levels of a goal-means function decomposition for a pressurized water reactor nuclear power plant (adapted from Roth, 1996).

The WPIS was designed to provide summary information on the status of these major engineered plant functions. Thus, the specific operational definition of *awareness of plant state and changes in plant state* that was used was awareness of the state of the major engineered plant functions that support the two high-level plant goals: maintaining plant safety and generating electrical power.

Figure 7.2 presents a schematic overview of the resulting WPIS design concept. The WPIS is composed of three panels. One panel focuses on plant safety functions; a second panel focuses on plant power generation functions; and the third panel includes general plant support functions, the task state/procedure overview, and the operator configurable area.

The plant state overview portion of the WPIS is organized around the different levels of the goal-means functional representation. At the top of the wall panels are areas devoted to each of the Level 2 functions. These areas alarm (indicated via color coding as well as an auditory tone) if one of these high-level functions is threatened. Below this section are areas corresponding to each of the major plant control functions (Level 4 functions). Specific criteria were developed to guide selection of plant parameters to be included under each plant function as well as selection of presentation format (e.g., whether a parameter would be presented as a digital value, a meter, or a trend display). The criteria reflected the monitoring and decision-making activities that the WPIS was intended to support.

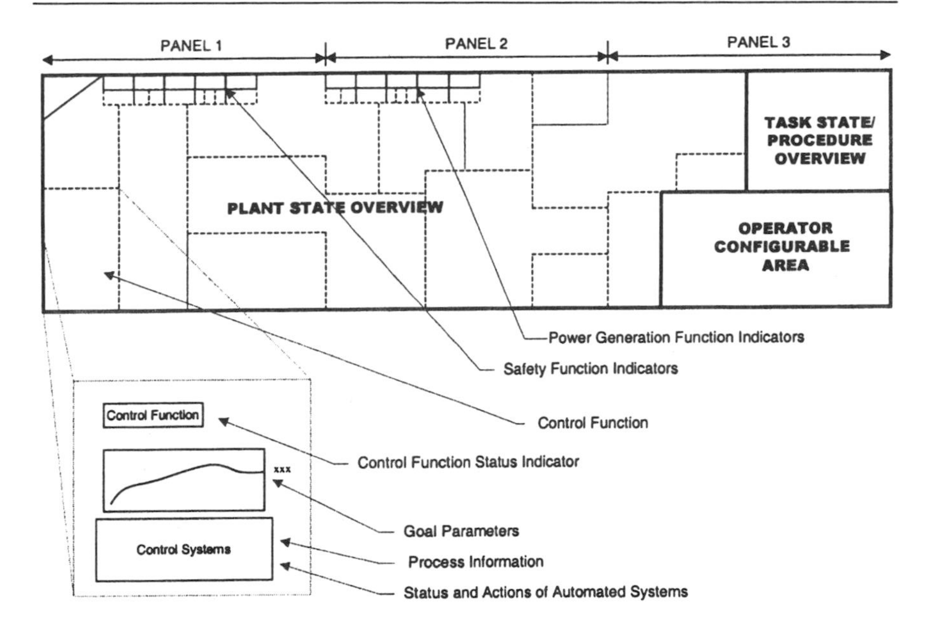

FIG 7.2 Schematic representation of different portions of the Wall Panel Information System (WPIS).

Criteria for selection of plant parameters included the need to enable operators to:

- Monitor whether the plant function goals are being satisfied and anticipate when goal violations are being approached
- Maintain awareness of the status of the processes that support the function goals
- Maintain awareness of the status of automatic process control systems affecting the plant function.

Thus each control function area includes:

- A control function status indicator that provides summary indication of whether the function is satisfied or being challenged (via color coding and an audio alarm)
- Plant parameter information that provides more specific information on goal achievement (e.g., a digital value, meter, or trend plot that provides indication of how well the goal is being achieved and whether limits are being approached to support projection into the future)
- Control system information that provides indication of the status of current processes, including automatic system actions that are affecting the goal parameter.

Figure 7.3 provides an illustration of the design concept. It depicts a rapid prototype of Wall Panel 1 that was generated as part of the design and test development cycle for the WPIS. Wall Panel 1 focuses on plant safety functions.

As illustrated in Fig. 7.3, a variety of different display formats were used to convey goal parameter information. The criteria used to determine what display format to employ for different goal parameters included potential rate of change of the parameter, the degree of historical perspective needed to interpret parameter behavior, and the need to correlate the behavior of the parameter in time in relation to the behavior of other parameters in the display. One of the outcomes of applying these criteria was that, in several cases, trend plots of key parameters were determined to be required to support operators in monitoring plant function goal achievement and anticipating future goal violations.

Another important design decision was to integrate graphically alarm and status messages into the plant overview portion of the WPIS. If a plant parameter exceeded an alarm threshold or resulted in an automatic system actuation, the graphic element representing that plant parameter (e.g., a meter or trend line) was color coded to indicate the alarm or change in status (e.g., it turned red). An audio signal was also provided to alert operators to the alarm condition.

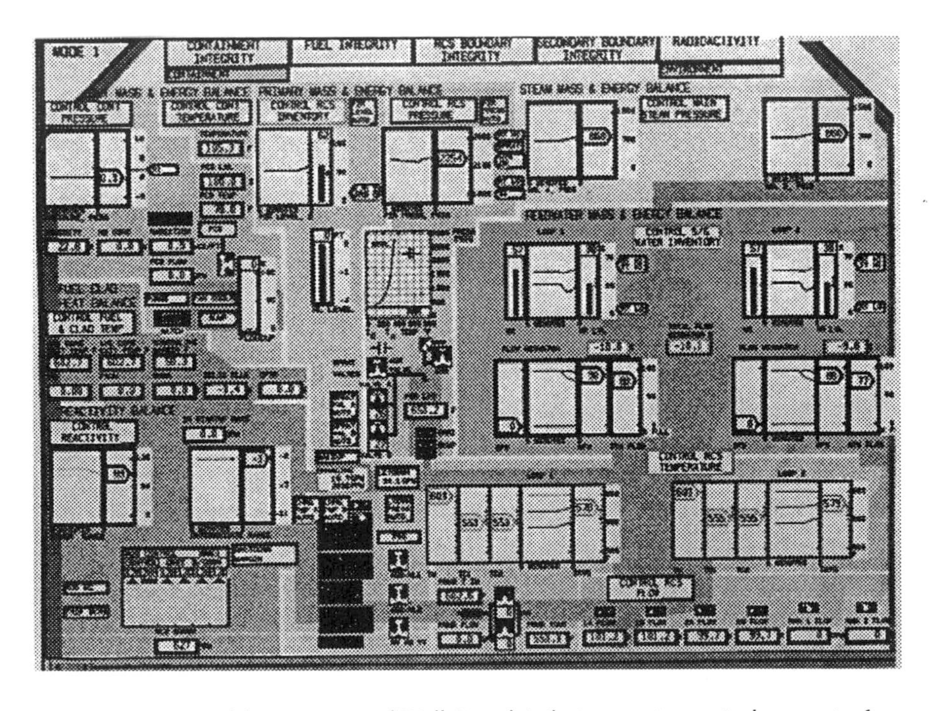

FIG. 7.3 A rapid prototype of Wall Panel 1 that was generated as part of the design and test development cycle of the WPIS. This wall panel focuses on plant safety functions.

The WPIS design includes a number of features that are not characteristic of typical plant status overview displays. Specifically:

- It is organized functionally to provide an overview of the status of major plant functions and the processes that support them
- It embeds alarms within the graphics of plant process state
- It includes extensive, detailed process state information, including presentation of trends.

In contrast to the WPIS design, traditional approaches to computer-based overview displays have tended to use physical mimics as the organizing principle. Typically, key plant parameters are presented as digital values organized around a physical mimic of the plant (Stubler & O'Hara, 1996). Figure 7.4 provides an example of an overview display that organizes plant parameter data around a physical mimic. As Fig. 7.4 illustrates, this type of display places more emphasis on depicting physical systems and their interconnections. However, these types of displays are less effective in communicating what are the major

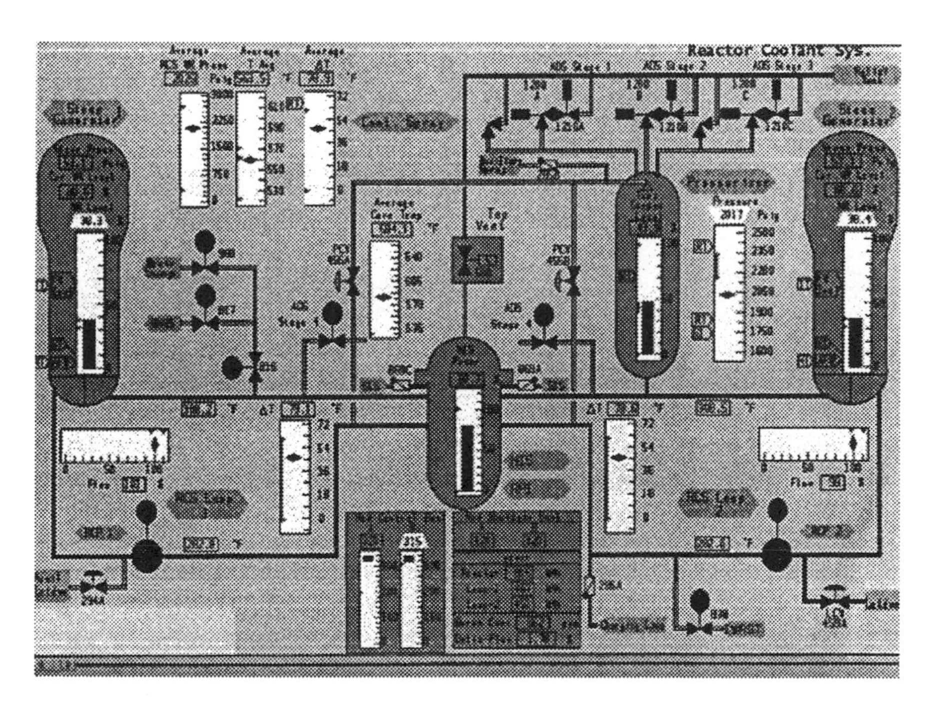

FIG. 7.4 An example of an overview display that uses a physical mimic as an organizing principle.

functions these systems are intended to achieve, whether the functions are being satisfied, and what alternative means are available to accomplish the same goal. Often the rationale for using a physical mimic is that it is more consistent with operators' mental models of the plant. As we mentioned earlier, part of our goal was to foster and reinforce goal-means mental models of the plant.

PHASED TEST PROGRAM

The WPIS display concept deviates extensively from conventional approaches to the design of group view displays. As a consequence, it was important to evaluate the effectiveness of the WPIS empirically in order to convince ourselves (the design team), our sponsors (a consortia of vendors and utilities), and government regulators of the industry of the viability of the design approach. A two-phased concept test program was performed as an explicit, preplanned, element of the first-of-a-kind concept development process.

Up until this point, the design concepts were developed based on a top-down analytic approach. The design concepts were guided by the human performance support requirements identified in the design basis and the results of the function-based cognitive task analysis. The operational perspective was represented through the inclusion of individuals with extensive operational background on the design team, but no attempt was made to directly elicit input of operators in determining the content and organization of the wall panel. At this point, the WPIS display concept was sufficiently concrete that it became possible to meaningfully involve plant operators. Plant operators were included in the concept test phase as test participants. We were able to examine the influence of the WPIS design on operator performance using objective performance measures and to elicit informed feedback from the operators on the general viability of the design approach as well as on specific features that could be improved on.

Two studies were conducted to evaluate the basic concepts underlying the WPIS design. The first WPIS study (Study 1) employed static displays and focused on the ability of the WPIS to provide a "big picture" overview of plant state and to support situation awareness of individual operators. The primary objective of this study was to establish that the design approach was generally on target. The second study (Study 2) employed a more tightly controlled experimental paradigm. Operator performance using the WPIS was compared with operator performance using an alternative overview display that was representative of traditional approaches to overview display design (see Fig. 7.4). Study 2 used dynamic displays driven by a high-fidelity plant simulator and focused on the ability of two-person teams to maintain broad situation awareness and coordinate work under dynamically evolving conditions.

Study 1

Study 1 was designed to assess whether the WPIS design concepts were on the right track before additional resources were devoted to further concept development and more rigorous testing. We briefly summarize the methods and results of the study as background for Study 2.

Method

Study 1 used a rapid prototype of the WPIS presented on a large 100-inch wall panel display unit. Nine individuals with operations experience served as participants. Participants were tested one at a time.

Two converging sets of measures were used to evaluate the WPIS concept. Objective measures were collected of how well the wall panel enabled individuals with operations experience to assess plant state when presented with a series of static WPIS displays that represented different points in time in an evolving plant scenario. In addition, subjective evaluations and specific suggestions for improvements of the WPIS were elicited from the test participants via a questionnaire and a final debrief session.

Each test participant was typically brought in for a 3-day period. The first day was devoted to training. The second day the participants were presented with three test scenarios. These consisted of a series of 5 to 7 static "snapshots" that represented different time points in an evolving plant transient. The test participant was asked to "think-aloud" as he or she tried to understand plant state. Follow-up probes were used to assess different aspects of situation awareness (e.g., What would you tell someone coming on shift about the current state of the plant—for example, regarding alarms, important parameters, active processes, trends, system availability? What should be the primary and secondary concerns of the operator at this time? If no operator actions are taken at this time, how would you expect the plant to change?).[2]

The third day the test participant filled out a written questionnaire. This was followed by an extensive verbal debriefing session where sections of the wall panel were systematically examined one at a time, and test participants were asked to comment on features they liked, did not like, think should be removed, or would like to see added. Typically, the debriefing session lasted approximately 4 hours. The scenario test presentation and debrief sessions were videotaped to provide a backup resource for analysis and documentation.

[2]The probes were asked immediately following the think-aloud portion for each snapshot. Probes whose answers were covered in the thinking-aloud portion were skipped.

Results

The results of Study 1 confirmed that the basic assumptions and approach being taken in designing the WPIS are on target. Analysis of performance data confirmed that the WPIS supported situation awareness of individual crew members. Participants were able to assess plant state from the wall panel displays in all the test scenarios. They were able to detect changes in plant state, assess the implications of those changes for achievement of plant safety and power generation goals, and project plant parameter behavior into the future. The results confirmed that the WPIS includes sufficient information to support situation awareness in a variety of normal and emergency conditions and that it presents the information in a way that can be readily extracted and understood by individuals with operational backgrounds.

Analysis of the rating questionnaire and debriefing results reinforced these conclusions. Debriefing session comments, although providing many specific suggestions for changes and improvements to the content of the wall panel, uniformly indicated approval of the basic approach we have taken in the design of the wall panels. Participants' questionnaire ratings and interview responses indicated that they liked the functional organization of the wall panel. They liked the integration of alarms that used color coding to direct attention to functional areas of concern. They also liked the fact that detailed information (including trend plots) was available on the wall panel to enable them to understand the nature of the problem without having to move to more specific displays.

It is interesting that several of the participants mentioned that their initial impression of the WPIS displays was that they were excessively dense. However, after going through the scenario exercises, they began to see that they could readily find the information they were looking for. They appreciated having detailed information available on the WPIS and no longer felt the displays were too dense.

Study 2

The first test confirmed the general viability of the design approach and provided a plethora of specific suggestions for improving the detailed design of the WPIS. A second design cycle ensued that incorporated the design suggestions.

At this point, a second, more rigorous evaluation was conducted. Study 2 examined the ability of the WPIS to support operator situation awareness and performance under real-time dynamic conditions, where operators had to actively control plant processes and diagnose and respond to plant malfunctions in real-time. A second difference is that Study 2 looked at performance of two-person teams rather than single individuals. Finally, Study 2 employed a more traditional experimental paradigm that compared performance of operators using the WPIS with performance of the same operators using an alternative display that was more representative of conventional approaches to overview display design.

Study 2 compared the effectiveness of the functionally-organized WPIS to an overview display that organized information around a physical mimic of the plant (*Conventional Overview Display Condition).* The displays used in the Conventional Overview Display Condition were slightly modified versions of overview displays currently being used in a nuclear power plant control room. Figure 7.4 presents one of the displays that were used in the Conventional Overview Display condition. This display is representative of a general class of overview displays that:

- Organize plant state information around a physical mimic of the plant
- Utilize digital values (and some meters) rather than trend plots to present plant parameter information
- Include minimal alarm information in the display
- Are less dense in terms of number of plant parameters presented on the displays than the WPIS.

Study 2 also examined the impact on performance of placing the WPIS on a large wall-mounted display unit. The study compared the effect of presenting the WPIS overview displays on large wall-mounted display units *(Large- Screen WPIS Condition)* versus presenting the same WPIS displays on VDUs at each individual operator's workstation (*Local WPIS Condition*). In the Local WPIS Condition, the three wall-mounted displays were turned off, and the WPIS displays were presented on three VDUs at each of the individual operator's workstations instead.

Method

Test Facility.

The test was conducted in a facility that included three large (100-inch) wall-mounted displays and two operator workstations, each with six VDUs (see Fig. 7.5). Three of the VDUs were used to display specific information and control displays that the operators used to control the plant. The remaining three VDUs were used in the Local WPIS condition to display the three WPIS displays. The operator workstations could be positioned one behind the other to serve as a control operator (CO) and a supervising operator (SO) workstation, or they could be placed side by side as two CO workstations. The wall panel displays and local workstations were driven by a real-time high-fidelity plant simulator.

Participants.

Twelve individuals with operations experience served as test participants. The participants were tested as six two-person crews. Each crew participated in all three test conditions: *large-screen WPIS*, *local WPIS*, and *Conventional Overview* condition. The order of the presentation of the three test conditions was counterbalanced across crews.

FIG. 7.5 Test facility where Study 2 was conducted. The facility included three 100 inch wall-mounted displays and two operator workstations each with multiple VDU.

Procedure. The study was conducted over a 5-day period. The first two days were devoted to training and practice. Test scenarios were presented on Days 3 and 4. Each crew participated in nine test scenarios driven by a high-fidelity training simulator (three scenarios in each of the three test conditions). The scenarios required them to understand plant state, take control action to change plant state (e.g., change power levels), detect disturbances in plant state and take control actions in response. Scenarios included normal operations maneuvers as well as disturbances that resulted in plant shutdowns and required use of Emergency Operating Procedures to respond to the transient.

In the first six scenarios that were run (two in each of the three test conditions), one test participant acted as a control operator (CO) who was responsible for taking control actions and the other as a supervising operator (SO) who had overview responsibility. In the last three scenarios (one in each of the three conditions), they acted as primary and secondary side control operators

At the end of Day 4, participants were given a written questionnaire that asked them to rate different aspects of the WPIS displays and to compare the different display conditions. On Day 5, a structured debriefing session was conducted, typically lasting 4 hours, to solicit detailed comments from the participants.

Performance Measures. The primary performance measures were (a) the ability of the crews to detect predefined *target events* (i.e., changes plant parameter values and/or alarms that provided an indication of a plant malfunction), and (b) the ability to diagnose *plant malfunction*s (of which the target events were symptoms). Crew detection of target events and diagnosis of plant malfunctions were assessed based on crew communication during the scenario. In addition all navigation key presses to access information and control displays were recorded. A final performance measure was subjective workload. Participants filled out a NASA task load index workload rating form after each test scenario (Hart & Staveland, 1988).

Results

Table 7.1 summarizes the performance of test participants in each of the three display conditions. As the table shows, test participants performed consistently better in the two WPIS conditions than in the Conventional Overview condition on all measures. Participants performed better with the WPIS than with the Conventional overview display in identifying target events (24% improvement) and in diagnosing plant disturbances (27% improvement). A chi-squared test established that these differences were statistically significant ($p < .05$).

Evidence of superiority of the WPIS in providing a broad overview of plant state also comes from examination of display navigation performance. The SO navigated to more than twice the number of detailed information displays in the Conventional Overview condition compared with the WPIS conditions. A paired *t* test found this difference in navigation to be statistically significant ($p < .04$). The difference in number of displays accessed provides converging evidence that the WPIS displays supported the SOs information needs better than the Conventional Overview.

Subjective operator workload also tended to be lower with the WPIS displays than with the alternative overview displays. In the case of the CO/SO scenarios, the difference was statistically significant ($p < .048$).

Operator comments during the debrief sessions and results of the questionnaire were consistent with the objective performance results. Test participants

TABLE 7.1
Summary of Operator Performance in Each of the Three Display Conditions

	Large-Screen WPIS	*Local WPIS*	*WPIS Conditions (Combined)*	*Conventional Overview*
Proportion correct detection of target events	.57	.65	.61	.50
Proportion correct diagnosis of malfunctions	.65	.67	.66	.52
Mean number of SO information navigations	1.25	2.50	1.88	4.42
Mean workload rating	.48	.47	.48	.54

provided positive feedback on the functional organization of information, the alarm organization and coding scheme, and the level of detail of the information displayed. Participants indicated that the WPIS displays were more effective than the Conventional Overview displays for maintaining broad situation awareness; being alerted to changes in equipment state, automatic system actuation, and parameter values; and assessing the nature and implications of abnormalities.

Collectively, the results of this test provide quantitative evidence of the superiority of the conceptual approach taken for the design of the WPIS overview display design, that is, functional organization, as compared to more traditional approaches, which organize plant state information purely around a physical mimic.

The results of WPIS test 2 also showed that the WPIS displays are effective overview displays whether they are presented on large wall-mounted display units or on local workstations. Objective performance on the test scenarios and workload ratings were similar in the WPIS Overview Condition and the Local Overview Condition, with no statistically significant differences. Furthermore, questionnaire ratings and final debrief comments indicated that test participants felt the WPIS would still be effective overview displays if they were made available on workstation CRTs instead of large, mounted display units.

At the same time, test participants generally felt that there were distinct advantages to large wall-mounted display units that provide a shared view to everyone in the control room. Overview displays at local workstations can be seen easily only by the operators sitting at their workstations. This limits the ability of new people entering the scene to get a quick overview of plant state. It also limits the ability of the operators to walk around and be able to see the plant overview from anywhere in the control room. The general consensus of test participants was that large wall-mounted displays should be included in an advanced control room.

DISCUSSION

The results support the basic design assumptions underlying the WPIS overview displays. Both objective performance measures and operator feedback via questionnaire and debrief session provided converging evidence that the WPIS overview displays were superior to the conventional overview displays in supporting situation awareness of plant state. Operators detected more targets and diagnosed more malfunctions with the WPIS overview display than with the conventional overview displays. In addition, workload ratings were lower with the WPIS displays than with the conventional overview displays. Finally, operator questionnaire ratings and debrief comments confirmed that operators strongly preferred the WPIS displays to the alternative overview displays.

There are several factors that contribute to the preference for, and improved performance with, the WPIS overview displays over the conventional overview displays. Questionnaire ratings and participant debrief comments clarified some of these contributing factors. They include the selection and density of information presented, the organization of information around plant functions as opposed to around a physical mimic, the integration of alarms into displays, and the display formats included in the WPIS—particularly the use of trend plots and integrated display elements.

IMPLICATIONS

In this chapter, we have presented a case study that traced the design process for development of a first-of-a-kind system—a group view display intended to support individual and team situation awareness for a power plant application. The process started with an analysis of the human performance issues and support requirements that was grounded in the existing research base and lessons learned across domains. This established the core elements that needed to be included in the group view display. A function-based cognitive task analysis was used to define the detailed information content and organization of the display. A key design decision was to create a plant overview that was organized around plant functions rather than around a physical mimic of the plant. Finally, a series of empirical studies were performed to establish the effectiveness of the displays in supporting individual and team situation awareness.

This case study has implications that extend beyond the specific WPIS design. Lessons can be drawn from the study for the design of group view displays in general as well as for the design of First-of-a-Kind Systems.

Implications for Design of Group View Displays

The study illustrates and provides support for the use of function-based task analysis to specify the content and organization of group view displays. The WPIS displays that were defined based on a function-based cognitive task analysis provided a richer and more effective set of parameters for maintaining broad awareness of plant state than the conventional overview display.

Similarly, the case study provides support for the use of a function-based representation as a framework for organizing plant overview information. The functionally organized WPIS was more effective and was preferred by operators over the conventional overview display that utilized a physical plant mimic as the organizational scheme. These results are consistent with the line of research on ecological interface design (Burns, in press; Vicente & Rasmussen, 1992; Vicente, 1999)

Finally, the studies illustrate that information dense displays that may appear cluttered on first inspection may in fact be more effective than sparser displays when used by experienced domain practitioners. On first exposure, the operators participating in the studies perceived the WPIS displays as being dense. However, after minimal experience using the displays operators expressed a clear preference for the more dense WPIS displays over the more data sparse Conventional Overview displays. The WPIS displays provide more parallel plant information and include integrated graphic elements such as trend displays that allow operators to more easily detect change and project plant parameter behavior into the future. These features allow operators to focus in more quickly on the causes of plant disturbances and their repercussions on achievement of plant safety and power production goals without having to navigate to lower level displays. Burns (in press) found similar findings in a more controlled laboratory setting.

Implication of Results for Design of First-of-a-Kind Systems

The WPIS study serves as a case study illustrating one approach to the design of first-of-a-kind systems that are envisioned to dramatically change the cognitive and collaborative activities entailed by the work environment. In those cases, traditional cognitive task analysis methods that rely on elicitation of expert knowledge and strategies are of limited value because much of the existent expertise is likely to reflect strategies and work-arounds intended to cope with limitations of the existing interfaces and technologies. The goal of first-of-a-kind systems is to change dramatically the work environment so as to eliminate the need for these work-around strategies. In those cases what is needed are cognitive task analysis methodologies that can identify the fundamental cognitive and collaborative demands of the work domain that transcend particular technologies and interfaces.

Function-based cognitive task analysis methods (Potter, Roth, Woods & Elm, 2000) and related cognitive work analysis methods (Rasmussen, 1986; Vicente, 1999), offer a principled approach to uncover and document the inherent cognitive and collaborative demands of the domain and the requirements for effective support systems.

REFERENCES

Burns, C. M. (in press). Putting it all together: Improving display integration. *Human Factors*.

Cannon-Bowers, J. A., & Salas, E. (Eds.). (1998). *Making decisions under stress: Implications for individual and team training*. Washington, DC: American Psychological Association.

Endsley, M. R. (1995). Situation awareness in dynamic human decision-making: Theory. *Human Factors, 37*, 32–64.

Hart, S. G., & Staveland, L. E. (1988). Development of NASA-TLX (Task Load Indexes): Results of Empirical and Theoretical Research. In P.A. Hancock & N. Meshkati (Eds.), *Human mental workload* (pp. 139–183). Amsterdam: North Holland.

Hogg, D. N., Folleso, K., Strand-Volden, F., & Torralba, B. (1995). Development of a situation awareness measure to evaluate advanced alarm systems in nuclear power plant control rooms. *Ergonomics, 11*, 2394–2413.

Hutchins, E. (1995). *Cognition in the wild.* Cambridge, MA: MIT Press.

Mumaw, R. J., Roth, E. M., Vicente, K. J., & Burns, C. M. (2000). There is more to monitoring a nuclear power plant than meets the eye. *Human Factors, 42*, 36–55.

Patterson, E. S., Watts-Perotti, J., & Woods, D. D. (1999). Voice loops as coordination aids in space shuttle mission control. *Computer Supported Cooperative Work, 8*, 353–371.

Patterson, E. S., & Woods, D. D. (1997). Shift changes, updates, and the on-call model in space shuttle mission control. In *Proceedings of the Human Factors and Ergonomics Society 41st Annual Meeting* (pp. 243-247). Santa Monica, CA: Human Factors and Ergonomics Society.

Potter, S. S., Roth, E. M., Woods, D. D., & Elm, W. C. (1998). A framework for integrating cognitive task analysis into the system development process. In *Proceedings of the Human Factors and Ergonomics Society 42nd Annual Meeting* (pp. 395–399). Santa Monica, CA: Human Factors and Ergonomics Society.

Potter, S. S., Roth, E. M., Woods, D. D., & Elm, W. (2000). Bootstrapping multiple converging cognitive task analysis techniques for system design. In J. M. C. Schraagen, S. F. Chipman, & V. L. Shalin (Eds.), *Cognitive task analysis* (pp. 317–340). Mahwah, NJ: Lawrence Erlbaum Associates.

Rasmussen, J. (1986). *Information processing and human-machine interaction: An approach to cognitive engineering.* New York: North Holland.

Roth, E. M. (1996). *Description of the westinghouse operator decision-making model and function based task analysis methodology* (Tech. Rep. WCAP-14695, Westinghouse Energy Systems). Pittsburgh, PA: Westinghouse Electric Corporation.

Roth, E. M. (1997). Analysis of decision-making in nuclear power plant emergencies: An investigation of aided decision making. In C. Zsambok & G. Klein (Eds.), *Naturalistic decision making* (pp. 175–182), Mahwah, NJ: Lawrence Erlbaum Associates.

Roth, E. M., & Mumaw, R. J. (1995). Using cognitive task analysis to define human interface requirements for first-of-a-kind systems. In *Proceedings of the Human Factors and Ergonomics Society 39th Annual Meeting* (pp. 520–524). Santa Monica, CA: Human Factors and Ergonomics Society.

Roth, E. M., Mumaw, R. J., & Lewis, P. M. (1994). *An empirical investigation of operator performance in cognitively demanding simulated emergencies.* (Tech. Rep. No. NUREG/CR-6208, Nuclear Regulatory Commission). Washington DC: U.S. Nuclear Regulatory Commission.

Roth, E. M., & O'Hara, J. (1999). Exploring the impact of advanced alarms, displays, and computerized procedures on teams. In *Proceedings of the Human Factors and Ergonomics Society 43rd Annual Meeting* (pp.158–162). Santa Monica, CA: Human Factors and Ergonomics Society.

Roth, E. M., & Woods, D. D (1988). Aiding human performance: I. Cognitive analysis. *Le Travail Humain, 51*, 39–64.

Rusnica, L. A., Kerch, S. P., Thomas, V. M., Kenney, S., Brockhoff, C. S., Morris, B. C., Roth, E. M., & Sugibayashi, N. (1999). Information Display System, U.S. Patent # 5,859, 885, Jan. 12, 1999.

Salas, E., Prince, C., Baker, D. P., & Shrestha, L. (1995). Situation awareness in team performance: Implications for measurement and training. *Human Factors, 37*, 123–136.

Serfaty, D., Entin, E. E., & Johnston, J. H. (1998). Team coordination training. In J. A. Cannon-Bowers and E. Salas (Eds.), *Making decisions under stress: Implications for individual and team training* (pp. 221–245). Washington, DC: American Psychological Association.

Stubler, W. F. & O'Hara, J. M. (1996). *Group-view displays: Functional characteristics and review criteria* (Tech. Rep. No. E2090-T-4-4-12/94, Brookhaven National Laboratory). Upton, New York: Brookhaven National Laboratory, Department of Advanced Technology.

Vicente, K. (1999). *Cognitive work analysis: Towards safe productive and healthy computer based work.* Hillsdale, NJ: Lawrence Erlbaum Associates.

Vicente, K. J., Mumaw, R. J., & Roth, E. M. (1997). *Cognitive functioning of control room operators - final phase* (Final Report AECB). Ottawa, Canada: Atomic Energy Control Board.

Vicente, K. J., & Rasmussen, J. (1992). Ecological interface design: Theoretical foundations. *IEEE Transactions on Systems, Man and Cybernetics, SMC-22*, 589–606.

III

Decision-Making Models

8

Self-Evaluation, Stress, and Performance: A Model of Decision Making Under Acute Stress

Michael R. Baumann
Janet A. Sniezek
Clayton A. Buerkle
University of Illinois, Urbana-Champaign

Many factors influence decision-making performance under acute stress. It is our belief that one of them is self-evaluation. This chapter proposes a model of how self-evaluation may affect performance through influencing motivation, anxiety, and self-regulatory processes. However, before proceeding, it is first necessary to define our terms.

Most of the work to date on performance under acute stress comes from the tradition of Naturalistic Decision Making. Therefore, we draw on that tradition in choosing our definition of acute stress. Salas, Driskell, and Hughes (1996) defined *acute stress* as a state that commonly occurs in situations involving high physiological arousal, the need to make multiple decisions rapidly, incomplete information on which to make these decisions (high uncertainty), and extreme consequences of failure—often, literally, life and death. For the stress to be acute, the event must cover a relatively small time span and involve focused threat (hence the term *acute*).

We have built a model of how various psychological processes relate to performance during the experience of acute stress. In defining the experience of acute stress, we build on the Salas, et. al. (1996) definition. To us, acute stress is a subjective state. For an individual to experience acute stress, four requirements must be met. First, at least some of the stressors Salas, et. al. (1996) employ in their definition must be present; second, the individual must be aware of the presence of these stressors; third, the individual must be motivated to attempt to resolve the situation, and fourth, the individual must be uncertain as to whether successful resolution of the situation is possible. The combination of these factors will lead to the subjective experience of acute stress.

For example, a situation might lead to high arousal, require that multiple decisions are made in a short time span with incomplete information, and have dire consequences of failure, without the individual experiencing stress. For the situation to induce acute stress, the individual must recognize that these factors are present. Even if the individual recognizes the arousal, requirements of the task, and consequences of failure, he or she still might not experience acute stress. The individual may value successful resolution of the situation, but consider success hopeless and resign him- or herself to suffering the consequences of failure instead of being motivated to resolve the situation. Or, perhaps the individual recognizes the situation, but is confident that he or she can successfully resolve it. Neither of these individuals would experience acute stress.

A PROPOSED MODEL

As stated earlier, we believe that four factors are necessary for the occurrence of acute stress: environmental demands, awareness of demands, performance goal, and uncertainty. However, our model of performance under acute stress can be understood in terms of the interrelationships of three factors: self-evaluation of performance, task experience, and anxiety (see Fig. 8.1 for a graphical representation of the model). We deal with the model in more detail next.

Self-Evaluation of Performance

For the purposes of the model, we define *self-evaluation of performance* as any process during which an individual expends any attention on evaluating his or her performance, performance-related behavior (such as strategies and goal attainment), or affective reactions to performance and related behaviors. Self-evaluation can occur before a task (*performance predictions*), during a task (*performance indications*), or after a task (*performance evaluations*). Performance predictions will affect final performance through their relationship to the individual's beliefs regarding the probability of successful performance (performance expectations). Performance indications involve comparing goal states to current states. As such, performance indications both can serve as the basis for self-regulatory processes and can function to update performance expectations. If self-evaluation reveals problems with performance, an individual may use this information to alter his or her strategies. If performance expectations change, the individual may alter the level of effort expended. Finally, although performance evaluation (posttask

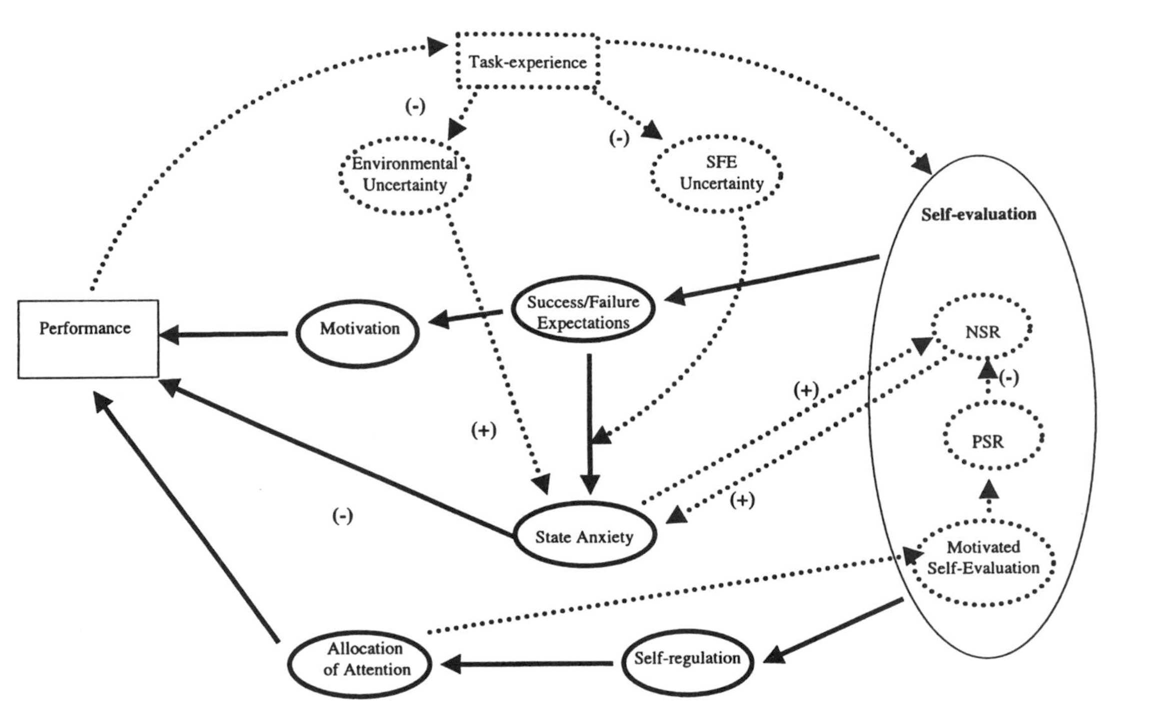

<u>Legend</u>. Due to the dynamic nature of our model, it cannot be drawn in a simple feed-forward manner. Proposed processes and states leading from self-evaluation to performance are denoted by solid arrows and shapes. All feedback processes occurring in the model are denoted by dotted arrows and shapes. For simple linear relationships, a (+) or (-) is used to represent positive and negative relationships, respectively.

FIG. 8.1. A proposed model for the relationships between self-evaluation, task experience, anxiety, and performance under acute stress.

self-evaluation) may serve many goals, in the current model its primary effect is to update the individual's task perceptions and performance expectations for future tasks.

Success/Failure Expectations. The first effect of self-evaluation of performance we propose is its effect on *success/failure expectations* (SFE). SFE reflect an individual's own beliefs about the probability of successfully accomplishing a task. As such, SFE include both a judgment of resources and uncertainty regarding that judgment. Essentially, SFE are a function of the difference between an individual's judgments of the personal resources available for the task and his or her judgments of the task demands. The larger the individual judges the difference between available resources and task demands to be, the further the probability of success differs from 50/50 (maximum uncertainty). In our model, performance expectations are fluid. When personal resources available are judged to exceed task demands, SFE are positive (lean toward success). When personal resources available are judged to be insufficient to meet task demands, SFE are negative (lean toward failure). Although initially established by performance prediction judgments, SFE can fluctuate over the course of the task as people gain task experiencc and re-evaluate their available resources relative to the demands of the task.

We propose that SFE will have two separate influences on performance. The first influence is due to a motivational effect. We postulate this effect due to the goal setting literature. A large body of research on goal setting suggests that people exhibit the highest level of performance when striving for a goal that they consider to be challenging, yet attainable (Latham & Locke, 1991). When people find the goal too easy, they do not put forth as much effort. The effort is not required, so why expend it? When people find the goal too difficult, they do not put forth as much effort. If the goal cannot be obtained, the effort is wasted. If the effort is wasted, then why not save the energy?

In any situation where acute stress is likely to occur, the high stakes will lead the decision makers to have the goal of successful resolution of the situation. If the decision makers do not believe that success is possible (SFE much less than zero), they are certain that they will fail to attain the goal and, therefore, are likely to expend less effort. This does not mean that people will never expend effort in the face of overwhelming odds against them. SFE are subjective judgments of the probability of success; they do not necessary reflect the probability of success calculated from external criteria. As such, it is possible that SFE will be positive (meaning success is judged to be possible) even when a more objective assessment suggests that success is impossible. Studies in self-evaluation (Radhakrishnan, Arrow, & Sniezek, 1996) and forecasting (Buckley & Sniezek, 1992) suggest people's desire for outcomes affects their judgments of the probability of those outcomes occurring. Therefore, if desire for goal attainment is sufficiently high, individuals may believe even an impossible goal is within their grasp. What we are saying is that unless people believe they can succeed, they will not put forth the effort required to do so.[1]

High SFE will also lead to reductions in the amount of effort expended. When judged personal resources available greatly exceed task demands, individuals may assume that they can successfully resolve the situation while expending only minimal effort. Although they may successfully resolve the situation, they might not perform as well as they are capable of performing. Individuals with SFE near zero will put forth the most effort. These individuals believe that the personal resources they can bring to bear on the problem are roughly equal to the demands of the task. In other words, they believe that success is possible, but only if they employ all of the resources available to them.

Proposition 1: The relationship between effort expended and SFE will be curvilinear, with the highest level of effort expended when judged resources available equal judged task demands.

The second way that SFE may affect performance is by affecting anxiety. Past research described stress as the relationship between perceived task demands and perceived personal resources (Lazarus, 1991; Maule & Hockey, 1993; Salas, et al., 1996; Sarason, 1980). Lazarus and Folkman (1984; Lazarus, 1991) view stress as a state that occurs when (a) an individual is motivated to successfully resolve a situation and (b) the individual perceives his or her personal resources to be barely adequate or inadequate to accomplish this. From this perspective, anxiety is essentially the result of a fear of the consequences of failure. Although we believe that this will explain part of the anxiety experienced under acute stress, it is not the whole story.

Uncertainty regarding the occurrence of unpleasant events has been shown to lead to anxiety (Monat, Averill, & Lazarus, 1972). Dawes (1988) argued that uncertainty does not have to involve a fear of loss to be unpleasant; uncertainty is an unpleasant state in and of itself. Dawes cited numerous examples of actions that people take to reduce uncertainty and feel in control of random events. This is true even when the expected outcomes of the events are positive, suggesting that uncertainty aversion is not due to fear of a loss or an unpleasant outcome. Research on training individuals for stressful situations suggests that the more preparatory information individuals receive before entering stressful situations, the less anxiety they experience (Driskell & Johnston, 1998). There are even

[1]Although there are a number of examples in military history where people face hopeless situations knowing success is impossible, we consider these situations to be rare exceptions and, therefore, we model only what we believe to be the more common case. However, our model would still hold if, in these "die trying" situations, the goal changes. For the Kamikaze of World War II, the goal was not survival, but to destroy enemy ships and make the U.S. advance too costly to continue.

mathematical models that incorporate people's tendency to avoid uncertainty into mainstream expected utility theory (Schmidt, 1998).

Therefore, we believe that uncertainty regarding task performance will affect anxiety and that this effect is separate from the anxiety stemming from fear of failure. If there is a large amount of uncertainty surrounding an SFE, then even individuals who estimate their available resources to exceed the task demands will not be certain of it. We assert that the uncertainty regarding performance and the fear of consequences of failure will separately contribute to the amount of anxiety experienced (see Fig. 8.2).

Consider a situation in which everyone has the same level of uncertainty in his or her judgments of personal resources available and task demands. There is a greater degree of uncertainty regarding performance for people who judge their personal resources available to be approximately equal to task demands than for those who judge there to be a large difference. When judging whether available personal resources exceed task demands, actual available resources may differ from estimated resources and actual task demands may differ from estimated task demands. As a result, when available personal resources are judged to be nearly

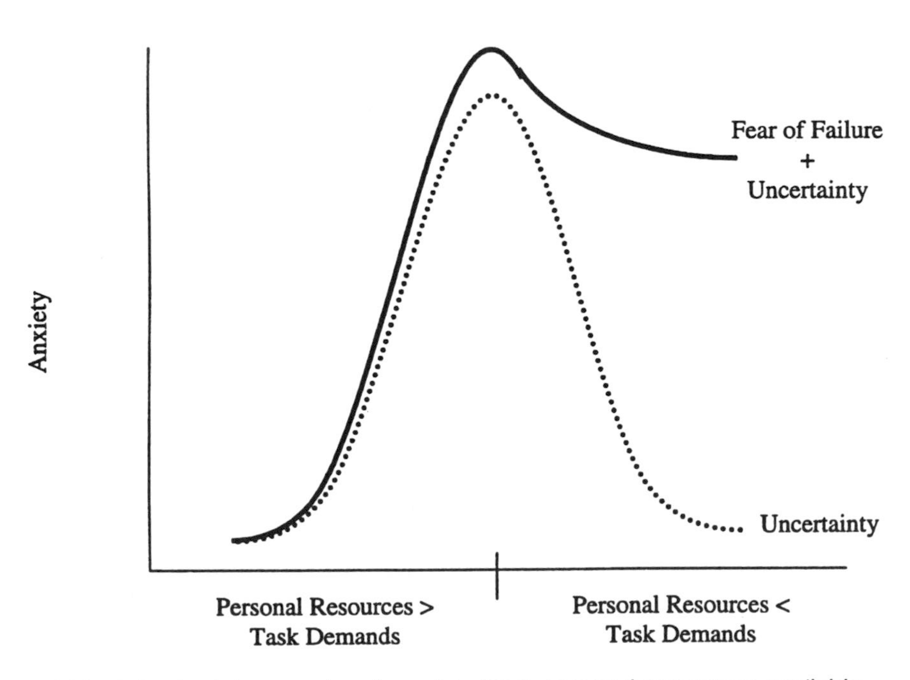

FIG. 8.2. Anxiety as a function of judged personal resources available and judged task demands.

equal to task demands, there is a large amount of uncertainty as to whether available personal resources exceed task demands or the reverse. For these people, there are two sources of anxiety: the fear of the consequences of failure and uncertainty regarding performance. We, therefore, predict that these individuals will experience more anxiety than either those who have very positive or very negative SFE.

When task demands greatly exceed available resources (very negative SFE), success is impossible (probability of success is much lower than 50%). There is no uncertainty regarding performance, so only fear of the consequences of failure will contribute to anxiety. Under such conditions, individuals may even reallocate their attention to accepting the consequences of failure. These individuals, therefore, should be less anxious than those for whom task demands and available resources are more nearly equal. However, unlike individuals who judge their available resources to greatly exceed the task demands (very positive SFE, probability of success much greater than 50%), individuals with very negative SFE still have anxiety stemming from their fear of the consequences of failure. Therefore, individuals with very negative SFE should still experience greater anxiety than those with very positive SFE. Figure 8.2 shows a graphical representation of our predictions.

Proposition 2: The relationship between anxiety and SFE will be curvilinear, with the highest level of anxiety occurring when judged resources available equal judged task demands. However, the curve will be nonsymmetrical.

Self-Regulation. There are two ways in which self-evaluation might serve self-regulatory purposes. The first of these is through a monitoring function and has been documented by other researchers (e.g., Kanfer & Ackerman, 1989). The primary task in many of these domains can be broken down into subtasks. Performance indications (self-evaluations of performance made during the task itself) can be used to monitor both overall performance and performance on specific subtasks. In so doing, self-evaluation can alert the individual when performance is unsatisfactory, allowing the individual to alter his or her allocation of attention.

We present as an example a domain we study in our own research—shipboard damage control. When a naval vessel takes damage in combat operations, a single individual (the Damage Control Assistant or DCA) is in charge of coordinating the containment and repair of the damage. There may be multiple types of damage to the ship (fire, flood, and hull ruptures), as well as damage in multiple areas. The DCA must distribute his or her attention across all of the various problems occurring in various parts of the ship, taking into account what systems are most critical for mission success. If the DCA is engaging in self-evaluation as a monitoring function, he or she might notice that all of the flooding is under control, but one of the fires keeps spreading. The DCA might then shift some of the attention he or she has been expending on coordinating flood containment to coordinating fire containment and extinguishment.

Proposition 3: Performance indications affect allocation of attention among subtasks.

The second way in which self-evaluation of performance may serve a self-regulatory purpose is through the use of motivated self-evaluation. We propose that motivated self-evaluation may reduce the amount of anxiety the individual experiences, thereby increasing performance. A common finding in anxiety research is that people experiencing anxiety are distracted by negative self-evaluations. Thoughts of failure, worries, and other "negative self-reactions" (NSR) are often reported by subjects under anxiety (e.g., Deffenbacher, 1980; Morris, Davis, & Hutchings, 1981; Wine, 1980). Research in the clinical literature suggests that if a person engages in a large amount of NSR, he or she will become more prone to perceiving events as negative (Meichenbaum, 1985). There is also some support for this notion in the literature on priming (Bylsma, et al., 1992). Engaging in NSR could prime other negative thoughts, which could affect task performance either directly by drawing attention away from the task or indirectly by increasing the individual's anxiety level. In short, NSR and anxiety could feed on each other in a vicious cycle, each continually increasing the other.

If, on the other hand, individuals employ motivated self-evaluation to generate positive self-reactions (PSR), they may be able to offset some of the effects of anxiety. By interjecting PSR when confronted with NSR, it may be possible to break the NSR-anxiety-NSR cycle. In the clinical literature, it has been shown that teaching people to engage in positive thoughts reduces their overall ratings of life stress (Hains & Szyjakowski, 1990; Meichenbaum, 1985) and in some cases even their levels of trait anxiety (Hains & Szyjakowski, 1990). In research on training for tactical decision making under stress, it has been shown that training people to evaluate the symptoms of stress as normal instead of something to worry about in their own right decreases the amount of anxiety experienced (Driskell & Johnston, 1998). In both cases, interjecting positive evaluations of the situation prevents anxiety from increasing. As long as motivated self-evaluation does not draw more attentional resources away from the task than does the NSR and anxiety it prevents, PSR should improve performance.[2]

Proposition 4: Using motivated self-evaluation to create PSR will decrease the amount of anxiety experienced in stressful situations.

[2]Although some research in the affect and cognition literature suggests that mood affects processing style (e.g., Bless et al., 1996; Bodenhausen, Sheppard, & Kramer, 1994), we do not believe that the use of PSR during acute stress will lead to similar effects. All individuals under acute stress are experiencing anxiety, regardless of whether they use PSR or not. In other words, they are all in the same "mood." The effects found in the affect and cognition literature only occur between moods, not within.

Task Experience

Expertise.

Task experience should lead to an increase in ability. The more experience an individual has with a task, the more likely it is that the participant has had the opportunity to develop strategies. Strategies for goal attainment often predict performance even better than the level of the goal itself (Chesney & Locke, 1991). These strategies may be strategies for coping with affective responses to the situation or for handling characteristics of the situation in addition to decision-making strategies necessary to resolve the situation.

Proposition 5: Performance will improve with task experience.

Uncertainty Reduction.

The more exposure individuals have had to the task, the more familiar they should be with the task. This should reduce two types of uncertainty: (a) general uncertainty about the environment, and (b) uncertainty regarding whether available personal resources exceed task demands. Reductions in uncertainty regarding the environment should reduce decision makers' anxiety by enabling them to anticipate likely events and avoid worrying about unlikely events. Their fear of the unknown is not reduced, but the amount that is unknown becomes smaller.

Proposition 6: Uncertainty will decrease with task experience

Proposition 7: The lower the uncertainty, the lower the level of anxiety that will be experienced under stress.

Reducing uncertainty regarding available personal resources and task demands will also decrease anxiety. In any SFE, there is a certain amount of uncertainty. If the uncertainty is large enough that successful resolution of the situation is not assured, then the decision maker should experience anxiety. As task experience increases, uncertainty regarding SFE should become smaller. Experimental support for this assertion has been reported (Radhakrishnan et al., 1996). As uncertainty is reduced, the amount of anxiety due to uncertainty is also reduced. This is demonstrated graphically in Fig. 8.3. Imagine the same individual, making the same SFE judgment at two different times. Both times, the individual judges his or her available personal resources to exceed the task demands by 2 units. Under small amounts of uncertainty, the individual will experience almost no anxiety—an SFE of +2 means the individual is certain of success (top panel of Fig. 8.3). However, when there are large amount of uncertainty regarding the SFE, even a +2 leaves a considerable amounts of uncertainty regarding whether success is possible (bottom panel of Fig. 8.3).

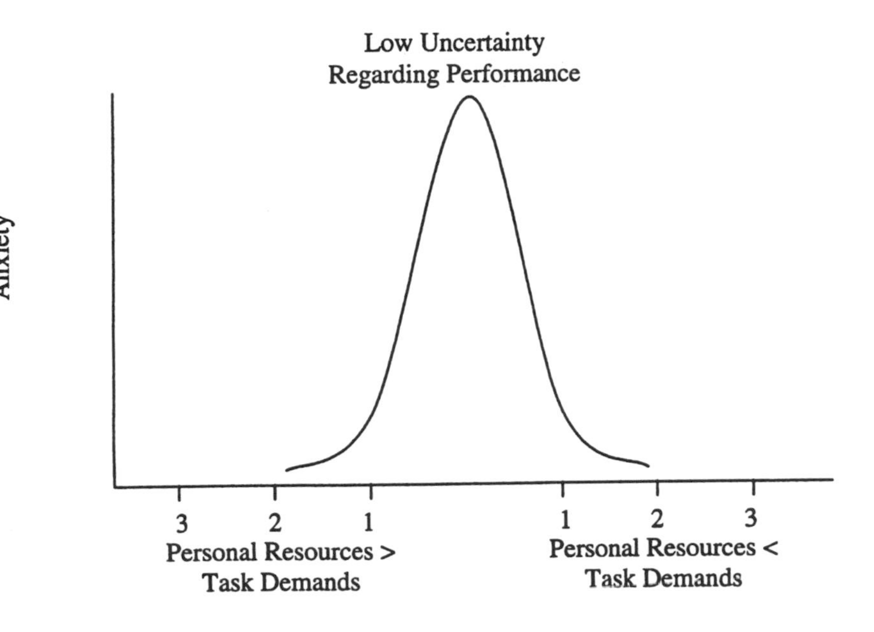

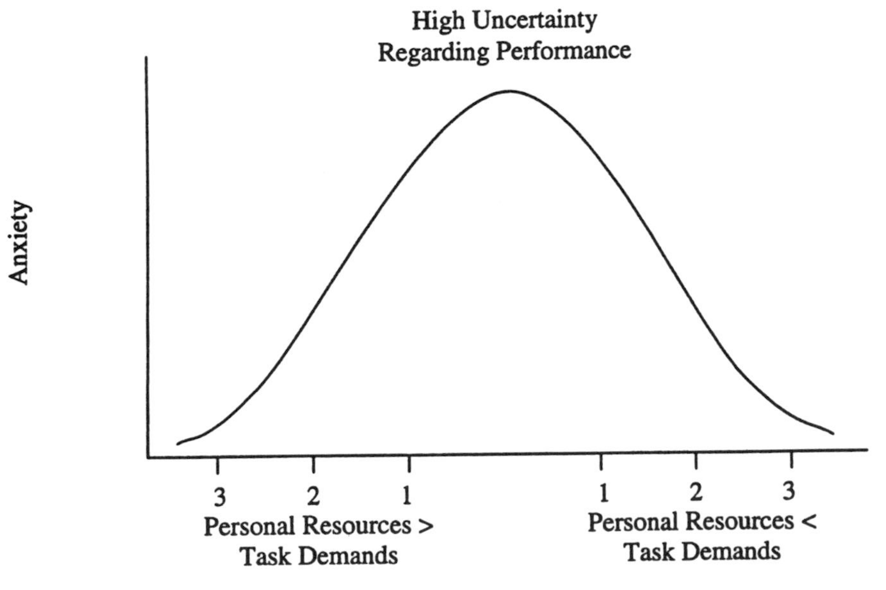

FIG. 8.3. Anxiety as a function of the difference between personal resources and task demands under different levels of uncertainty.

Anxiety

We believe that the subjective experience of acute stress involves anxiety. Many of the stressors involved in situations that typically induce acute stress also have been shown to lead to anxiety. For instance, time pressure (Leon & Revelle, 1985), uncertainty (Monat et al., 1972), and threat of bodily harm (Orasanu & Backer, 1996). It has been argued that the presence of stressors, per se, does not necessarily lead to decreases in performance (Klein, 1996), and we do not dispute this. However, we believe the anxiety typically induced by these stressors will impair performance. Across a wide range of tasks, from the almost entirely physical (Burton, 1988) to the almost entirely cognitive (Leon & Revelle, 1985; Wright, 1974), and places in between (Hardy & Parfitt, 1991), anxiety has been found to have negative effects on performance. There are two specific effects of anxiety that are of interest to the current chapter.

Proposition 8: Anxiety will be negatively related to performance.

Narrowing of Focus. A number of studies have shown that anxiety leads to a narrowing of attentional focus (see Geen, 1980, for a review). In other words, anxiety leads to "tunnel vision." In simple decision making, people attend to fewer cues (Wright, 1974). The more complex the stimulus field, the greater the performance disadvantage of those under higher levels of anxiety (Leon & Revelle, 1985). Research with dual-processing tasks has found that, when placed under stress, participants tend to ignore one of the tasks and focus on the other (e.g., Bacon, 1974). This suggests that on complex tasks, anxiety might lead individuals to tunnel in on a specific subtask and forget about what else is going on around them. Klein pointed out that this narrowing of focus is not necessarily a problem and may in fact even be adaptive because it would lead the decision maker to focus only on important information (1996). However, because people focus on only what seems important to them, at the time of the narrowing, the narrowing may prevent people from noticing when something else becomes important. According to our definition of acute stress, a number of decisions must be made rapidly. Most acute stress domains involve complex tasks. If the individual becomes overfocused on a specific subtask, he or she may not notice when another subtask becomes more important.

There is some evidence that, at least at a metacognitive level, participants are aware of the onset of tunnel vision (Baumann & Sniezek, 1999). If so, individuals using self-evaluation as a monitoring function may be able to recognize the onset of tunnel vision and deliberately switch their focus of attention.

Negative Self-Reactions. As mentioned in the section on self-evaluation, a common occurrence in anxiety is the experience of negative self-reactions (NSR). In the anxiety literature, these thoughts are usually referred to as either "worry" or as "off-task thinking" (e.g., Deffenbacher, 1980; Morris et al., 1981). There are two ways in which NSR may negatively affect performance. First, situations associated with acute stress tend to require the decision maker's full attention. NSR pulls attention away from doing the task and places it on worries about failing to do the task (e.g., Sarason, 1980; Wine, 1980). As such, it pulls attentional resources away from a task where attention is already stretched to the limit or beyond. By so doing, NSR reduces the decision maker's ability to attend to and process all of the information available. Given that the information available is typically incomplete, being unable to even process the information that is available is likely to result in the decision maker having very little information upon which to base a decision. This should lead to poor decisions, resulting in higher NSR, further reducing the available attentional resources.

Proposition 9: NSR will interfere with performance by consuming attentional resources needed to resolve the situation.

The other way in which NSR may lead to a decrease in performance is by actually increasing the level of anxiety. Engaging in NSR may prime individuals to view events in a more negative manner (Byslma, et al., 1992). A minor setback is suddenly seen as "the end of the world." As more and more decisions and events are viewed as having gone badly, the individual becomes more uncertain about his or her ability to successfully resolve the situation (Deffenbacher, 1980). This increase in uncertainty leads to an increase in anxiety. The increase in anxiety then leads to a further increase in NSR, and the cycle spirals upward.

Proposition 10: Anxiety and NSR are related in a positive feedback loop.

We believe that it is possible to overcome both of these possible effects of NSR. Individuals who effectively use self-evaluation for self-regulatory purposes may engage in positive self-reactions (PSR) to prevent NSR from cycling with and increasing anxiety. Using PSR will not directly reduce the attentional resources drawn away from the task—thinking positive thoughts is likely to draw off just as much attention as thinking negative thoughts. However, it is likely to keep the level of NSR in check by breaking the NSR-anxiety cycle. By preventing NSR from increasing, engaging in PSR prevents the drain on attentional resources from increasing. By preventing NSR from leading to increasingly negative evaluations of events, PSR prevents NSR from leading to higher levels of anxiety. There is some evidence that engaging in positive thinking even reduces anxiety (Hains & Szyjakowski, 1990; Meichenbaum, 1985; Morris & Fulmer, 1976). The end result is that engaging in PSR may indirectly reduce the attentional drain due to NSR, as well as preventing NSR from leading to further anxiety.

METHODOLOGY FOR TESTING THE MODEL

There are several difficulties associated with studying acute stress. First, situations involving acute stress occur only rarely in the real world. Second, when these situations do occur, the stakes are too high to risk interference for research purposes. Third, there are no psychometrically validated measures of acute stress. A successful methodology must account for all three of these difficulties.

Solutions to the first two problems can be found in the literature on training for acute stress. Because hands-on practice is not feasible, many organizations have turned to the use of simulator training as an alternative. Simulation is widely used for training in both commercial and tactical aviation (Salas, Bowers, & Rhodenizer, 1998), as well as training for space flight, armored vehicles, and even locomotives (Flexman & Stark, 1987). There is some evidence that these simulators are useful in at least some domains (see Bell & Waag, 1998, for a review of simulator effectiveness in air combat training).

The key to an effective training simulator is psychological realism (Dennis & Harris, 1998; Kantowitz, 1992). A good training simulator requires participants to make decisions similar to those required in the task for which they are training, and these decisions must be made under similar psychological conditions. For acute stress, this means that the simulator must capture the time pressure, the anxiety, the uncertainty, and the high stakes involved in an actual acute stress situation. Research in the tradition of Naturalistic Decision Making suggests that simulators with moderate levels of physical fidelity involving stressors introduced at levels only moderately close to those encountered under acute stress are sufficiently real psychologically to lead to improved performance (Driskell & Johnston, 1998). If they are sufficiently real to be effective training tools, they should be sufficiently real to provide an experimental environment for the more general study of performance under acute stress (Donovan, Sniezek, & Baumann, 1996). In short, turning to a simulator methodology solves the first two problems mentioned above.

To solve the third problem, the lack of psychometrically validated measures of acute stress, we take a multivariate approach. To measure the subjective state of stress, we choose to employ a measure of anxiety. It has been shown in the past that the stress produced by many of the stressors involved in the Salas, et al., definition of acute stress can be detected using measures of anxiety (e.g., Leon & Revelle, 1985 on time pressure; Monat et al., 1972 on uncertainty). Therefore, we believe that anxiety measures will serve as a useful indicator of the overall level of acute stress experienced.

However, anxiety and acute stress may not have a one-to-one mapping. Therefore, we also employ measures of subjective time pressure, subjective difficulty, motivation, and uncertainty to determine whether the simulation fulfilled the requirements of our definition of acute stress sufficiently well to be used in research. We manipulated the environmental stressors to insure they occurred (Requirement 1). The measures of subjective time pressure and difficulty were employed to determine whether participants recognized the stressors as present (Requirement 2). Motivation was measured to determine whether successful task completion was a valued goal (Requirement 3). Finally, uncertainty regarding performance was measured to determine the amount of doubt participants had regarding whether or not successful performance was possible (Requirement 4)

EMPIRICAL SUPPORT FOR THE PROPOSED MODEL

Although the research is still ongoing, we have now conducted several studies and have tested several of the links in our model using a simulator methodology and quantifying stress in terms of anxiety. So far, the tests have been favorable. We describe the methods and preliminary results of the most recent study as an example, but all of our simulator studies have used roughly the same methodology.

In our most recent study, 58 DCA candidates at the U.S. Navy's Surface Warfare Officer's School (SWOS) served as participants. Each participant was assigned four trials on an immersive, multimedia, ship combat crisis simulator currently under development as a training tool. Participants issued orders through a computer console. The same console assailed them with reports of floods, fires, explosions, and system failures. All reports were presented in audio, video, and text format. In case the reports alone were not sufficient to lead to information overload, klaxons, sirens, and shouting were constantly played at a background level. To simulate the uncertainty involved in actual damage control situations, probabilistic elements were included in the simulation of damage propagation and the effectiveness of damage control efforts. Performing the correct actions did not always result in successful damage control. Successful task completion required approximately 25 minutes, and unsuccessful task resolution (i.e., losing the ship) required significantly less time.

In keeping with SWOS's training practices, DCA candidates were organized into 15 groups of three and 2 groups of four ahead of time by SWOS instructors and assigned to simulator time. The participants within each group rotated their turn as either DCA, plotter (an officer who aids the DCA by keeping track of ship status on a large dry-erase ship diagram), or observer (for instructional purposes). Time pressure, information overload, decision making under conditions of incomplete information (uncertainty), and distractor noises played in the background (sirens, klaxons, etc.) were used to create a situation favorable to the experience of acute stress. Simulator trials occurred during the last week of the training course and were portrayed by SWOS instructors as relevant to at-sea damage control performance. Therefore, we expected participants to consider good performance on the simulator a valued goal, making the situation even more conducive to the experience of acute stress. Pre- and posttest self-report questionnaire data were collected in an entry room next to each simulator room.

The variables measured in this particular study were SFE, anxiety, subjective workload, expected performance, and uncertainty regarding estimates of performance. SFE was measured with a newly developed subjective workload scale called the Resource/Demand Index (RDX; Buerkle, Baumann, & Sniezek, 1998). On each of 10 different dimensions, participants are given a 10 centimeter line with hash marks every centimeter, but no numbers. The line is labeled at one end with *available resources greatly exceed task demands*, in the center with *available resources equal task demands*, and at the far end with *task demands greatly exceed available resources*. Participants circle the hash mark that most closely corresponds to the difference between their available personal resources and task demands. Participants are then asked to rate the importance (from 0 to 10) of each dimension. A composite score is calculated as the sum of the judged differences between available resources and task demands for each dimension, weighted by the rated importance of each dimension. For the purposes of this chapter, *available resources greatly exceed task demands* was assigned a point value of 5, *available resources equal task demands*, a value of 0, and *task demands greatly exceed available resources*, a value of -5.

Anxiety was used as a measure of stress. The scale was a short version of Spielberger, Gorsuch, and Lushene's State-Trait Anxiety Inventory (STAI; 1970, 1983) developed by Marteau and Bekker (1992). Subjective workload was measured by the NASA Task Load Index (TLX; Hart & Staveland, 1988). The TLX was chosen both for its ease of administration (Hill et al., 1992) and the particular dimensions it includes (temporal demand, mental demand, and effort demand). We obtained participants' estimates of their performance (both pre- and postperformance) using a simple 0 to 100 scale where 0 was anchored with *no correct actions taken* and 100 with *perfect human performance*. Ninety-five percent credible intervals were obtained for each estimate of performance as a measure of uncertainty regarding performance expectations. Self-report measures were used for NSR and PSR (Baumann, Donovan, & Sniezek, 1998).

In terms of task characteristics, the simulator scores above the midpoint of the scales for time pressure, mental demand, and effort expended on all trials (all $t > 2.2, p < .03$). Averaged across all trials, participants' state anxiety exceeded their baseline levels by 20%, a statistically significant amount ($t = 3.2, p < .001$). In short, the task was cognitively taxing, and our participants were anxious, time pressured, and putting forth effort. Consistent with the work of Baumann, Sniezek, Donovan, and Wilkins (1996) on an earlier version of the simulator, the major psychological elements of acute stress appear to have been present. It is, therefore, worthwhile to examine our predictions.

Our predictions regarding the effects of self-evaluation as mediated by SFE (Proposition 1) received at least initial support. To determine whether SFE had a curvilinear relationship with motivation, each participant's SFE were squared, and these squared scores were then correlated with reported effort expended. Averaged across trials using Fisher's r to z transformation, the correlation was significant (average $r = .32$, $p < .04$). The greater the difference between perceived resources available and perceived demand, the greater the reported effort expended. From these results, it appears that self-evaluation as mediated by SFE does affect motivation in the manner predicted

The same procedure used to examine the relationship between SFE and effort was used to examine the relationship between SFE and anxiety (Proposition 2). When transformed and averaged across trials, the correlation between SFE squared and anxiety was nonsignificant (average $r = -.11$, $p > .44$), although the relationship was in the predicted direction. The strongest individual trial relationship between the squared SFE score and anxiety was $r = -.27$ ($p < .07$). These results do not rule out the possibility that SFE and anxiety are related in the manner predicted, but they do fail to provide support for the predicted relationship.

Support for other portions of our model has been previously presented. In two separate samples, we found support for the idea that motivated self-evaluation can be used to mitigate the effects of anxiety (Proposition 4). As predicted in Proposition 10, we found anxiety and NSR to be positively correlated in multiple samples (Baumann, Donovan, & Sniezek, 1998; Baumann, Buerkle, & Sniezek, 1998). If PSR is not being used as a self-regulatory form of self-evaluation, then one would expect PSR and NSR to be negatively correlated. People who report more negative thoughts about their performance should also report fewer positive thoughts about their performance. However, we have found PSR and NSR to be positively related to each other across individuals in one sample of DCAs (Baumann, Donovan, & Sniezek, 1998) and within individuals in a second sample of DCAs (Baumann, Buerkle, & Sniezek, 1998). The fact that we can obtain this effect within participants in addition to across participants suggests that the relationship is not due merely to individual differences in reactivity or in scale usage. Although the data are only correlational in nature, they suggest that some people do use PSR to keep NSR in check.

Perhaps more interesting is that the amount of PSR participants engage in is more highly correlated with anxiety on previous trials than with anxiety on the current trial. PSR is also highly correlated with trait anxiety—more so in fact than is NSR (Baumann, Donovan, & Sniezek, 1998). This evidence suggests that expectations regarding anxiety led to increases in PSR. Those who previously experienced high anxiety or who typically experience high anxiety are more likely to engage in PSR during a stressful task. Again, results so far are merely correlational, but intriguing and consistent with our model.

The effects of experience have also been largely as predicted. Participants exhibit narrower credible intervals in their pretrial predictions of performance on later trials than on earlier trials (Buerkle et al., 1998), suggesting a reduction in uncertainty regarding task performance (Proposition 6). When participants are doing repeated trials of a single scenario, anxiety decreases across trials (Proposition 7; Baumann et al., 1996). However, when each trial involves a different scenario, uncertainty regarding performance does decrease across trials, but anxiety does not decrease across trials (Sniezek, Baumann, & Buerkle, 1997). Taken together with the quadratic relationship between SFE and anxiety suggesting that greater uncertainty regarding performance is associated with higher levels of anxiety, these findings suggest that uncertainty regarding performance and uncertainty regarding the task act separately on anxiety.

As of yet, we have been able to test few of the proposed links involving performance. Because of the general support already in the literature for the relationships between motivation and performance, allocation of attention and performance, and anxiety and performance, tests of these links in the model have been lower priority than the other less intuitive links. One of the few such tests performed examined the link between anxiety and performance (Proposition 8). Consistent with the literature, anxiety was negatively related to performance within participants (Baumann et al., 1996).

SUMMARY

We believe that acute stress must be measured as a state. To define it only in terms of the events likely to cause it is to repeat the mistakes of other areas of research. Just as it is possible to have time-critical situations where individuals do not experience time pressure and to have set goals where individuals do not engage in goal-oriented behavior, it is possible to have situations involving all of the external factors commonly associated with acute stress that do not lead to the experience of acute stress. Until reliable measures of acute stress are developed, a useful method of measuring acute stress may be to employ anxiety measures.

We assert that self-evaluation, task experience, and anxiety all affect performance in acute stress domains. Self-evaluation affects SFE and self-regulatory activities. SFE affects performance both by affecting motivation and by affecting anxiety. Self-regulatory activities affect performance both by influencing the task solution strategies employed and the use of motivated self-evaluation of performance to mitigate the negative effects of anxiety.

Task experience also affects performance in multiple ways. The first is a simple practice effect. People learn what strategies do or do not work when they have had opportunities to practice the task. Task experience also affects self-evaluations. With task experience, the uncertainty surrounding SFE should be reduced, leading to lower anxiety. Finally, task experience should affect anxiety by reducing the amount of uncertainty regarding the environment.

Anxiety also affects performance under acute stress. Anxiety narrows focus of attention, often resulting in overfocusing or tunnel vision. There is some evidence that people can recognize when this is occurring. We believe that, if people do use self-evaluation for self-regulatory purposes, people may be able to compensate for tunnel vision by deliberately stepping back and taking time to survey the entire situation before returning to a specific part of the problem.

Anxiety may also affect performance by leading to NSR. NSR may lead people to both expect more negative events and interpret events more negatively. Either of these could lead to further anxiety. Therefore, NSR and anxiety may be parts of a positive feedback loop, each increasing the other. People engaging in self-regulatory self-evaluation may deliberately engage in PSR to break this anxiety-NSR spiral. Although we do not expect PSR to reduce existing levels of anxiety, we do believe that it prevents NSR from feeding back into anxiety and thereby keeps anxiety in check.

Many portions of the model are as yet untested. Of those that have been tested, the results are generally favorable. We have some evidence that motivation and anxiety are related to task efficacy in the manner predicted by our model. PSR and NSR are related to each other in a positive manner, and the amount of PSR that a particular individual engages in appears to be related to his or her previous experience with anxiety. Task experience appears to reduce uncertainty regarding performance as well as anxiety. And finally, anxiety appears to be negatively related to performance.

ACKNOWLEDGMENTS

The authors would like to thank Carol L. Gohm for comments on previous drafts. The work presented in this chapter was supported in part by grants from the Office of Naval Research and Naval Research Laboratories (grants N00014-95-0749 and N00014-97-C-2061, respectively).

REFERENCES

Bacon, S. J. (1974). Arousal and the range of cue utilization. *Journal of Experimental Psychology, 102*(1), 81–87.

Baumann, M. R., Buerkle, C. A., & Sniezek, J. A. (1998, May). *Cognitive activities and anxiety in a high stress decision-making task.* Poster presented at the 4th Conference on Naturalistic Decision Making, Arlie Center, Warrenton, VA.

Baumann, M. R., & Sniezek, J. A. (1999). *The effects of time pressure on awareness of others: Implications for group coordination.* Manuscript in review.

Baumann, M. R., Donovan, M. A., & Sniezek, J. A. (1998, April). *Can "Thinking Positive" help performance? Cognitive resource allocation on a complex task under time pressure.* Poster presented at the 70th annual meeting of the Midwestern Psychological Association, Chicago, IL.

Baumann, M. R., Sniezek, J. A., Donovan, M. A., & Wilkins, D. C. (1996). *Training effectiveness of an immersive multimedia trainer for acute stress domains: Ship damage control* (University of Illinois Tech. Rep. No. UIUC-BI-KBS-96008). Urbana, IL: Beckman Institute.

Bell, H. H., & Waag, W. L. (1998). Evaluating the effectiveness of flight simulators for training combat skills: A review. *The International Journal of Aviation Psychology, 8*(3), 223–242.

Bless, H., Clore, G., Schwarz, N., Golisano, V., Rabe, C., & Wolk, M. (1996). Mood and the use of scripts: Does a happy mood really lead to mindlessness? *Journal of Personality & Social Psychology, 71*(4), 665–679.

Bodenhausen, G. V., Sheppard, L. A., & Kramper, G. P. (1994). Negative affect and social judgment: The differential impact of anger and sadness. *European Journal of Social Psychology, 24*, 45–62.

Buckley, T., & Sniezek, J. A. (1992). Passion, preference, and predictability in judgmental forecasting. *Psychological Reports, 70*(3), 1022.

Buerkle, C. A., Baumann, M. R., & Snizek, J. A. (1998, May). *Expertise and self-regulatory processes in decision-making under stress*. Poster presented at the 4th Conference on Naturalistic Decision Making, Arlie Center, Warrenton, VA.

Burton, D. (1988). Do anxious swimmers swim slower? Reexamining the elusive anxiety performance relationship. *Journal of Sport Psychology, 10*, 45–61.

Bylsma, W. H., Tomaka, J., Luhtanen, R., Crocker, J., & Major, B. (1992). Response latency as an index of temporary self-evaluation. *Personality & Social Psychology Bulletin, 18*(1), 60–67.

Chesney, A. A., & Locke, E. A. (1991). Relationships among goal difficulty, business strategies, and performance on a complex management simulation task. *Academy of Management Journal, 34*(2), 4000–4024.

Dawes, R. M. (1988). *Rational choice in an uncertain world*. New York: Harcourt Brace Jovanovich.

Deffenbacher, J. L. (1980). Worry and emotionality in test anxiety. In I. G. Sarason (Ed.), *Test Anxiety: Theory, Research, and Applications* (pp. 111–128). Hillsdale, NJ: Lawrence Erlbaum Associates.

Dennis, K. A., & Harris, D. (1998). Computer-based simulation as an adjunct to ab initio flight training. *The International Journal of Aviation Psychology, 8*(3), 261–276.

Donovan, M. A, Sniezek, J. A., & Baumann, M. R. (1996, August). *Learning from unusual events: Decision making in crises*. Symposium sponsored by Managerial and Organizational Cognition and Organizational Behavior Divisions of the Academy of Management, Cincinnati, Ohio.

Driskell, J. E., & Johnston, J. H. (1998). Stress exposure training. In J. A. Cannon-Bowers & E. Salas (Eds.), *Making decisions under stress: Implications for individual and team training* (pp. 191–218). Washington, DC: American Psychological Association.

Flexman, R. P., & Stark, E. A. (1987). Training simulators. In G. Salvendy (Ed.), *Handbook of human factors* (pp. 1012–1037). New York: Wiley.

Geen, R. G. (1980). Test anxiety and cue utilization. In I.G. Sarason (Ed.), *Test anxiety: Theory, research, and applications* (pp. 43–62). Hillsdale, NJ: Lawrence Erlbaum Associates.

Hains, A. A., & Szyjakowski, M. (1990). A cognitive stress-reduction intervention program for adolescents. *Journal of Counseling Psychology, 37*(1), 79–84.

Hardy, L., & Parfitt, G. (1991). A catastrophe model of anxiety and performance. *British Journal of Psychology, 82*(2), 163–178.

Hart, S. G., & Staveland, L. E. (1988). Development of NASA-TLX (task load index): Results and empirical and theoretical research. In P. A. Hancock & N. Meshkati (Eds.), *Human mental workload* (pp. 139–183). Amsterdam: North Holland.

Hill, S. G., Iavecchia, H. P., Byers, J. C., Bittner, A. C., Zaklad, A. L., & Christ, R. E. (1992). Comparison of four subjective workload rating scales. *Human Factors, 34*(4), 429–439.

Kanfer, R., & Ackerman, P. L. (1989). Motivation and cognitive abilities: An integrative/aptitude-treatment interaction approach to skill acquisition. *Journal of Applied Psychology, 74*(4), 657–690.

Kantowitz, B. H. (1992). Selecting measures for human factors research. *Human Factors, 34*(4), 387–398.

Klein, G. (1996). The effects of acute stressors on decision making. In J. E. Driskell & E. Salas (Eds.), *Stress and human performance* (pp. 49–88). Mahwah, NJ: Lawrence Erlbaum Associates.

Latham, G. P., & Locke, E. A. (1991). Self-regulation through goal setting. *Organizational Behavior & Human Decision Processes, 50*(2), 212–247.

Lazarus, R. S. (1991). Psychological stress in the workplace. *Journal of Social Behavior and Personality, 6*(7), 1–13.

Lazarus, R. S., & Folkman, S. (1984). *Stress, appraisal and coping*. New York: Springer.

Leon, M. R., & Revelle, W. (1985). The effects of anxiety on analogical reasoning: A test of three theoretical models. *Journal of Personality and Social Psychology, 49*(5), 1302–1315.

Marteau, T. M., & Bekker, H. (1992). The development of a six-item short-form of the state scale of the Spielberger State-Trait Anxiety Inventory (STAI). *British Journal of Clinical Psychology, 31*(3), 301–306.

Maule, A. J., & Hockey, G. R. J. (1993). State, stress, and time pressure. In O. Svenson & A. J. Maule (Eds.), *Time pressure and stress in human judgment and decision making* (pp. 83–101). New York: Plenum Press.

Meichenbaum, D. (1985). *Stress innoculation training*. New York: Pergammon.

Monat, A., Averill, J. R., & Lazurus, R. S. (1972). Anticipatory stress and coping reactions under various conditions of uncertainty. *Journal of Personality & Social Psychology, 24*(2), 237–253.

Morris, L. W., Davis, M. A., & Hutchings, C. H. (1981). Cognitive and emotional components of anxiety: Literature review and a revised worry-emotionality scale. *Journal of Educational Psychology, 73*(4), 541–555.

Morris, L. W., & Fulmer, R. S. (1976). Test anxiety (worry and emotionality) changes during academic testing as a function of feedback and test importance. *Journal of Educational Psychology, 68*(6), 817–824.

Orasanu, J. M., & Backer, P. (1996). Stress and military performance. In J. E. Driskell & E. Salas (Eds.), *Stress and Human Performance* (pp. 89–125). Mahwah, NJ: Lawrence Erlbaum Associates.

Radhakrishnan, P., Arrow, H., & Sniezek, J. A. (1996). Hoping, performing, learning, and predicting: Changes in the accuracy of self-evaluation of performance. *Human Performance, 9*(1), 23–49.

Salas, E., Bowers, C. A., & Rhodenizer, L. (1998). It is not how much you have but how you use it: Towards a rational use of simulation to support aviation training. *The International Journal of Aviation Psychology, 8*(3), 197–208.

Salas, E., Driskell, J. E., & Hughes, S. (1996). Introduction: The study of stress and human performance. In J. E. Driskell and E. Salas (Eds.), *Stress and Human Performance* (pp. 1–45). Mahwah, NJ: Lawrence Erlbaum Associates.

Sarason, I. G. (1980). Introduction to the study of test anxiety. In I. G. Sarason (Ed.), *Test anxiety: Theory, research, and applications* (pp. 3–14). Hillsdale, NJ: Lawrence Erlbaum Associates.

Schmidt, U. (1998). A measurement of the certainty effect. *Journal of Mathematical Psychology, 42*(1), 32–47.

Sniezek, J. A., Baumann, M. R., & Buerkle, C. A. (1997, March). Raw data from simulator validation studies at Surface Warfare Officer School.

Spielberger, C. D., Gorsuch, R. L., & Lushene, R. E. (1970). *Manual for the state-trait anxiety inventory*. Palo Alto, CA: Consulting Psychologists Press.

Spielberger, C. D., Gorsuch, R. L., Lushene, R., Vagg, P. R., & Jacobs, G. A. (1983). *State-trait anxiety inventory for adults: Sampler set, manual, test, scoring key*. Palo Alto, CA: Mind Garden.

Wine, J. D. (1980). Cognitive attentional theory of test anxiety. In I. G. Sarason (Ed.), *Test anxiety: Theory, research, and applications* (pp. 349–386). Hillsdale, NJ: Lawrence Erlbaum Associates.

Wright, P. (1974). The harassed decision maker: Time pressures, distractions, and the use of evidence. *Journal of Applied Psychology, 59*(5), 555–561.

9

Reflective Versus Nonreflective Thinking: Motivated Cognition in Naturalistic Decision Making

Henry Montgomery
Stockholm University

There are two psychological traditions that are concerned with explaining the dynamics behind human behavior. Motivational psychology is focused on how "hot" factors, such as needs, emotions, and different kinds of motivation, lead to particular behavioral patterns. By contrast, judgment and decision-making (JDM) research is concerned with how various "cold" information processes may explain how people choose between different courses of action. Another way of contrasting the two traditions is in terms of locus of control. Thus, motivational psychology views behavior as controlled by factors inside the individual whereas JDM researchers attempt to understand how people's understanding of the world outside themselves explains their behavior.

In recent years, the study of naturalistic decision making (NDM), that is, the study of how professionals make decisions in field settings, has become an established research field (Klein, 1997). The relationship between NDM and JDM has been discussed extensively (e. g., Klein, 1998). By contrast, I have not found any discussion in the NDM literature of how NDM relates to motivational psychology. In this chapter, I first present a model of judgment and decision making—the so called perspective model (Montgomery, 1994, 1997)—that may be used as a point of departure for describing how motivational and information processing factors interact in professionals' decision making, but also in other kinds of decisions. It may be noted that the present description of the perspective model on some points updates and clarifies the earlier accounts of the model. I discuss how experience (a key notion in naturalistic decision making models) enters into the perspective model. Thereafter, I relate the model to the notion that thinking may be driven by different kinds of motivation. I show how the resulting perspective–motivational model allows a distinction between reflective and nonreflective decision making. The distinction is illustrated by means of data from a naturalistic study of financial credit decisions. In a final section, the distinction between reflective and nonreflective decision making is related to JDM and NDM models.

THE PERSPECTIVE MODEL OF JUDGMENT AND DECISION MAKING

The perspective model develops the idea that decision makers tend to view the decision situation in such a way that it supports the decision to be made (Montgomery, 1993). The model differs from most JDM models by assuming that there is a close analogue between evaluative judgments and visual perception. Evaluative judgments are assumed to follow from what an observer—the subject—sees in a mental object. Thus, evaluations are not computed in a more or less arbitrary way from given information, but follow directly from how the subject sees given characteristics of an object from a given viewing position. The perspective model implies that evaluative judgments of fairly complex objects can be made quickly, in one step, and also that different viewing positions will result in gestalt switches in the evaluation of the same object.

The central idea of the perspective model is that an observer—the subject—views an object with positive (attractive) and negative (unattractive) sides from a given position in a mental space. The subject is positioned outside the object facing either its positive side or its negative side. Figure 9.1A shows a case in which two subjects—a stock investor and a credit manager—view a company (the object) from its positive and negative side, respectively.

The sides are assumed to be positioned in parallel to each other (see the upper and lower line in Fig. 9.1A). On each side either positive or negative aspects of the object come together. Each aspect corresponds to a certain amount of a given quantity (e. g., amount of potential gain). The positive and negative aspects are linked in pairs via the same descriptive aspect that is positioned between the positive and negative side. For example, the positive aspect "amount of potential gain" (see black section of the upper line in Fig. 9.1A) and the negative aspect "amount of insecurity" (see black section of the lower line in Fig. 9.1A) are both linked to the descriptive aspect "amount of expansiveness" (see black section of the central line in Fig. 9.1A). In these links, the descriptive contents are assumed to be the same whereas the evaluative contents (positive or negative) are different.

The aspects on a given evaluative side are assumed to be located on bipolar dimensions where each dimension contains a certain amount of a given aspect and a certain amount of another aspect which is assumed to have trade-off relationship to the first aspect. In Fig. 9.1A "potential gain" and "security" are assumed to have such a trade-off relationship on the positive side. On the negative side, the trade-off relationship holds between "insecurity" and "missed potential gain." On the descriptive axis, the trade-off relationship holds between "expansiveness" and "restrictiveness." It may be noted that bipolar dimensions could also be defined by the diagonal corners in Fig. 9.1. These dimensions go between opposites that logically exclude each other, that is, from "potential gain" to "missed potential gain" and from "security" to "insecurity."

In Fig. 9.1A, it is assumed that the amount of expansiveness is about twice as great as the amount of restrictiveness (as reflected in the length of the black section of the central line as compared with the grey section of the central line). This

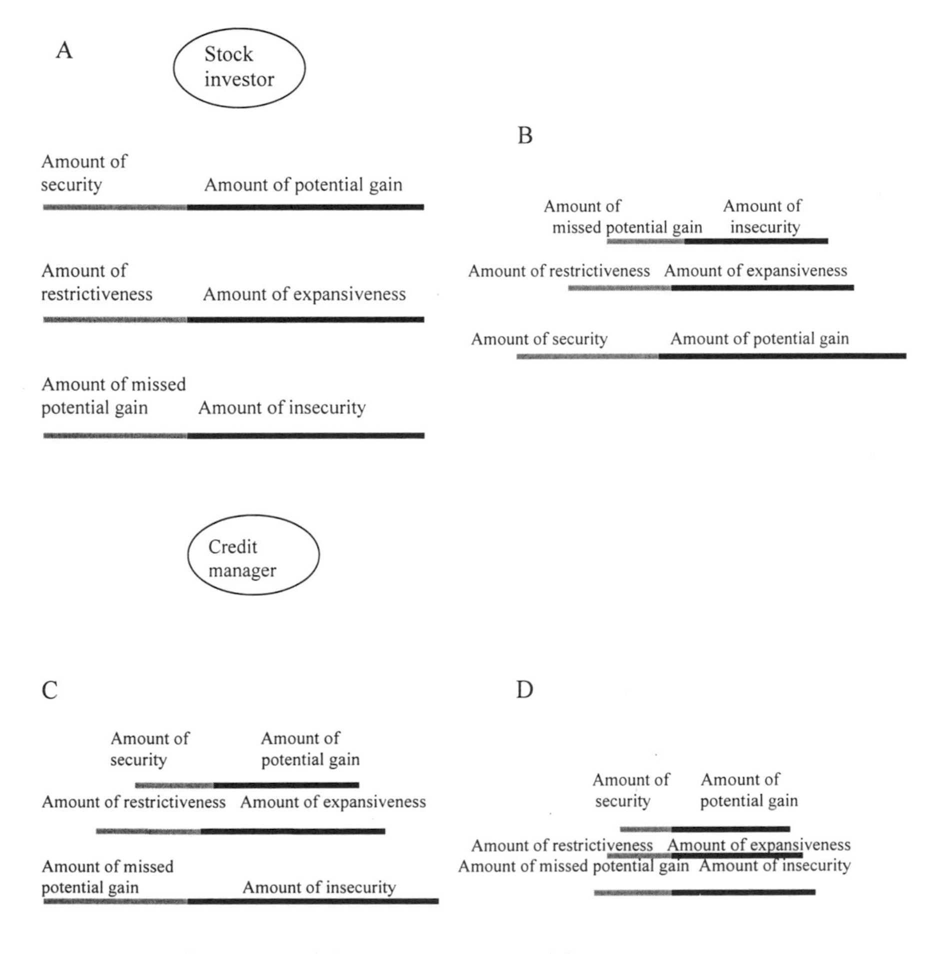

FIG. 9.1. Illustration of the perspective model.

relationship is paralleled by about twice as great an amount of potential gain than security on the positive side and by about twice as great an amount of insecurity than missed potential gain on the negative side (compare the lengths of the sections and the grey sections of the lines on the positive—upper—and negative—lower—side). However, it is not necessarily true according to the perspective model that the positive and negative aspects are "objectively" equal as has been assumed in Fig. 9.1A. What is necessarily true according to the model is that in some sense there exists an objective reality, independent of specific perspectives, with fixed amounts of positive and negative aspects (in this case assumed to be equal). However, depending on the viewing position, different aspects of the object will appear with different sizes, the general rule being that the closer the aspect is to the viewer the larger it appears. This implies that if the object is seen from its positive side (see position of stock investor in Fig. 9.1A) the positive aspects will tend to be seen as larger than the negative aspects (see Fig. 9.1B, which shows what the stock investor sees from his or her perspective) and vice versa if the object is seen from its negative side (see the position of credit manager in Fig. 9.1A and what the credit manager sees from this position in Fig. 9.1C). More specifically, in our example (Fig. 9.1) the stock investor will experience the object as having more potential gain than insecurity and also as having more security than missed potential gain (compare the black or grey sections of the upper and lower lines in Fig. 9.1B), whereas the opposite will be true for the credit manager (Fig. 9.1C). Moreover, the larger the distance is to the object, the smaller the difference will be between positive and negative aspects. That is, a longer viewing distance will yield a more objective view of the object (see Fig. 9.1D).

The adoption of a given viewing position is assumed to imply that the subject identifies himself or herself with an entity of some sort. This entity is denoted as a subject identification and may correspond to a role, a person, or a group of persons. Each subject identification is located in one or several positions in the mental space surrounding the object. A subject identification is associated with certain interests that determine how it is located in relation to the object. In general, the subject identification will be located close to the positive side of the object, if the object is experienced as being congruent with its interests, and on its negative side, if the object is experienced as incongruent with its interests. In this example, the object is more expansive than restrictive. This is congruent with the stock investor's interests to gain much money, implying that he or she will be located on the positive side of the object (the company), and it is incongruent with the credit manager's interests of not losing credited money implying that he or she will be located on the negative side of the object.

Figure 9.1 also implies that "gestalt switches" will occur if the subject shifts his or her attention as a result of adopting a new subject identification. For example, if the stock investor in Fig. 9.1 identifies with the credit manager (e. g., because of friendship), he or she may experience the company as insecure rather than leading to potential gains.

THE ROLE OF EXPERIENCE IN THE PERSPECTIVE MODEL

Research on naturalistic decision making is focused on how people use their experience to make decisions in field settings (Klein, 1997). How does experience enter into the perspective model? More precisely, how is experience related to perspective-based "seeing"? Consider again the stock investor who prefers to buy stocks in expansive (but risky) companies. His or her "seeing" presumably is based on personal experiences that associate expansiveness with gains rather than with losses. He or she may also have experiences of managing losses in a successful way. It may be speculated that such experiences reinforce the investor's interest in taking risks to get great gains, which in turn means that it will be natural to continue to adopt the subject identification of being a gain-seeking stock investor. On the other hand, the investor may also have been in contact with cases of unsuccessful investments that however are less salient, for example, because they are not personal but associated with other people. If the stock investor identifies with a credit manager, these negative experiences instead may be salient because, for a credit manager, other people's (i. e., clients') economic failures or successes are more relevant than the outcomes of one's own investments. To conclude, the subject identification influence which experiences will be salient and reinforcing.

At the same time, experience influences what subject identification will be adopted. If a policy associated with an adopted subject association leads to a disaster, a radical gestalt shift may occur. However, subject identifications may not be automatically controlled by reinforcements. There may also room for deliberate choices among different subject identifications. This possibility is discussed in the following section.

REFLECTIVE AND NONREFLECTIVE THINKING

Traditionally, motivation has been linked to external behavior. However, recent research has stressed that there is a motivational basis also for internal mental processes, that is, thinking in some sense. Typically, a distinction is made between two kinds of motivation for thinking: (a) thinking that aims at acquiring an accurate picture of the external world and (b) thinking that has goals other than accuracy. Nonaccuracy goals include a desire to defend preestablished ideas, values, or interests (defense motivation), or goals that are primarily focused on the interpersonal consequences associated with expressing a given judgment in a particular setting (impression formation; Chaiken, Wood, & Eagly, 1996). Various terms have been used for these two types of thinking. Type (a) thinking has been labeled as accuracy-oriented (Chaiken et al., 1996), open-minded (Rokeach, 1960), vigilant (Janis & Mann, 1977), or as being analytical (Epstein, Pacini, Denes-Raj, & Heier, 1996). Type (b) thinking is sometimes labeled by one comprehensive term, such as close-minded (Rokeach, 1960) or, as exemplified above, is subdivided into different types of nonaccuracy orientations, such as defense-oriented and impression-oriented thinking.

In the present chapter, the two types of thinking are characterized and interpreted as reflective and nonreflective thinking, respectively. *Reflective thinking* aims at becoming aware of (reflecting) facts external to the thinking person. The reflective thinker is curious. He or she wants to find the truth, and nothing but the truth. This implies that the thinker also wishes to uncover factors that may bias the thinking process. In *nonreflective thinking* there is the reverse motive of not becoming aware of all factors that influence the thinker. More precisely, the thinker is not (and does not want to be) fully aware of those nonaccuracy motives that bias his or her thinking (cf. Epstein's, 1989, notion of implicit values). This is because awareness of nonaccuracy motives may weaken their impact on the thinking and, hence, counteract the attainment of these motives. The reason for this, in turn, is that nonreflective thinking also has a reflective component in the sense that it is oriented to external facts, but only to those facts that may help the thinker to attain his or her nonaccuracy motives. In other words, when a person is involved in nonreflective thinking, he or she believes that he or she increases his or her understanding of the external world (believed accuracy orientation) although in fact in his or her beliefs he or she uses certain aspects of the external world to serve interests beyond accuracy. In other words, there is always an ingredient of self-deception in what I have called nonreflective thinking.

PERSPECTIVES AND REFLECTIVE VERSUS NONREFLECTIVE THINKING

I am now ready to relate the perspective model to the two kinds of motives in human thinking described earlier. As emphasized by Mead (1959) and many others, thinking always is oriented toward the external world from a given perspective. In the same vein, the perspective model assumes that people view the world from different subject identifications, each of which occupies a certain position in relation to an observed object.

Let us now examine in more detail implications of the perspective model for how reflective thinking works in a situation where an individual wants to understand and evaluate an object. For example, the object may be an alternative in a decision situation. Basically, in reflective thinking the individual is interested in getting a balanced and a comprehensive view of the object. To do so, it will be necessary to see the object from different sides and, in particular, from both its positive and negative sides. This goal may be attained by means of the following mental moves.

First, the individual may adopt a subject identification that is associated with interests that invite seeing the object from different sides. That is, in this case, the individual moves the same subject identification into different viewing positions. For example, a businessperson who identifies with the company's interests may aspire to view an economic prospect from both its positive and negative sides.

Second, the individual may also move between different subject identifications that occupy certain positions in relation to the object. For example, besides identifying with the company, a businessperson may also identify with the interests of other economic actors.

Third, in order to get an overview of the object the individual will be interested in finding a suitable viewing distance. To attain this goal, the individual may move an adopted subject identification or find a new subject identification that has the right distance to the object.

Fourth, the reflective individual will be interested in seeing the object from a short distance in order to see the details of the object. All viewing positions that may add new information are of interests for the reflective individual.

Fifth, the reflective individual will always be interested in focusing on the interests associated with different subjective identifications in order to avoid a biased understanding of the object. Put differently, he or she wants to understand and control his or her perspectives. To do so, he or she will not only focus on the to-be-evaluated objected but also on the interests associated with different subject identifications that may be adopted for viewing the object. However, as is true for all mental focusing according to the perspective model, this focusing also will always be done from a certain subject identification, which makes it difficult or perhaps impossible to be completely unbiased when examining the interests associated with a given subject identification. In any case, the reflective individual is assumed to have a genuine ambition of becoming aware of how the interests associated with an adopted subject identification justifies his or her moving in the space surrounding the object.

Sixth, as a final step the reflective individual tries to synthesize what he or she has seen from different positions, where each position is associated with particular interests. In this step, the individual tries to find a balanced overview of the object and its relationships to interests associated with different subject identifications.

Let us now examine how nonreflective thinking works in terms of the perspective model. Generally, the nonreflective thinker tends to view the object from just one evaluative perspective, which implies that it will contrast to reflective thinking with respect to most of the characteristics of this kind of thinking listed earlier.

In particular, the nonreflective individual is primarily interested in seeing the object from either its positive or its negative side. He or she has no genuine interest to balance out positive and negative features in order to get an accurate overview of the object. His or her primary interest instead is to defend and consolidate his or her positive or negative evaluation of the object.

As discussed, nonreflective thinking is assumed to involve a "blind spot," that is, to follow from interests which the thinker is not aware of and does not want to be aware of. According to the perspective model, these interests must be based on a subject identification. Thus, nonreflective thinking is assumed to be driven by subject identifications that are not clearly acknowledged by the thinker. Let us label such identifications as implicit subject identifications (cf. Epstein, 1989, notion of implicit values). This influence is masked by having access to an additional explicit subject identification that justifies the evaluation. For example, a credit manager may grant a large credit to a client because he or she identifies himself or herself with the client's needs (implicit subject identification), but he or she justifies the credit decision in terms of external facts that are compatible with the company's interests (explicit subject identification).

EMPIRICAL ILLUSTRATION OF REFLECTIVE AND NONREFLECTIVE DECISION MAKING

Using experiences from an ongoing study of decision making concerning financial credits (Montgomery & Lundin, 1998), I now present a typology of different kinds of reflective and nonreflective decision making. The study involved interviews with credit evaluators (credit manager and salesmen) and clients related to a large Swedish wholesale business in plumbing. Five cases of credit decisions were investigated. All cases were selected to exemplify difficult decisions about whether to reject or accept a credit request.

In the data, I found two basic types of subject identifications, which could be at work in both reflective and nonreflective decision making. On the one hand, the credit evaluator may adopt an ego identification, where the evaluator's interests are seen as separate from the client's interests. On the other hand, the credit evaluator may adopt a "we" identification, implying that the evaluator and the client are seen as having common interests.

Both types of subject identifications could be explicit or implicit. In explicit subject identifications, the decision maker is aware of how his or her interests determine what is seen. Moreover, because the reflective decision maker is concerned with getting accurate information, it is important for him or her to start out from explicit interests in order to distinguish between these interests and what he or she focuses on as a consequence of having these interests. In the context of credit decisions, an explicit ego identification can be assumed to be associated with the interest of being an accurate credit evaluator. As a rule, this identification will coincide with the selling company's interests. In an explicit "we" identification, the credit evaluator is interested not only in getting accurate information about the client's financial position, but also in possibilities of satisfying common economic interests that are shared by him or her and the client.

Implicit ego identifications concern the decision maker's private fears and hopes, which are unrelated to or even threatened by information that is focused from an explicit subject identification. For example, a credit evaluator may wish to avoid a client because of his or her rude manners (private fear) and, as a consequence, stress problems in the client's financial position, although from a truly explicit subject identification these financial problems would be seen as being balanced by other more positive features of the client's financial position. An implicit "we" identification implies that the credit evaluator's perception and actions are driven by a wish to collaborate with the client, even if this behavior from a reflective perspective may be seen as threatening the credit evaluator's economic interests. It may be noted that in both types of implicit subject identification, the credit evaluator's evaluation of the client's financial position is affected by how he or she perceives the social climate in the relationship to the client.

Let us now relate the different types of subject identification to reflective and nonreflective decision making, respectively, in the data from the Montgomery and Lundin (1998) study. In 3 out of 5 cases, I found clear evidence that the credit evaluator (credit manager or salesman) was influenced by the social relationship to the client. In two cases, which ended with credit losses and bankruptcy, there had been a preceding story of good social relationships between credit evaluator and client. Moreover, the economic transactions between the selling company and the client had worked well. As a consequence, the credit evaluator focused on positive economic aspects and, apparently, failed to take warning signals seriously when they occurred. In both cases an implicit "we" identification may have pushed the credit evaluator, when explicitly being in the role of credit evaluator, to adopt a perspective that brought positive economic aspects to the foreground and relegated negative economic aspects to the background.

In the third case, a negative social relationship seems to have influenced the credit evaluator to reject credit, although it later turned out that the client would have been able to pay back the credit. Here, a frustrated implicit ego or "we" identification seem to have been at work and influenced the evaluator. It should be noted, however, that there indeed were negative economic signals in this case, although the economic situation apparently was still worse for the cases that were granted credit.

Let us now consider two cases where I found evidence of reflective decision making. One case (Case 4) may be seen as being based on an explicit "we" identification. Here, an old friend of the credit evaluator, a lawyer who invested money in the firm, took care of a firm in a bad economic shape. As a consequence, the credit evaluator reevaluated the economic prospects of the firm. The credit evaluator believed in the lawyer's competence and honesty, and he appreciated the fact that the lawyer invested money in the firm. He was aware that his evaluation might have been influenced by his good relationship to the lawyer. He took a risk, but a calculated risk based on reflective thinking that prepared him for the possibility of being mistaken.

The final case (Case 5) exemplifies an explicit ego identification leading to a decision to deny credit. In this case, there were both negative and positive economic indicators, but with a perceived overweight for the negative indicators (estimated to be 70/30 or 60/40). There were no signs of close social relationships between credit evaluator and client.

Summarizing the results across the five cases, the following observations can be made. In four cases (Cases 1 through 4), the personal relationship between credit evaluator and client influenced the credit decision. This seems to have been particularly true for the salesmen who tended to have a closer relationship to the client than was the case for the credit manager. In three cases (Cases 1 through 3), the credit decisions appear to have been based on nonreflective thinking inasmuch as the credit evaluator's economic judgments seems to have been pushed in a certain direction by his implicit subject identifications. In the remaining two cases (Cases 4 and 5), the evaluator was more clearly aware of his subject identifications, and he deliberately tried to get a balanced view of the client's situation.

The data also illustrate how an object will be seen from its positive or negative side depending on whether it satisfies the interests associated with the adopted subject identification. In Cases 1 and 2, the positive side was in focus, due to the good fit of the implicit "we" identifications, and in Case 3, the negative side dominated, at least partly because of a frustrated ego or "we" identification. Case 4 illustrates a case that started with a negative view but changed into looking at the object from its positive side as a result of new identification possibilities. In Case 5, the dominance of negative aspects seen from an explicit ego identification led to a focus on the negative side of the object.

OTHER ACCOUNTS OF DECISION MAKING

How does the distinction between reflective and nonreflective decision making relate to other accounts of human decision making? First, it should be noted that the present distinction has a motivational rather than cognitive basis. In this sense, the distinction is close to the distinction between vigilant and various variants of nonvigilant decision making (Janis & Mann, 1977), action versus state-oriented thinking (Kuhl & Beckman, 1985), implemental and deliberative mindset (Taylor & Gollwitzer, 1995), and need for closure versus need to avoid closure (Kruglanski & Webster, 1996). Common to all these distinctions is that they contrast an action-oriented mode of thinking (i. e., thinking that aims at facilitating ensuing action) to a more passive, deliberative mode of thinking. The present distinction instead contrasts an ambition to get a maximal knowledge (reflective thinking) versus an ambition to avoid such knowledge that will threaten the attainment of certain interests (nonreflective thinking). Both types of thinking may be conceived as being action oriented or not. Although reflective thinking seems to be close to deliberative, nonaction-oriented thinking, the data exemplify how reflective thinking also may serve the reverse role of preparing the individual to perform actions that are grounded in more or less uncomfortable facts (Case 5). In the same vein, nonreflective thinking may serve the role of precluding action (e.g., to avoid an uncomfortable negative credit decision) as was true in several of the cases, but also serve the role of facilitating action after the decision maker has made up his or her mind (Case 3).

It may now be asked how this distinction relates to various cognitive decision-making biases, such as the representativeness and availability heuristics, anchoring effects, framing effects, and so forth.

Possibly, several researchers in this area would claim that cognitive biases might occur both in reflective and nonreflective thinking because these biases are due to cognitive limitations rather than being motivationally controlled (Kahneman, Slovic, & Tversky, 1982). On the other hand, it is tempting to assume that nonreflective thinking is more vulnerable to cognitive biases than is true for reflective thinking because a reflective thinker would be expected to get rid of all kinds of biases. In particular, framing and anchoring effects may be motivationally controlled because these effects imply that one of several possible perspectives, in some sense, is picked out as the salient perspective. Apparently, people can switch between different frames or anchorings (Maule, 1989). To the extent that this done in order to eliminate one-sidedness and increase awareness of possible biases, reflective thinking will be a means to reduce the impact of cognitive biases in an understanding of the world.

Finally, I consider how the reflective–nonreflective distinction relates to how decisions are conceived within research in naturalistic decision making. This research typically examines decisions made by professionals using their experience in situations calling for rapid action. Researchers within this tradition often stress the adaptive and reality-oriented nature of naturalistic decision making. To the extent that this picture is true, I would say that naturalistic decision making is reflective. *Reflective*, as this term is used here, does not necessarily mean elaborate thought processes, but it means simply to be reality-oriented, that is, a motivation to reflect reality. On the other hand, professionals also may depart from reality in their decision making, being misled by their implicit subject identifications, as was exemplified in the Montgomery and Lundin (1998) study of credit decisions. Thus, in order to improve professionals' decision making, it may be important to find out how nonreflective thinking following from implicit subject identifications may be in the way of making adaptive and reality-oriented decisions.

REFERENCES

Chaiken, S., Wood, W., & Eagly, A. H. (1996). Principles of persuasion. In E. T. Higgins & A. W. Kruglanski (Eds.), *Social psychology: Handbook of basic principles* (pp. 702–742). New York: Guildford.

Epstein, S. (1989). Values from the perspective of Cognitive Experiential Self-Theory. In N. Eisemberg, J. Reykowsky, & E. Staub (Eds.). *Social and moral values* (pp. 3–22). Hillsdale, NJ: Lawrence Erlbaum Associates.

Epstein, S., Pacini, R., Denes-Raj, V., & Heier, H. (1996). Individual differences in intuitive-experiential and analytical-rational thinking styles. *Journal of Personality and Social Psychology, 71*, 390–405.

Janis, I. L., & Mann, L. (1977). *Decision making*. New York: The Free Press.

Kahneman, D., Slovic, P., & Tversky, A. (Eds.). (1982). *Judgment under uncertainty*. New York: Cambridge University Press.

Klein, G. (1997). The current status of the naturalistic decision making framework. In R. Flin, E. Salas, M. Strub, & L. Martin (Eds.), *Decision making under stress: Emerging themes and applications* (pp. 11–28). Aldershot, England: Ashgate.

Klein, G. (1998). *Sources of power: How people make decisions*. Cambridge, MA: MIT Press.

Kruglanski, A. W., & Webster, D. M. (1996). Motivating closing of the mind: "Seizing" and "freezing." *Psychological Review, 103*, 263–283.

Kuhl, J., & Beckman, J. (Eds.). (1985). *Action control: From cognition to behavior*. New York: Springer.

Maule, A. J. (1989). Positive and negative decision frames: A verbal protocol analysis of the Asian disease problem of Tversky and Kahneman. In H. Montgomery & O. Svenson (Eds.), *Process and structure in human decision making* (pp. 163–180). Chichester, England: Wiley.

Mead, G. H. (1959). *Mind, self and society*. Chicago: University of Chicago Press.

Montgomery, H. (1993). The search for a dominance structure in decision making: Examining the evidence. In G. Klein, J. Orasanau, & R. Calderwood (Eds.), *Decision making in action* (pp. 182–187). Norwood, NJ: Ablex.

Montgomery, H. (1994). Towards a perspective theory of decision making and judgment. *Acta Psychologica, 87*, 155–178.

Montgomery, H. (1997). Surrender at Perevolochna: A case study of perspective taking and action control in decision making under stress. In R. Flin, E. Salas, M. Strub, & L. Martin (Eds.), *Decision making under stress: Emerging themes and applications* (pp. 193–204). Aldershot, England: Ashgate.

Montgomery, H., & Lundin, R. (1998). [Interviews with credit evaluators on credit decisions]. Unpublished raw data.

Rokeach, M. (1960). *The open and closed mind*. New York: Basic Books.

Taylor S. E., & Gollwitzer, P. M. (1995). Effects of mind sets on positive illusions. *Journal of Personality and Social Psychology, 69*, 213–226.

10

The Management of Temporal Constraints in Naturalistic Decision Making: The Case of Anesthesia

Véronique De Keyser
Anne-Sophie Nyssen
University of Liege

In naturalistic situations and, more precisely, in dynamic risk environments, one of the major competences expected from operators is *situativeness*: "the tendency to take the particularity of the situation into account" (Hukki & Norros, 1998, p. 317; Hutchins, 1995; Suchman, 1987). However, training people to acquire such competence in their natural environment is difficult for obvious risk and economic reasons. The use of simulators is becoming more and more widespread. Up to now, the emphasis has mainly been put on (a) acquisition by the trainees of procedural knowledge and problem solving skills to control rare and potentially serious incidents, (b) acquisition of crisis management skills (Gabba, Fish, & Horward, 1994), and (c) methodological support for instructors (Hukki & Norros, 1998). Less research has been done in the United States on the temporal features of naturalistic situations that should be preserved in simulators, for they are part of the situativeness difficulty. In Europe, a strong trend of field and experimental studies, mainly in process control, has pinpointed this problem and stressed the "temporal errors" linked to these situational characteristics (De Keyser, 1995). Regarding time and situation, we have to fight two common views. The first restricts the problem to temporal pressure and dynamic decision making under stress. Based on field studies and microworld experiments in dynamic environments, Hoc (1989), Brehmer (1990a, 1990b), and Van Daele (1993) clearly demonstrated that many temporal characteristics other than temporal pressure shape the operator's behavior: time scales of actions, delay of feedback, slow/rapid tempo of process responses, and so forth. In fact, these characteristics determine, among other factors, the mode of control of the situation: from an anticipatory mode, where people plan their actions in advance and can predict the evolution of the process to a reactive one, where people react immediately to any change in the situation. Sometimes "scrambled" modes of control (Hollnagel, 1998) or cognitive breakdowns (Amalberti, 1996) emerge when people do not succeed in managing the situation. The other reductionist view is to think that the problem will be solved by increasing technical and functional validity of simulators. To a certain extent this is true. If we compare full-scale and screen-based simulators, the former preserves the temporal characteristics of the situation better than the latter, but it does not preserve all of them. Screen-based simulators could be useful training for some aspects of situativeness. In fact, what we need to assess is their psychological validity. This is the extent to which simulated situations generate similar psychological conditions of action to naturalistic situations (Baker & Marshall, 1989; Grau, Doireau, & Poisson, 1998; Leplat, 1989). Psychological validity stems from similarity between the conditions of action rather than similarity between material or technical characteristics per se. But conditions of action cannot be exclusively derived from formal task constraints; they depend strongly on the context in which the activity takes place. What has to be kept in simulators directly depends on the analysis of the activity in naturalistic situations. This activity analysis aims to identify the relationships between the contextual task constraints and the performance demands.

Training to cope with temporal constraints is a rather delicate question. At a very general level, time can be classified into three classes of problems: duration, order, and temporal perspectives (Block, 1990; De Keyser, d'Ydewalle, & Vandierendonck, 1998; Sougné, Nyssen, & De Keyser, 1993). In dynamic environments, *duration* constraints are omnipresent: action time, feedback latency, lag between operations or events, and so forth. Being slightly behind or ahead of schedule can dramatically change the course of actions and can have quality, safety, contextual, and economic consequences. However, duration estimation does not exclusively rely on psychological time (Boltz, 1998). For example, "The doctor came too late" possibly refers to different temporal reference systems (Javaux, 1996): "too late" because the doctor said he was coming at noon and it was already 4 p.m., "too late" because the next door drugstore was closed, or "too late" because the patient's state had deteriorated so much that the doctor could not save her. Clock, contextual, and action time: all participate in the estimation of duration in work environments and can be in conflict (De Keyser, 1996). Training about duration does not imply training people to keep deadlines, but rather is about how to increase their temporal flexibility and their knowledge about evolution and duration. Training must also enhance their competence in making satisfactory compromises between diverse time reference systems, taking into account economic, safety, and technical constraints.

Order is another facet of time. It generally suggests a linear and sequential approach to time, for instance, a sequence of actions has to be followed according to a plan. But order is not always that simple. Complex order can be derived from temporal logic (Allen, 1984) which defines seven basic temporal relations between time intervals: before, equals, started-by, during, finished-by, overlapped-by, and met. This logic enables very complex plan structures to be described, close to those found in naturalistic situations (Javaux, 1996; cf. Figs. 10.1 and 10.2).

In the naturalistic context, complex plans have to be organized, memorized, executed, and, very often, modified in an opportunistic way (Seifert, Patalano, Hammond, & Converse, 1997), and this is possibly a matter of training. However, simulators exhibit constraints that make this type of training difficult. Indeed,

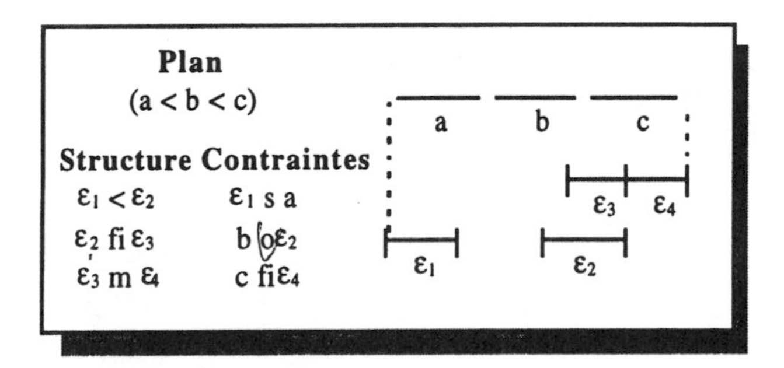

FIG. 10.1. Complex order in plans (Javaux, 1996). The description of a complex temporal structure based on intervals is possible with the aid of Allen's relations: before (<), equals (=), started-by (si), during (di), finished-by (fi), overlapped-by (oi), met (mi). A synchronization constraint (c fi ϵ) is expressed between element c of the plan (a < b < c) and the temporal element ϵ.

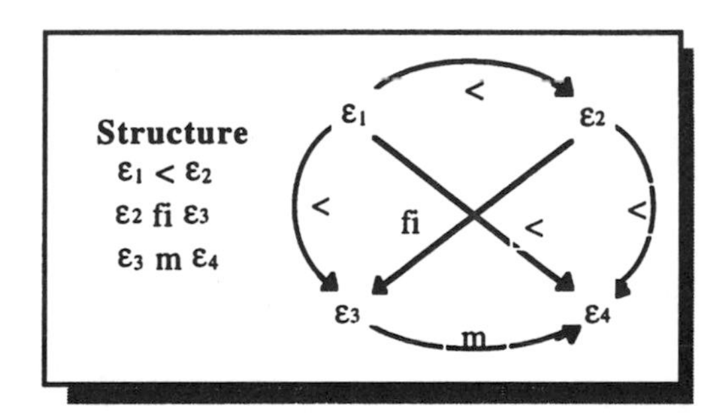

FIG. 10.2. It imposes a specific temporal relation between c and ϵ: "c finished by ϵ." It means that the termination of action c must coincide with the end of the temporal element. The recognition by the operator of the regularity of some temporal structures enables predictions. In its most general form, a prediction on a regular temporal structure corresponds to a subset of possible futures underway in the present.

explicit learning of such complex temporal structures competes with the working memory span and consumes resources from a cognitive point of view. But implicit learning necessitates a very frequent exposure to these structures: this condition can generally not be met in full-scale simulators, which are rather expensive. Finally, in simulators, temporal perspectives are, by definition, truncated. Even if in the briefing session instructors try to simulate the history of the case, this attempt is severely limited and does not cover the history of the instruments, of previous similar cases, of the participants themselves, and so forth. The same for future: there is no future after the case, no consequences, and no stakes. The training session is a theater play with its classical constraints of location, time, and action. Does it means that no temporal expertise can be acquired using simulators? This chapter addresses this issue and focuses on three types of competence: dynamic diagnosis, anticipation and planification, and synchronization. All of them are important for situativeness and rely on duration, order, and temporal perspectives. All of them are potential source of errors and accidents. The domain covered by this chapter is anesthesia.

NATURALISTIC SITUATIONS AND TEMPORAL COMPETENCES

Temporal competences and their possible relations with human errors have to be studied in context—before any attempt to teach them using simulators. The following results are based on a large field study, carried out by a multidisciplinary team of psychologists and anesthetists in a Belgium hospital (Nyssen, 1997). The data have been gathered by diverse analysis techniques: observations in operating rooms, questionnaires, and quality conferences. The anesthetist's cognitive task analysis sketched in Fig. 10.3 is a simplified version of the result of more than 200 hours of observation in operating rooms (Nyssen, 1997; De Keyser & Nyssen, 1993).

The map illustrates the complexity of the task and stresses the importance of its temporal demands. The anesthetist must continuously adjust his behavior to the contextual conditions, with dynamic diagnosis, anticipation and planification, and with synchronization.

Dynamic Diagnosis

Dynamic diagnosis includes monitoring patient condition and the course of actions taken by the team in order to maintain an up-to-date representation of patient state during surgery. Although anesthesia is usually referred to as a procedure for putting the patient into an unconscious state so that surgery is possible, this procedure is highly contextual depending on the patient's responses and

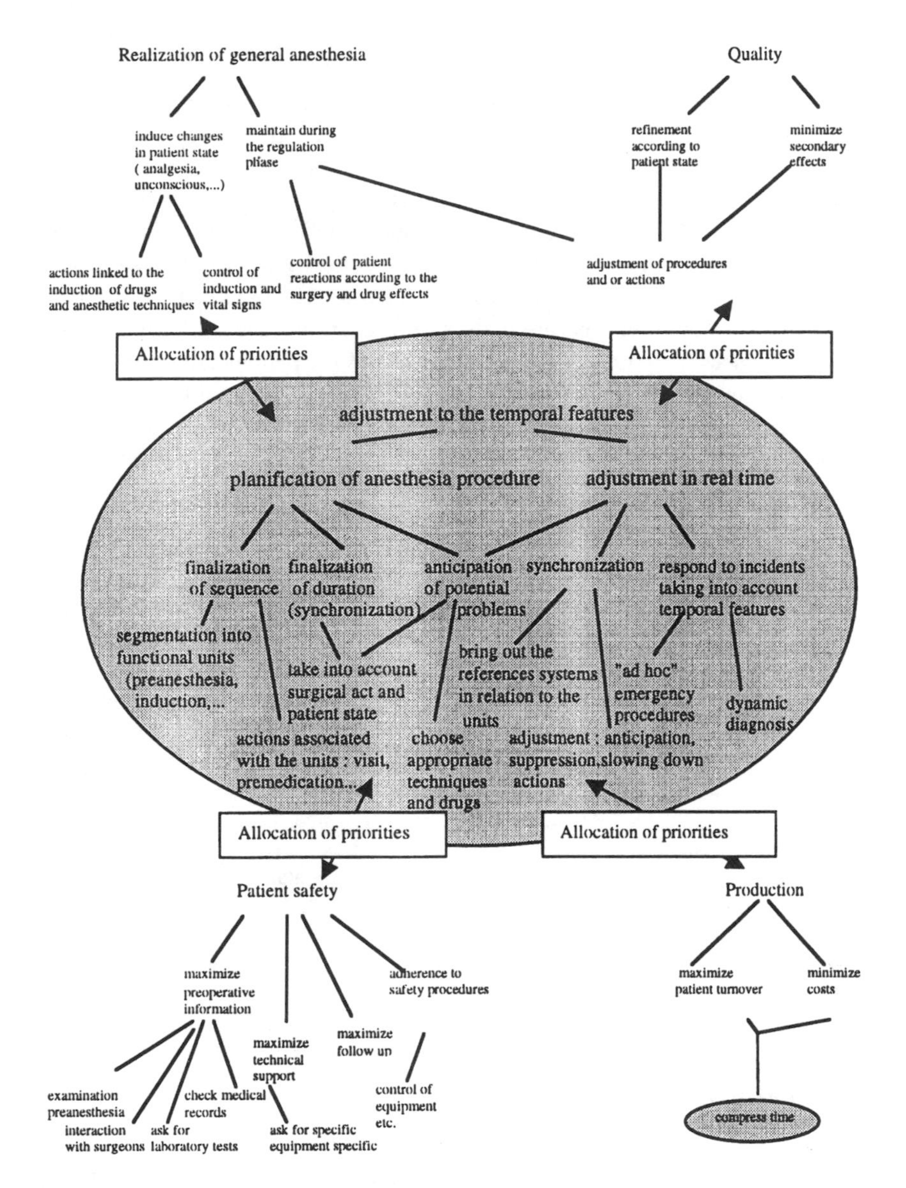

FIG. 10.3. Simplified cognitive task analysis in anesthesiology emphasizing the temporal task demands. The goal structure includes realization of anesthesia, patient safety, quality, and production as well as the temporal adjustment to the temporal features of the task situation. These abstract goals can be in conflict and lead to failures. Note that the temporal elements are surrounded with gray. This part of the structure can be seen as a metatask whose goal is to mediate and allocate priorities between safety, quality, production, and realization of anesthesia.

Note: From "Les Erreurs humaines en anesthésie, by V. De Keyser and A. S. Nyssen, 1993, *Le Travail Humaine, 56* (2–3) p. 243–266. Copyright 1993 by A. S. Nyssen. Adapted with permission.

the surgeon's activity. Consequently, the anesthetist must continuously update his or her mental representation of the patient's state and adapt strategy to the current situation. Any delay before diagnosis is crucial, but less than the way the anesthetist recovers the incident. Sometimes, priority is given to the recovery process and diagnosis is delayed. In naturalistic situations, when a trainee does not succeed in controlling an incident, procedure requires the trainee to call a supervisor. Our field studies and error collections show that trainees fail to estimate correctly how long they should wait before calling and do so too late: they overestimate their competences, they underestimate the speed of the patient's deterioration, or, sometimes, they do not want to lose face. This leads to a delay in both the treatment and diagnosis processes. Failure to reassess the current situation can lead to fixation errors (De Keyser & Woods, 1990) as illustrated in the following case story collected in the field (De Keyser & Nyssen, 1993, p. 258):

> The anesthetist was in his third year of training; the senior was in the neighboring operating room. The patient, a man in his 70s with chronic bronchitis, was in generally poor health and was being operated on for a facial tumor. It was a matter of comfort surgery because the tumor was too advanced to be cured. The duration estimated in the schedule was 3 hours. The case was prolonged for more than 8 hours without the anesthetist alerting the surgeons to the evolution and increased risk for the patient. The intervention required the patient to be turned from his initial position on his back to a side position. While the patient was being moved, the endotracheal tube slipped downward and the patient was only ventilated on one side. Moreover, the patient's face was bleeding profusely, and the blood saturated the connections between the tracheal tube and the respirator, which made them less airtight. At the end of the afternoon, the electrocardiogram alarm sounded just as the anesthetist was discussing the case with his supervisor. Both hurried in and discovered respiratory troubles with bradycardia. Resuscitation procedures were started, and they succeeded in recovering the heart.

Anticipation and Planification

The speed of the deterioration process can limit the amount of time available to respond and diagnose a problem, encouraging planification and anticipation. This is true for the three phases characterizing an anesthesia: the preoperatory phase, the peroperatory phase and the postoperatory phase. Before the intervention, the anesthetist must anticipate any difficulties by taking into account the patient's information and the surgical act and must choose the best anesthesia procedure. During the surgery, the anesthetist can anticipate, to some extent, the evolution of the situation taking into account knowledge about the patient state, the drug latencies, and the surgical act. This is also true for the postoperatory phase for which one must predict potential incidents and take safety measures. Failure to anticipate becomes dramatic when the anesthetist is confronted with a problem situation that evolves rapidly with a very short time interval in which to respond. Anticipating and planning illustrate how temporal competences are closely related to causal and procedural knowledge—and not exclusively to time estimation. On the other hand, the earlier case illustrates the importance of temporal perspectives that cannot be simulated easily: the preoperatory phase, for instance, is crucial not because of the risk involved but because it is during this early phase that the anesthetist learns about the patient from questions and physical examination as well as from laboratory tests and questions to the surgeon. All these conditions involving dynamic interaction between multiple actors are difficult to create in simulators.

Synchronization

There are a number of actions or events that must be synchronized when conducting an anesthesia, notably because of the collective nature of the work situation. For instance, extubation must be carried out as soon as the surgical act is finished, when autonomous respiratory function is completely recovered; otherwise, the risk of an incident is increased. We observed that synchronization is a difficult process, in particular for trainees who do not estimate the duration of the drugs or the duration of the surgery acts well enough. This leads to synchronization failures, which can have dramatic consequences for the patient if they are not recovered in time, as in the following case (De Keyser & Nyssen, 1993, p. 260):

> The anesthetist on duty was scheduled to be on call that night. It was the beginning of the afternoon. The intervention was carried out on a 2-year-old child. The anesthesia was administered without any problem, and the surgical act finished. Hurried by the hospital planning, the anesthetist extubed a bit too fast at the first sign of awakening (cough in the tube). The child went into a laryngal spasm (the reflex closing of vocal chords causing a complete or partial glottal obstruction). The anesthetist succeeded in quickly reintubing the patient and recovered the incident. Remarks: small children are particularly vulnerable to these complications: they have been known to go back to sleep after the first signs of awakening. The anesthetist must then wait for this phase to pass before extubing the child.

Along this line of field observations, the question is: Can these temporal competences be taught using simulators, knowing the simulators limitations? When asking this question, we must be aware that, up to now, at least in Belgium, simulator training programs have concentrated on intraoperative management of critical incidents. Thus, it was not possible to build a clean experimental setting, focusing exclusively on the temporal aspects. However, taking advantage of the tight collaboration between psychologists and anesthetists and the presence of the researcher in all the training sessions organized for the past 3 years with audio and video records, it was at least possible to observe the trainees' performance and to make some assumptions.

SIMULATORS AND TEMPORAL COMPETENCES

Full-Scale Anesthesia Simulator

The full-scale simulator (CASE Series: Comprehensive Anesthesia Simulation Environment, built by CAE LINK Corporation, NY, USA) is presented in the form of a conventional operating room equipped with all the necessary equipment for an operation except that the patient has been replaced by a mannequin best representing the clinical reality. The mannequin provides electronically generated sounds and breathing and dynamically changeable airway anatomy. Full hemodynamic monitoring is supported. The cardiovascular model can generate and model the hemodynamic consequences of a variety of incidents (dysrhythmias, . . .). The simulation session is directed by a simulation director (senior anesthetist) who can adjust in real time the different physiological and pharmacological parameters according to the reactions of the trainee. The training simulation follows, to some extent, the segmentation of the anesthetist's task in naturalistic situations and includes three phases:

1. A briefing during which information is given concerning the patient's records (approximate duration, 15 min). This phase attempts to create the conditions necessary for anticipation and planification of the anesthesia procedure based on the preanesthesia visit, but there is no patient examination.
2. A simulation session for which the duration varies from 30 to 60 minutes, according to the type of scenario. The session is recorded on video. In naturalistic situations, the duration of problem situations can last much longer than 60 minutes but simulating this type of session is difficult for economic and human resource reasons.
3. A debriefing during which the performance is discussed in collaboration with the instructor and the psychologist, using the videotape recording (approximate duration, 35 min).

Screen-Based Anesthesia Simulator

The screen-based simulator (Anesthesia Simulator Consultant (ASC); Swchid, 1996) consists of a standard personal computer with one video monitor and a mouse. The system provides graphical representation of the mock monitoring displays and clinical equipment and photographs to display the patient and actions taken on the patient. It uses physiological and pharmacological models to provide dynamic clinical data in real-time. If the trainee has difficulties with one problem, an expert system consultant can outline the most important therapeutic interventions or take over the case management completely. The system tracks the drug levels and effects of more than 80 drugs. The subject responds to the simulated problem situation using a straightforward menu system. The user can control the variables describing the patient's state and act on the airway, ventilation, fluids, and drugs using the mouse. This technique allows the anesthetist to act very rapidly on the system. Even if drug latencies and action feedback are included in the simulator, all the intermediary steps, such as the preparation of the syringe, are not. In naturalistic situations, carrying out these steps can be the origin of many difficulties, which in turn can delay the diagnosis and treatment processes.

In the next section, we describe the data collection methods from three of our studies to validate these two types of simulated situations on three temporal demands emphasized by the task analysis: (a) dynamic diagnosis, (b) planification and anticipation, and (c) synchronization.

Dynamic Diagnosis

We compared the diagnosis performance of 20 anesthetist trainees (10 novices and 10 more experienced trainees) confronted with two problem situations (an anaphylactic shock[1] and a malignant hyperthermia[2]) in the two types of simulators. These two scenarios have different temporal characteristics or tempo (Cook, Woods, & Mc Donald, 1991). The first scenario (AS) arises suddenly and evolves rapidly. The second (MH) evolves slowly with an exponential breakdown. These tempos affect the diagnosis and recovering process differently. If the patient's state changes rapidly and the short-term consequences of these changes are negative, the anesthetist must decide to act quickly. There is no time for knowledge-based reasoning. Adjustment to the dynamics must be done in real-time, preferentially based on automatisms. In the second scenario (MH), the anesthetist has more time available to diagnose and plan the recovery process. He or she can decide to consult a senior without any risk for the patient. In any case, the diagnosis process involves, upstream, an awareness of the change of the situation. When the tempo is slow, awareness of changes appear to be more difficult (Jones & Boltz, 1989). Moreover, in such tempo, the value of information changes over time, making the diagnosis conditional to the reassessment of the situation.

The experimental setting used to compare diagnosis performance was inspired by Chopra et al. (1994; cf. Fig. 10.4). The performance of the trainees was

[1]Immediate allergic reactions involving a generalized response to certain medications used by the anesthetist.

[2]Lethal disorder of skeletal muscle metabolism triggered by volatile anesthetics or muscle relaxants. The symptomology is perverse: accumulated increase of CO_2 production precedes the increase in temperature that characterizes the incident.

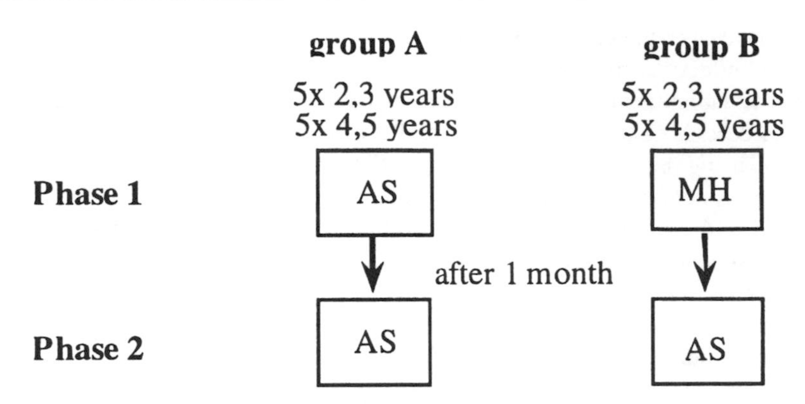

FIG. 10.4. Schematic representation of the study design. The subjects are divided into two groups, and the experiment unfolds in two phases. During the first phase, Group 1 is confronted with the anaphylactic shock, whereas Group 2 is facing the malignant hyperthermia. During the second phase, both groups have to solve the anaphylactic shock. The performance of the two groups is evaluated following a grid elaborated by two anesthetist experts. Diagnostic scores and delays in diagnosis are compared for the two groups.

assessed by two experienced instructors based on a list of treatment actions (from 0 to 100; see Nyssen, 1997).

No statistically significant differences were found, in terms of precision and time of diagnosis, between the two types of simulators. We found a learning effect between Phase 1 and Phase 2 for Group A in both simulators ($p < .01$). However, observation of the subjects' strategies and the debriefing discussion suggest that the increase in performance on the screen simulator is based more on the knowledge of the software than on the understanding of the anesthetic process. The subjects play with the ASC just as with a videogame. Analysis of similarity between the time dimension and the associated cognitive strategies in the simulator and in the task environment show important discrepancies. We noted an artificially high speed of the process in the screen simulator. The use of the mouse to act on the system accelerates the process. The high tempo of the deterioration makes knowledge-based reasoning difficult and favors a reactive diagnosis strategy. Additionally, only one predetermined treatment procedure is included in the software. This is far from naturalistic situations, in which there are more than one solution to any problem. In the full-scale simulator, the instructor can adapt the scenario to the trainee's behavior during the simulation and focus the debriefing on the optimal treatment strategy, taking into account the contextual conditions. This is not the case in the screen-based simulator, in which only one treatment procedure is outlined by the expert system consultant.

Planification and Anticipation

In the full-scale simulator, five scenarios have been investigated with 6 novices and 6 experts each (Nyssen & De Keyser, 1998) to clarify and amplify available data on planification and anticipation competencies as part as the cognitive task analysis. In all scenarios, diagnosis times are shorter in the more experienced group ($p < .01$). We can intuitively connect this reduction in the time of diagnosis to the anticipation competencies observed during the briefing among the more experienced trainees. What was more puzzling was that even when novices anticipated the potential for incident based on the patient's information during the briefing, it did not help them to plan the recovery process during the simulation. They got lost in routines, procedures, and equipment checks that were not yet automatic. For instance, in one scenario, heart rate troubles on the electrocardiogram were detected by both the novices and the experts. While the experts explicitly decided to interrupt the intubation procedure (delay of cross-checking verification) and shift to managing the disturbances, the novices stuck to the intubation procedure, wasting precious time. In another scenario, all the anesthetists anticipated the risk of ischemia during the briefing but this did not orient novices' attention during the simulation session. They were overwhelmed by practical operational problems. Consequently, the difference in performance between the two groups should not be analyzed in terms of anticipation competences, but rather in terms of limited attentional resources (Khaneman, 1973). Everything happened as if the task underway mobilized an important set of attentional resources in the novices. The disturbance is perceived but successively processed at the central and decisional level, which in turn increases the time of diagnosis and response.

These data show how an in-depth analysis of performance in full-scale simulators conducted by a psychologist can help one to understand the relevant task demands better. Results show that besides medical knowledge and diagnosis competency taught in the screen-based simulator, there is also a need for various other competences that strongly influence performance on the job. Full-scale simulators create the conditions to acquire some of these competences. Yet, without psychological research, there is a risk that the sources of errors will be misunderstood, leading to unadapted training. Based on our analyses it appears, for instance, that explicit anticipation training in simulators might not be the only direction to pursue. Trainees, including novices, are capable of anticipating some incidents from the patient's examination during low workload periods such as the preanesthesia visit or briefing. But, this competency does not help novices to manage the case better when the problem occurs. Following this, we can expect an indirect effect of the training of procedures in simulators: the acquisition of automatisms in such environments will release the attentional resources for higher level information processing such as planification and anticipation. The learning of such automatisms is not possible in the screen-based simulator. On the contrary, in the full-scale simulator, the instructor, together with the psychologist, can include numerous contextual variables and interferences encouraging parallel processing. During the debriefing, they can emphasize the need for planification and time-saving strategies, which release more available resources in case of problems. Even if the complexity of the simulated situations is more complex than on the screen-based simulator, the hospital and the history are still missing. These are difficult to simulate. Next, we show how the work organization influences performance with synchronization demands.

Synchronization and Conflicting Temporal Reference Systems

Field observations are very important to gain access to the temporal structure of the task sketched in Fig. 10.5 (Nyssen & Javaux, 1996). Previously, we described the synchronization demands inherent in the task. Additionally, the anesthetist is simultaneously involved in other processes of synchronization. The hospital schedule introduces, up to a certain point, clock time as a constraint of synchronization, sometimes leading the anesthetist to take certain risks. In order to meet that constraint, we observed that the experienced anesthetist does not wait until the end of the surgical act to decrease the drugs before extubing the patient, but anticipates it. With experience, he or she has acquired knowledge about duration. Taking into account the drugs administered and an estimation of the probable duration of the operation, the anesthetist decreases then stops injecting the drugs, starting from the moment when he or she estimates their duration of effect to be identical to the time necessary for the end of the operation. In this way, the anesthetist reduces the awakening time and ensures the synchronization constraint linked to the hospital schedule is met. The success of this synchronization strategy depends on a low variability of the work situation; an unusual patient condition can make this acquired strategy inappropriate and lie at the source of synchronization failures (see Fig. 10.5).

Teaching such competency implies, as we said before, an improvement in the subject's temporal flexibility and the elaboration of good compromises between the different time frames (clock, contextual, and action time), which can be in competition for the anesthetist. In the full-scale simulator, the instructor can include some social aspects, such as conflictual relationships between anesthetists and surgeons, in order to create the learning conditions for compromise. But economic, personal, or organizational aspects of work are difficult to simulate. In naturalistic situations, these aspects highly influence performance and can lead to temporal failures which, in turn, can have dramatic consequences. We present a case that leads to the death of the patient (The "Soir" Journal, 09/22/99):

> The anesthetist was in his fourth year of training; his senior was in the neighboring operating room and was supervising three operating rooms. The patient, a 9-year-old girl, was being operated on for multiple knee fractures. The operation was planned to be long but without any high risk for the patient. After 10 hours of intervention, the

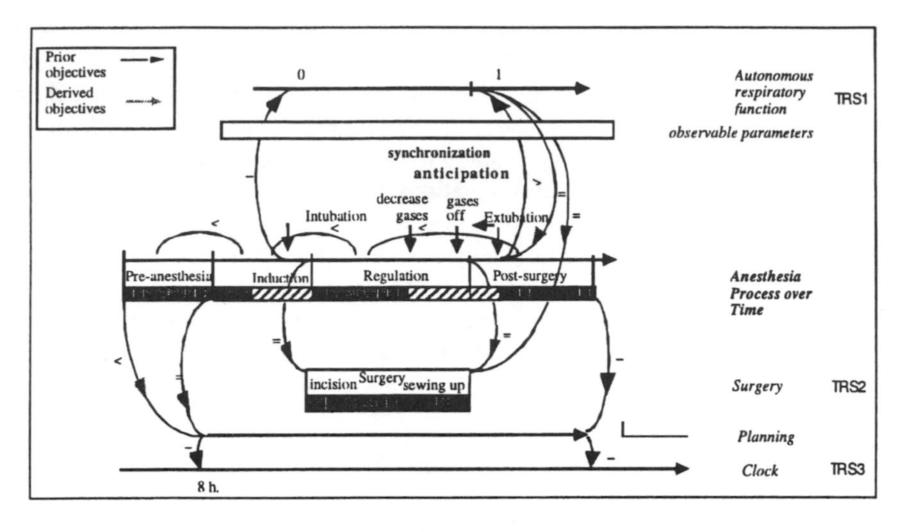

FIG. 10.5. A formal description of the temporal constraints an anesthetist negotiates during an operation. The anesthetist has to synchronize the decision process and the interventions (drugs, gas, intubation, and extubation, etc.) with: the autonomous respiratory function (from 0 to 1), the sequence of the surgical act, and hospital planning.

blood pressure dropped just as the anesthetist was discussing the case with his supervisor. Both hurried in, discovered respiratory troubles with bradycardia and cardiac arrest. Resuscitation procedures were started but without success. The patient died from hemorrhagic shock.

The anesthetist trainee committed suicide after the accident. The surgeon and the senior anesthetist were sued. The state report acquitted the surgeon but found the senior anesthetist guilty. He was blamed for accepting to supervise three operating rooms at the same time, which affected his potential to anticipate the problem or, in any case, to treat it in emergency.

This dramatic case illustrates how work organization creates new temporal demands, which do not take only the form of temporal pressure, as has been largely debated (see Cannon-Bowers & Salas, 1998). Here, the form taken is more indirect. Three anesthesia plans must be supervised in parallel, creating a new form of synchronization requirement for the supervisor. The dynamics of the event (tempo of symptoms and delayed effects on the patient) are other important factors making it difficult for the trainee and supervisor to realize the possibility of a problem. Training simulators can force the subject to think about the temporal evolution of incidents. However, training does not appear as the most appropriate prevention measure. Another approach would be to link performance failure to work organization. Following a study on stress we carried out among the Belgium anesthetist community, the lack of control over time management (dealing with tied schedules, deadlines, and quantitative overload) appears as the first factor generating stressful situations (François, 1999). This perceived lack of control can make subjects doubt the adequacy of their own resources with which they think they have to cope to deal with the situation. The conclusion drawn from this kind of research is that an individual in a position to control and regulate his or her activity can cope better with stressful situations (Arronsson, 1989; Hansez & De Keyser, 1997). Improving job control in a workplace, particularly control over time management in our case, is highly relevant for successful management of dynamic problem situations. On this point of work organization, the questions of prevention must be studied at the collective level and involve the hospital authorities and government with a view to developing general health policies.

CONCLUSIONS

If we refer to the three facets of time described in the introduction—duration, order, and temporal perspectives—training simulators exhibit severe limits:

Duration. All the absolute durations of naturalistic situations are distorted. Moreover, their relative time intervals are usually not preserved. For instance, in anesthesia simulators, the simulated surgical act is generally much shorter than in similar real operations with some of the anesthetist acts having a quasi-immediate effect while others have a realistic duration. These distorsions are much more important in screen simulators than in full-scale simulators. This has different consequences on training. First, it prevents any internalization of durations. It means that in simulators, time adaptation requires attentional resources and actions have to be planned and executed in a conscious way—looking for external temporal reference systems. In naturalistic situations, regular durations are internalized, which spares cognitive resources for unexpected events and contextual changes. Second, higher dynamicity in simulated situations, especially in screen simulators, entails a specific mode of control of the situation that is more reactive than anticipatory. This means a change in temporal perspectives, a reduction of the anticipation span usually associated with having fewer patient variables to take into account.

Order. Usually, simple order is preserved. This is immediately related to the causal model underlying the simulation. For instance, a sequence such as "action A precedes B and has an effect on C" referring to the relations between patient variables and anesthetist actions is found in naturalistic situations as well as in simulators. But complex order including multiple and conflicting constraints, parallel plans to follow with interferences, and interruptions found in naturalistic situations is never simulated. In fact, simulators do not teach the complexity and the variability of the real world; they use regular predictable problem situations. Sources of variability, such as patient variability and medical team variability, can be introduced in the simulated situations but they are far removed from the complexity found in naturalistic situations.

Temporal perspectives. By definition, a training session is a kind of classical theater play with unity of time, location, and action. Past, present, and future are shrunk into a very short time interval. The briefing tries to reconstitute the history of the case, but it concerns only the patient, and does not cover the monitoring systems, the medical team, or previous cases this team has already encountered in the past. Even though the present is rather realistic, there is no future beyond the training session. It removes any consequences or social stakes associated with the case. This is excellent from a training point of view, for it gives trainees the opportunity to learn from errors, without any risk for the patient. But, of course, this is not the real world.

Simulators are far from incorporating the real-life conditions of action as they introduce either temporal distorsion and/or temporal simplification. While simplification generally supports situativeness by helping trainees to focus on specific situation aspects (without being distracted or overwhelmed by the complexity of the world), this is not the case with distorsion. It significantly changes the strategies and mode of control of the process. This is reflected in dynamic diagnosis. This competence can be trained using screen-based and full-scale simulators. Trainees learn to improve the accuracy and the speed of their diagnosis in both simulators. But, the diagnosis strategy they develop on the screen simulator cannot be transferred to any naturalistic situation. It is a reactive and skill-based strategy, well adapted to fast videogames but inadequate in some real-life situations. In a full-scale simulator, durations are still distorted, but at least trainees have to manipulate instruments and action time is better preserved. This increases the psychological validity of the exercise. For example, periodic reassessment of the patient state proves to be difficult in full-scale simulators as well as in naturalistic situations, and many fixation errors have been recorded. These errors are analyzed and exploited in the debriefing sessions. The fact that temporal perspectives are simplified does not prevent full-scale simulators from being an excellent training support for anticipation and planning. Errors committed by trainees show that, even if they anticipate the problem correctly during the briefing, the patient evolution, the drug manipulations, and complex the procedures are enough to create realistic learning conditions of action. This is not the case for synchronization which is totally missing in simulator training programs despite the difficulties it creates in naturalistic situations. Synchronization is generally an implicit requirement. It implies the introduction of coherency and order in a complex world and relies on two aspects that are not well preserved in simulators: complex order, which is highly simplified, and duration accuracy, which is distorted. In naturalistic situations, people have to adjust their actions to different plans simultaneously, to take into account multiple temporal constraints, to perform opportunistic and reactive planning, and to meet deadlines without taking risk: this is real life. Yet, is it really necessary to train people to cope with all these constraints? Judiciously, simulators remind us how unnatural naturalistic situations are. In working environments, they are pure human inventions and, as such, must be continuously questioned, criticized, and modified. If they are too incoherent and chaotic, the problem has to be tackled at a work organization level and not at the training level.

The Role of the Psychologist in Simulator-Based Training Programs

The value of a psychologist in a training simulator is in helping to see the cognitive demands imposed by a task situation and then in promoting the learning process. Training simulators too often "assume" rather than "observe" decomposition of naturalistic situations. This can lead to a premature focalization of performance, which can create an inappropriate prioritization of the training issues. From our point of view, a key element for the psychologist is to collect and compare data on human performance from various sources, from the field, and from simulators (Nyssen, 2000). The goal is not to improve the functional or physical validity of the simulators. Rather, the juxtaposition of data allows better understanding of the cognitive demands of the task situation, the sources of errors, and the potential of preventive measures, either training or other improvements of the man–machine system. At the same time, the results of such a research process, swinging from field situations to simulated situations, can serve to guide the ongoing evaluation of the simulator-based training programs. It can also be part of the improvement of the learning process as well as that of training the instructors. Too often, simulator instructors are domain experts who do not exploit recent scientific findings in the training, learning, and cognitive areas (Salas, Bowers, & Rhodenizer, 1998). The psychologist can help the instructor to assess the trainee's state of knowledge and skill, to create appropriate simulated situations, to provide extrinsic feedback, and to draw attention to the learner's self-monitoring behavior during the debriefing. The reciprocal relationship between psychologist and instructor will, in turn, improve the effectiveness of the training program.

ACKNOWLEDGMENTS

This work was supported by the Belgian programme on Interuniversity Poles of Attraction (I.A.P.), initiated by the Belgian Federal Government, Prime Minister's Office, Science Policy Programming and the SSTC Programme.

REFERENCES

Allen, J. F. (1984). Towards a general theory of action and time. *Artificial Intelligence, 23,* (pp. 123–154).

Amalberti, R. (1996). *La conduite des systèmes à risque* [Management of Risk Systems]. Paris: Presses Universitaires de France.

Arronsson, G. (1989). Dimensions of control as related to work organization, stress and health. *International Journal of Health Services, 19*(3), 459–468.

Baker, S., & Marshall, E. (1989). Simulators for training and the evaluation of operator performance. In L. Bainbridge & S.A.R. Quintanilla (Eds.), *Developing skills with information technology* (pp. 293–314). Chichester, England: Wiley.

Block, R. A. (1990). Models of psychological time. In R. A. Block (Ed.), *Cognitive Models of Psychological Time* (pp. 1–35). Hillsdale, NJ: Lawrence Erlbaum Associates.

Boltz, M. G. (1998). The relationship between internal and external determinants of time estimation behavior. In V. De Keyser, G. d'Ydewalle, & A. Vandierendonck (Eds.), *Time and the dynamic control of behavior* (pp. 109–122). Germany: Hogrefe & Huber.

Brehmer, B. (1990a). Strategies in real-time, dynamic decision making. In R. Hogarth (Ed.), *Insights in decision making* (pp. 262–279). Chicago: University of Chicago Press.

Brehmer, B. (1990b). Towards a taxonomy for microworlds. In J. Rasmussen, B. Brehmer, M. de Montmollin, & J. Leplat (Eds.), *Taxonomy for analysis of work domains*. Roskilde, Denmark: Riso National Laboratory.

Cannon-Bowers J. A., & Salas, E. (1998). *Making decisions under stress*. Washington, DC: The American Psychological Association.

Chopra, V., Gesing, B. J., De Jong, J., Bovill, J. G., Spierdijk, J., & Brand, R. P. (1994). Does training an anesthesia simulator lead to improvements in performance? *British Journal of Psychology, 73,* (pp. 293–297).

Cook, R., Woods, D., & McDonald, J. (1991). *Human Performance in anesthesia: A corpus of cases. Report to the Anesthesia Patient Safety Foundation.* (Cognitive Systems Engineering Laboratory Tech. Rep. No. 91-TR-03). Ohio State University, Department of Industrial and Systems Engineering.

De Keyser, V., (1995). Time in ergonomics research. *Ergonomics, 38*(8), 1639–1660.

De Keyser, V. (1996). Les erreurs temporelles et les aides techniques [Temporal Errors and technical aids]. In J.-M. Cellier, V. De Keyser, & C. Valot (Eds.). *Gestion du temps dans les environnements dynamiques* [Time management in dynamic environments]. (pp. 287–304). Paris: Presses Universitaires de France.

De Keyser, V., & Nyssen, A.S. (1993). Les erreurs humaines en anesthésie [Human Errors in Anesthesia]. *Le Travail Humain, 56*(2–3), 243–266.

De Keyser, V., & Woods, D. D. (1990). Fixation errors in dynamics and complex systems. In A. G. Colombo & A. Saiz de Bustamente (Eds.), *Systems reliability assessment*. Dordrecht, The Netherlands: Kluwer Academic.

De Keyser, V., d'Ydewalle, G., & Vandierendonck, A. (1998). *Time and the dynamic control of behavior*. Germany: Hogrefe & Huber Publishers.

François, J. C. (1999). *Evaluation de risques psycho-sociaux chez les anesthésistes* [Assessment of psycho-social risks in anesthetists]. Unpublished manuscript, Université de Liège, Liège, Belgium.

Gaba, D., Fish, K., & Horward, S. (1994). *Crisis management in anesthesiology*. New York: Churchill Livingstone.

Grau, J., Doireau, P., & Poisson, R. (1998). Conception et utilisation de la simulation pour la formation: Pratiques actuelles dans le domaine militaire [Design and use of simulation for training: Lessons drawn from present military use]. *Le Travail Humain, 61*(4), 361–385.

Hansez, I., & De Keyser, V. (1997, October). *Stress at work: A methodology for diagnosis and collective intervention. The case of public services workers.* Paper presented at the 5th European Conference on Organizational Psychology and Health Care, Utrecht, The Netherlands.

Hoc, J. M. (1989). Strategies in controlling a continuous process with long response latencies: Needs for computer support to diagnosis. *International Journal of Man-Machine Studies, 30*, 47–67.

Hollnagel, E. (1998). *Cognitive reliability and error analysis method.* New York: Elsevier.

Hukki, K., & Norros, L. (1998). Subject-centred and systemic conceptualisation as a tool of simulator training. *Le Travail Humain, 61* (4), 313–331.

Hutchins, E. (1995). *Cognition in the wild.* Cambridge, MA: MIT Press

Javaux, D. (1996). La formalisation des tâches temporelles [Formalism for temporal tasks]. In J.-M. Cellier, V. De Keyser, & C. Valot (Eds.), *Gestion du temps dans les environnements dynamiques* [Time management in dynamic environments] (pp. 122–155). Paris: Presses Universitaires de France.

Jones, M. R., & Boltz, M. (1989). Dynamic attending and responses to time. *Psychological Review, 96*(3), 459–491.

Khaneman, D. (1973). *Attention and effort.* London: Prentice-Hall.

Leplat, J. (1989). Simulation and simulators in training. In L. Bainbridge & S.A.R. Quintanilla (Eds.), *Developing skills with information technology* (pp. 277–291). Chichester, England: Wiley.

Nyssen, A. S. (1997). *Vers une nouvelle approche de l'erreur humaine dans les systèmes complexes. Exploration des mécanismes de l'erreur humaine en anesthésie* [Towards a new approach of human errors in complex systems: Exploration of human error mechanisms in anesthesia]. Unpublished doctoral dissertation, Université of Liège, Liège, Belgium.

Nyssen, A. S. (2000). Analysis of human errors in anesthesia: Our methodological approach from general observations to targeted studies in simulator. In C. Vincent (Ed.), *Safety in medicine.* (pp. 43–63). New York: Elsevier.

Nyssen, A. S., & De Keyser, V. (1998). Improving training in problem solving skills: Analysis of anesthetist's performance in simulated problem situations. *Le Travail Humain, 61*(4), 387–401.

Nyssen, A. S., & Javaux, D. (1996). Analysis of synchronization constraints and associated errors in collective work environments. *Ergonomics, 39*, 1249–1264.

Salas, E. , Bowers, C. A., & Rhodenizer, L. (1998). It is not how much you have but how you use it: Towards a rationale use of simulation to support aviation training. *The International Journal of Aviation Psychology, 8*(3),197–208.

Schwid, M. D. (1996). Graphical anesthesia simulators gain widespread use. *APSF Newsletter, Fall*, 32–34.

Seifert, C. M., Patalano, A. L., Hammond, K., & Converse, T. M. (1997). Experience and expertise: The role of memory in planning for opportunities. In P. J. Feltovich, K. M. Ford, & R. R. Hoffman (Eds.), *Expertise in context* (pp. 102–123). Cambridge, MA: MIT Press.

Sougné, J., Nyssen, A. S, & De Keyser, V. (1993). Temporal reasoning and reasoning theories: A case study in anaesthesiology. *Psychologica Belgica, 33*, 311–328.

Suchman, L. (1987). *Plans and situated actions: The problem of human-machine interaction.* Cambridge, England: Cambridge University Press.

Van Daele, A. (1993). *La réduction de la complexité par les opérateurs dans le contrôle des processus continus* [Decrease of complexity by operators in continuous processes]. Unpublished doctoral dissertation, Université of Liège, Liège, Belgium.

11

Analyzing Submarine Decision Making: A Question of Levels

Susan S. Kirschenbaum
Naval Undersea Warfare Center Division, Newport

Most sciences begin in an observe-and-describe mode (e.g., biological taxonomies, early chemistry, etc.). Mechanistic explanations, prediction, and control follow later. Thus, Naturalistic Decision Making (NDM) research has described the problem and decision makers' strategies and responses. Many readers are familiar with some of the "classic" NDM stories, such as the rescue team leader who used mental simulation to solve the problem of how to successfully use a harness (Klein, 1998) and the fire ground commander who recognized the laundry chute fire had spread (Klein, 1993). These decision makers in the field used recognition and mental simulation to evaluate their options, not elimination-by-aspect (Tversky, 1972) or anchoring-and-adjustment (Einhorn & Hogarth, 1985). These are the stories that led researchers away from the normative and descriptive models of classical decision-making research. Now is the time to dig deeper in the existing data and to collect new data that support the more detailed analysis characteristic of more mature science.

The goal of this chapter is to demonstrate the contribution to NDM theory that accrues when process-tracing data are analyzed at a more detailed level than is customarily reported in the NDM literature. Just as a detailed Goals, Operators, Methods, and Selection Rules (Card, Moran, & Newell, 1983) analysis has proved useful for gaining insight into user-computer interactions at the device level (Gray, John, & Atwood, 1993), I believe that a detailed analysis of the interaction between decision maker and information at the cognitive level can yield equally useful insights into the decision process. With all the understood complications of uncertainty, ambiguity, and event dynamics, NDM requires these insights to progress from a descriptive to a predictive science.

This chapter begins with a discussion of the concept of levels of analysis and then describes how the submarine problem and the task of the submarine Approach Officer (AO) fit within the NDM framework. It then delves into naturalistic submarine decision making, with an emphasis on information-gathering strategies and processes. Finally, it shows how phenomena, at the detailed level, inform understanding at higher levels.

LEVEL OF ANALYSIS: MULTIPLE POSSIBILITIES

A major problem in analyzing data from experts performing military problem solving/decision making is that of establishing the appropriate level of analysis (Gray & Kirschenbaum, 2000). The concept of *level of analysis* has been discussed in detail elsewhere (Anderson, 1990; Marr, 1982). One of the more general formulations is Newell and Card's (1985) analysis that showed how human actions can be analyzed in terms of timescales ranging from decades to milliseconds. A modified and truncated version of this formulation (see Table 11.1) shows the timescale of decision actions. Just as human actions can be described at many timescales or levels, data on human actions can be collected and analyzed at many of these same levels. Any single event can be analyzed as part of a social

TABLE 11.1
Levels of Analysis (after Newell's) of the Timescale of Human Actions

Time		*Action*	*AO's Task*	*Theory*
(sec)	*(common units)*			*(world)*
10^6	(years)	System	Development	Social
10^7	months	Planning	Training	&
10^8	weeks	Crew	Mission	Organization
10^5	days	Crew	Mission	
10^4	hours	Task	Scenario	Bounded
10^3	10 min	Task	SituationAssessment	Rationality
10^2	minutes	Task	Goals	
10	10 sec	Unit task	Subgoals	
10^0	1 sec	Operator	Operators	Cognitive Band
10^{-1}	100 ms	Cycle Time	Buffers	
10^{-2}	10 ms	Signal	Integration	
10^{-3}	1 ms	Pulse	Summation	Biological Band

interaction, an instance of situation assessment for decision making, the supporting goals, a string of subgoals (unit tasks) and operators (actions to accomplish the subgoals), or a string of neural impulses. Behavior at one level is often explained (and even predicted) by analysis at the next lower (more detailed) level.[1] Thus, situation assessment that takes place over 10s of minutes is explained by goals (minutes) and subgoals (10s of seconds).

Often the same data can be analyzed at more than one level. For example, in a large-scale military command and control simulation experiment, Pascual and Henderson collected verbal and behavioral protocol data, including video and audio recordings, behavioral and event records, and subjective judgments of the situation.[2] These verbal and behavioral protocols provide data suitable for analysis within the Social, Bounded Rationality, or Cognitive bands. This wealth of data poses a challenge to the researcher, both in terms of quantity and complexity. Furthermore, as noted by Pascual and Henderson, complex military data are especially difficult to encode. Thus, although these data could support a more detailed analysis (goals, subgoals, and operators), they chose to encode the data entirely at the level of Scenario and Situation Assessment. Their encoding focused on the strategies used for these scenario and situation assessment decisions. They found that 60% of the encoded utterances reflected a Recognition Primed Decision-Making (RPD) strategy. Of the eight other strategies identified, none accounted for more than 11% of the encoded utterances.

Pascual and Henderson hypothesized that RPD dominated because the participants were experienced and the locale was familiar. This is an explanation at the next higher level of analysis, reflecting training and mission knowledge. It explains neither why the RPD strategy was so dominant nor how that strategy supported decision making. The analysis techniques they used were not sufficient to tie decision strategy to information-gathering behavior or information content. Their results provided interesting and useful information, but they were limited in explanatory power by the level of analysis.

In part, research goals determine the appropriate level of analysis. Often, however, those goals are well served by a detailed analysis that can do more than describe the phenomena—that can explain the mechanisms that cause and control the phenomena. The Pascual and Henderson example is just one of many where data could support detailed analysis and answer the questions of why and how the phenomena described in the NDM literature differ from those described in other segments of the judgment and decision-making literature (Endsley & Smith, 1996; Flin, Slaven, & Stewart, 1996; Klein, 1993). For example, a more detailed

[1]The same analysis is true in the so-called hard sciences, such as chemistry. Behavior of elements at the macroscopic level is explained by properties of atoms at the microscopic level, and so forth.

[2]They also collected workload assessments and individual differences data, but these are irrelevant to the current discussion.

analysis might explain how decision makers using a RPD strategy differ in their use of information from those using another strategy. Payne and colleagues (Payne, Bettman, & Johnson, 1993) used just such an operator-level analysis to show that decision makers adaptively change strategies as constraints change (time, available information, etc.). Understanding of NDM phenomena could benefit from such a detailed analysis of decision-maker behavior in complex, dynamic, and uncertain situations.

SUBMARINE PROBLEM SOLVING AS NATURALISTIC DECISION MAKING

The list of situations that fit within the NDM family has always included many military situations (Howell, 1993; Kaempf, Klein, Thordsen, & Wolf, 1996). That list continues to grow, still, as submarine command is added to it. Each addition helps refine the definition. Submarining has a strong family resemblance to the others, but has some unique characteristics of its own.

Uncertainty

Among of the hallmarks of NDM are ill-structured problems, uncertainty, dynamic, and changing decision environments, action feedback loops, time stress, and high stakes (Klein, 1993; Orasanu & Connolly, 1993). As noted by several authors, traditional decision literature is unsuited to this problem space (Beach & Lipshitz, 1993; Orasanu & Connolly, 1993). Traditional decision models assume (at least implicitly) that the decision maker can gather all the relevant information and review it to make a rational choice or judgment, based on all the facts. They also assume that the information sources are reliable and the information is valid for some reasonable period of time.

In contrast, many decisions take place in environments where none of these assumptions are valid. When working in the field, rather than behind a desk, information is often incomplete, ambiguous, unreliable, and difficult or expensive (in time and effort) to obtain. It arrives—or does not—by questionable or undependable means. In some cases, such as the military (Kirschenbaum, Gray, & Ehret, 1997), the information may be deliberately distorted to hide or mislead. In other cases, such as a crisis (Flin et al., 1996), vital information can fall victim to the situation and never reach the key decision maker. Lastly, information can be lost by old or inappropriate technology or be impossible to extract from noisy data given current science and technology.

All of these conditions apply to submarining. Perhaps the most significant is the uncertain and noisy quality of even the most basic information. The submarine's primary sensors are sonar hydrophones that listen passively for sounds that indicate the presence of targets (also called contacts) in the ocean. This is an exercise in ambiguity. Underwater sound transmission is bent, reflected, scattered, and refracted by changes in temperature and salinity, and by the action of waves, currents, and different bottom types (Urick, 1983). It is masked by the noise of waves, schools of fish, whales, and noisy watercraft of all types. Thus, locating these auditory contacts is like locating the source of the image among fun house mirrors. Classifying an auditory contact in the noisy ocean environment is difficult. Localizing (establishing course, speed, and range) the contact is even more difficult (Kirschenbaum, 1990).

In virtually every other domain, the decision maker has at least some direct access to vital information and has multiple sources of information. She or he can query on-site participants, observe vital signs, scan reports, read gauges, and so forth. In contrast, when submerged the submarine decision maker has only one source of information, his[3] sonar system. As noted, the transmission characteristics of the underwater environment pose challenges, even for this information source. Moreover, while tracking the direction of the noise source, passive sonar cannot determine the range (distance) to the contact. The submariner has only two measurements, time and bearing (direction), from which to compute the three unknowns of course, speed, and range (see Fig. 11.1). Because it is mathematically underconstrained, solving the localization problem requires a combination of waiting for data to accumulate and maneuvering to get data from another point of view, an action-feedback loop. The solution emerges gradually, not suddenly. The best solution is the one with the most stable data whose elements are independently credible and mutually supporting.

Dynamic Change

Dynamically changing conditions also influence the validity and reliability of available information. When the situation is dynamic, information that is slow in arriving may no longer be valid by the time the decision maker sees it. On the other hand, the information that is available often cannot be validated under the time pressure of dynamic situations. Thus, dynamic situations pose special problems for the decision maker, even if the situation changes relatively slowly. Situations can change both due to the actions of the decision team and due to circumstances outside of their control. While many of the incidents reported in the NDM literature are emergencies (see Flin et al., 1996 for an example), there are many other situations that also fit the framework, but are not crises (for example, Roth, Mumaw, Vicente, & Burns, 1997).

In the world of the submarine decision maker, even routine decision making is uncertain and dynamic. The problem does not move as quickly as a fire, but it is

[3]All submariners are male as of 2000.

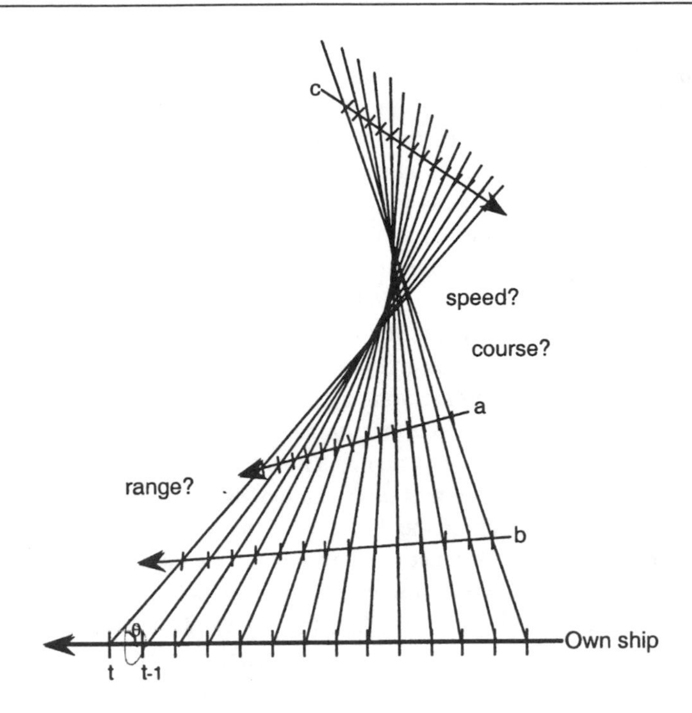

FIG. 11.1. One leg of the submarine problem. While OS travels at a constant course and speed, it tracks a contact with bearing θ at time t. The tick marks indicate equal time intervals. Assuming the contact went at a constant course and speed during the interval, many course, speed, and range combinations fit the data (e.g., lines a, b, or c).

continually changing. Submarines depend on sound as their principal information source and sonar as their principal sensor. Both the receiver, own ship (OS), and the transmitters, the sound sources, are moving. This movement can either be constant or changing. OS and contacts may also be changing direction, speed, and/or depth. Each action, by either of the ships, can dramatically change the situation and the availability and quality of information. Thus, the decision maker must solve a dynamic, multivariable problem with noisy, ambiguous data (see Kirschenbaum, 1994).

Expertise and the Role of Schema

Another hallmark of NDM research is the role that expertise and schema-driven problem solving play. The *schema* is the expert's memory structure for storing and retrieving relevant experience (Bartlett, 1932/1995; Ericsson & Kintsch, 1995). One can think of the schema structure as an organized set (or, possibly, a network) of attribute-value pairs. One of the principal decision strategies found in NDM research, the Recognition Primed Decision Making (RPD) strategy (Klein, 1989) is grounded in schema-driven, expert problem solving. (Arguably, decision making is a special case of problem solving.) The problem schema includes information about the class of problems and about their solutions. Like VanLehn's (1989) more general description of schema-based problem solving, RPD depends on the expert's ability to connect given information to the correct situational schema.

The three stages in VanLehn's (1989) schema-driven problem solving model are schema selection, schema instantiation, and action (see Fig. 11.2). During the first stage, schema selection, the actor selects the appropriate class of problem. As with decision making (specifically, RPD), schema selection is often key to solving the problem correctly. Thus, the physician determines if the presenting symptoms represent a case of heartburn (caused by gastric disease or food-borne bacteria) or heart attack (caused by heart disease or malformation). Diagnostic tests serve two purposes. They distinguish between the two possible schemas, and they begin the schema instantiation process (see later). However, there are many possible diagnostic tests, and they can be costly in time, money, and patient discomfort. Physicians (and insurance companies) order only those tests that they think are relevant to the likely diagnosis. VanLehn noted that the selection of a single (test) schema selection is triggered by information obtained very early in

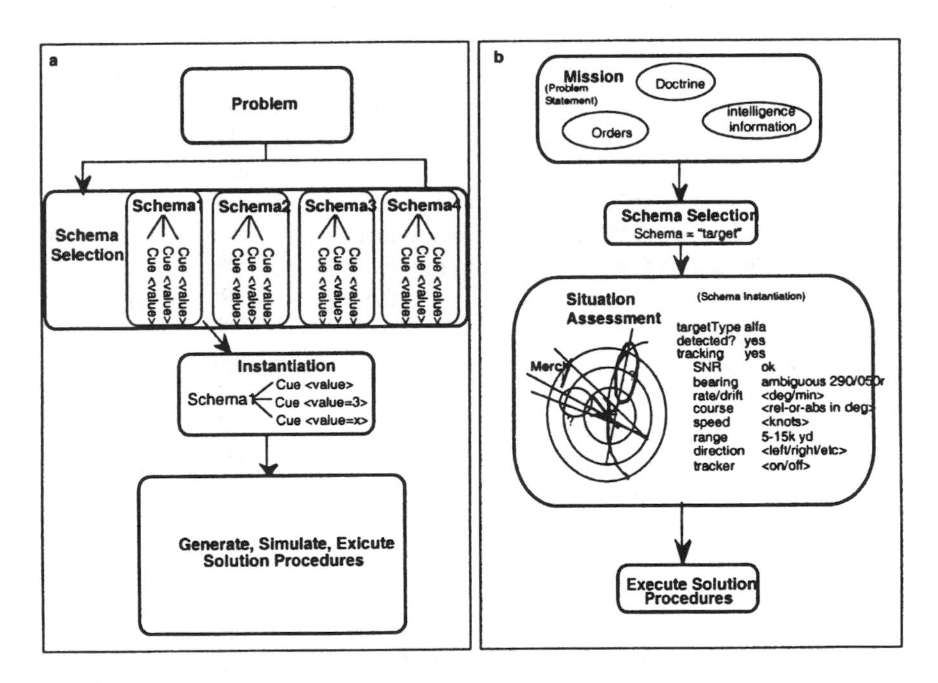

FIG. 11.2. Situation assessment as a process of schema instantiation. Note how VanLehn's process (a) and the submarine process (b) differ. (The graphic representation in (b) is typical of that drawn by submarine officers.)

the problem-solving process. This phenomenon has been found in naturalistic decision making as well. As in RPD, problem solvers proceed to attempt a solution using the selected schema. Only when the solution method fails do they reconsider the schema. Marshall (1995) showed how error at the schema selection stage leads to failure in subsequent stages.

The second stage is schema instantiation in which specific values are attached to the attributes in the schema. After the heart attack schema is selected, diagnostic tests enable the physician to attach specific values to cholesterol and other blood chemical levels, blood pressure, heart structural features, blood vessel blockage, patient factors (age, gender, and general health), and so forth. These serve to distinguish this instance of heart attack from the generic schema.

The final stage, action, is based on the diagnosis and the instantiated values of the schema. Thus, based on the instantiated heart attack schema, treatment decisions would differ for a high cholesterol reading versus a high blood pressure reading (or a combination of the two).

A similar three-stage process was expected in the submariner's decision making. (RPD was even posited in proposing the program of research.) Research was focused on predecisional information gathering in a search for the diagnostic triggers for schema selection. Contrary to expectations, results indicated that the submariners have only a single, generic "target/contact" schema (Gray & Kirschenbaum, 2000). Thus, while the emphasis in other NDM problem domains is on recognizing the correct situational schema, the problem in submarining is instantiating the contact schema. Where, for example, the physician has many disease schema from which to choose, the submariner has only the generic target/contact schema (Fig. 11.3[4]). This schema is composed of leaf nodes and relationships among these nodes. The leaf nodes are attribute-value pairs for such attributes as sensor, contact type (class), bearing (by), bearing rate (dby), signal strength (SNR), and so forth. Establishing the value for each attribute is what is meant by the term schema instantiation, and is the key phase in submarine decision making.

Because there is only a single schema structure, the initial information that is required to trigger schema selection in, for example, the medical domain, can bypass that step and become the first attribute values to be instantiated. Note that the contact schema must be instantiated for each contact. In classic RPD, identifying the appropriate schema leads to decision-making actions (or generation,

[4]The structure of the submarine Approach Officers' schema was obtained from unpublished data collected in my laboratory. In this study, 10 experienced submarine officers sorted 35 nouns and noun phrases on the basis of similarity. The phrases were the most frequently used terms from 20 verbal protocols, edited to eliminating overlapping terms. Cluster analysis revealed the schema structure as shown in Fig. 11.3. The generic contact schema is composed of leaf nodes and relationships among these nodes. The leaf nodes are attribute-value pairs for such attributes as sensor, target type (class), bearing, bearing rate, and signal strength (SNR).

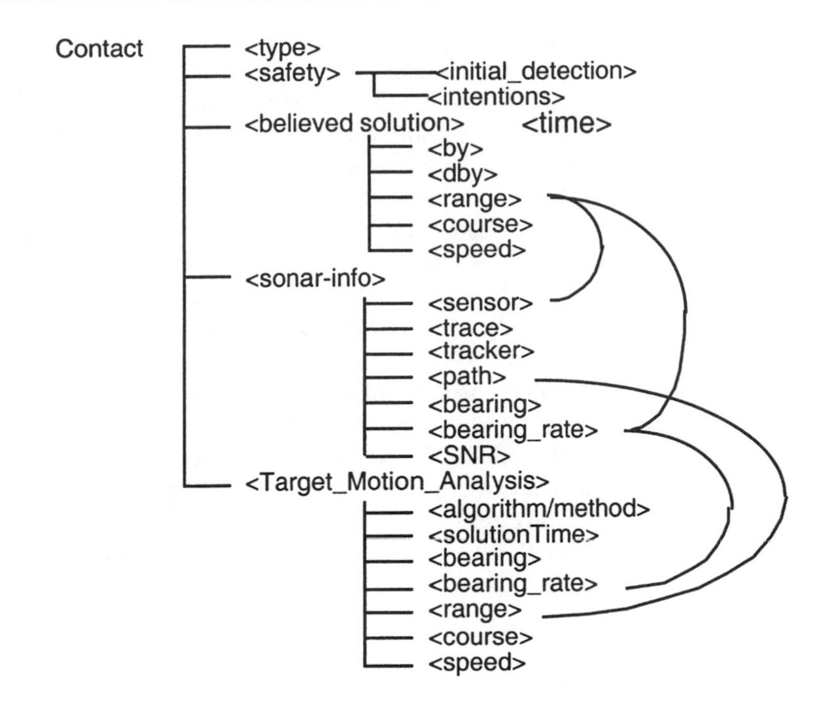

FIG. 11.3. Contact schema. Items enclosed in carets, <xxx>, are variables that must be instantiated to understand a specific contact.

mental simulation and evaluation, of possible actions). Meanwhile, the submariner spends the bulk of his time in the schema instantiation phase. The prime concern is to generate a "solution" (to solve the localization problem) for each contact. As noted, the problem is underconstrained, and there can be several credible solutions. After the solution attributes have been instantiated, the decision action is largely determined by doctrine. This functions in much the same way that the RPD decision is dictated by the schema that was selected.

SO WHAT?

Thus far, I have shown how the submarine problem fits within the NDM family of decision situations. The family resemblance is strongest in the uncertain, dynamic nature of the problem environment, in the role of an action-feedback loop, and in the key part played by the expert's knowledge schema. There are differences as well. The submariner's expertise lies not so much in recognizing the schema that best fits the available information, but in interrogating and interpreting the information to complete the schema, that is, schema instantiation, not selection. Inclusion of this stage within the NDM definition extends the family of NDM domains.

At this point, the reader has every right to ask "so what?" Let's dig a bit deeper. My research goal has always been to understand the information-gathering behavior of submarine officers in order to inform the building of better information displays for these decision makers. To accomplish this requires understanding both the behavior and the reasons for the behavior, the "whys" and "whens" that control information gathering. Thus, we needed to dig down from the rational analysis band into the cognitive band of processes and events to learn more about the relationship between schema knowledge in the head and information in the world. Taking inspiration from Newell and Simon (1972; Ericsson & Simon, 1984) and from the Exploratory Analysis of Sequential Data (ESDA; Sanderson & Fisher, 1994), a process-tracing procedure was employed to further the investigation. This is the focus of the remainder of this chapter. Two sets of research findings serve as examples to illustrate the value added for the added time and effort. In the first (Kirschenbaum, 1992), I observed the use of a small set of strategies to guide information gathering by experts. In the second experiment (Kirschenbaum et al., 1997), my collaborators and I tried to understand why and how the strategies are applied. For clarity, I shall call these the "MouseLab" study and the "CSEAL" study, after the two simulations used.

THE EXPERIMENTS

To illustrate the earlier argument, I explore the submarine officer's information-gathering behavior in support of decision making. From a close analysis of this behavior, I can extract both common strategies and an adaptive subgoaling procedure for staying aware of the changes in the dynamic situation.

The MouseLab Study

The data to support the strategies observation come from a study that was only partly naturalistic (Kirschenbaum, 1992). Although the participants were submarine officers at three levels of expertise and the scenario was derived from an at-sea exercise, the display was artificial. Data were presented in a computerized information board-like matrix (MouseLab; Johnson, Payne, Schkade, & Bettman, 1986). MouseLab allowed me to present a matrix of alphanumeric values for eight parameters, displayed over seven time slices for each of three "legs."[5] In order to track the information-gathering behavior, the values are only visible while the cursor is in the cell. The resultant data stream is a detailed action protocol consisting of the duration, latency, and order of information selection. The program also records interim and final decisions. Follow-up questions evaluated both recall and situation understanding.

[5]A leg is a period of constant course and speed. Submarine change course in order to resolve elements of the limited and uncertain target data.

Common Strategies. These data facilitated analysis at the level of tenths of seconds (latencies) and aggregation at the level of tens of seconds to 1 to 2 minutes (duration of clusters). The key result, for the purposes of this chapter, was that the action protocols of the most experienced officers showed structured strategies. These strategies included examination of data history (history), comparison between measured (raw) data and processed "solution" data (comparison), complete survey of data types[6] (survey), and systematic chunking of certain sets of data (sets). In comparison, novices looked at more data, more often reexamined previously seen cells, and more often used a strategy of closeness (examining adjacent cells). These strategies were discovered by analyzing the latencies between cells. Short latencies are characteristic of members of the same data set. Longer latencies indicate that the participant needed to decide what to do next.

These strategies served to guide information-gathering behavior in support of situation assessment. They also provide insight into the schema structure itself (see again Fig. 11.3). For example, the history strategy is a strong indicator of the importance of time and data trend in the knowledge schema. The sets and comparison strategies are evident in the links between attributes in the schema.

Strategies take place at the level of goals or subgoals, and each strategy instance is composed of two to five operators. As with most good experiment, the results resolved some questions and raised others. It was still not clear when and why each of these strategies was employed. What controls strategy selection? What is the relationship between expert schema knowledge and the information collected with the strategies? How do the dynamic changes in the world interact with the expert's in-the-head knowledge? Lastly, would these results hold up in a more natural experimental environment?

The CSEAL Study

Although my colleagues and I aimed for a more naturalistic setting for the next study, the inaccessible (limited space, classified displays, at-sea environment, etc.) and uncontrollable nature of the natural submarine task environment set a practical limited on naturalism (Ehret, Gray, & Kirschenbaum, 2000; Kirschenbaum, in press). CSEAL, a high-fidelity simulation,[7] proved an acceptable compromise. It includes all the components of a generic submarine command center and Navy-standard models of the ocean environment and of participating ships (OS and contacts). Participants interacted with the simulation

[6]Novices actually looked at more cells and reexamined more previously seen cells.

[7]This simulation, Combat System Engineering and Analysis Laboratory (CSEAL), is also used to develop submarine decision support systems.

through an operator. This had three advantages: it reduced training time, it mimicked onboard conditions, and it facilitated audio/video recording of concurrent verbal protocols because all requests for information were verbalized to the operator. Each time that the participant ordered an OS maneuver and at the conclusion of each scenario, the participants reported their current understanding of the situation. Ten experienced submarine officers (Executive Officers or Commanding Officers) spent 2 hours engaged in typical submarine scenarios. All scenarios had two contacts, a merchant ship and a hostile submarine. The mission was to protect commercial shipping from hostile attack.

Although data (primarily verbal and action protocols) can be described at many levels, the goal is to find the optimal explanatory level for the phenomena observed. Taking NDM behaviors for granted, I was interested in explaining why and how they occurred and in accounting for the strategies observed earlier. The pursuit of such explanations again forces us into the range of minutes to seconds. Because we are interested in explaining "why" the NDM behaviors that emerge at the 10 minute level occur, my colleagues and I focused our encoding on goals that emerge over minutes, subgoals that emerge over 3 to 30 sec, and operators that emerge over 300 milliseconds to 3 seconds. The explanation for "why" something happens the way it does at one level is often found in the level below it. The "whys" of the situation assessment level at which Pascual & Henderson (1997) worked (hours to 10 min) can be found at the goal (min) and subgoal (10s of seconds) levels. These, in turn, are grounded in operators (elementary units of our analysis) that take 300 milliseconds to 3 seconds.

The verbalizations and screen events were transcribed, segmented, and encoded as one of nine operators. Three of these are information operators and four are action operators. The remaining two operators are *instrumentation* and *NA*. The information operators include one request, *query*, for information and two identifying the receipt of information, *receive* and *derive*. (Derive is defined as an operator that combines currently visible data with information in memory to form new information.) These information operators most often follow the sequence of a *query* followed by one or more *receive*, possibly followed by a *derive*. The *receives* do the work of carrying information from the world to the head of the expert, filling some slot in the current instantiation of the schema. The action operators include setting a tracker to follow a moving contact, maneuvering OS to "see" the contact from another angle, or taking some action on the solution data values. They were interspersed among the information operators, based on either the passage of time or on received information such as the detection of a contact.

All goals and operators could be observed directly or deduced from verbalizations. This assertion is validated by very high interrater reliabilities as measured by Cohen's Kappa, K (corrects for chance agreement). The mean agreement for operators was 75%, $K = 0.67$, $p < .001$. The operators were encoded into five level-1 goals, thirteen level-2 subgoals, and one level-3 sub-subgoal (agreement = 44%, $K = 0.31$, $p < .001$).

The low-level and generic operators are in service of goals and subgoals specific to the submariners' task of seeking and finding an enemy submarine. The five level-1 goals deal with the entities with which the submariners are concerned; OS, hostile contact, and merchant contact. The 13 level-2 goals represent the basic level (Rosch, Mervis, Johnson, & Boyes-Braem, 1976) of submariner information processing.

Differences Between the Goals. As with the MouseLab data, the interesting results are found in the process, not just the raw tallies. For these data, the answers to the experimental questions are in the interplay between the layers, that is, the search for information to instantiate the MERCHANT and SUBMARINE versions of the contact schema. The comparison between the two principal goals, LOCATE-SUB (LOC-SUB) and LOCATE-MERCHANT (LOC-MERC), demonstrate what can be learned from this detailed analysis and refine our understanding of the expert's information gathering strategies.

The two goals occurred at roughly equal frequency, although the submarine was the higher priority contact (LOC-MERC, mean = 3.67, *SD* = 1.94; LOC-SUB, mean = 4.89, *SD* = 1.62[8]). Examining timelines shows that the participants interleave submarine and merchant goals. Where the two goals differ is in the duration of each episode. Specifically, the number of subgoals per goal accounts for the difference in duration. There was very little variance in the number of subgoals per LOCATE-MERCHANT goal, but a large variance in the number of subgoals per LOCATE-SUB goal (see Fig. 11.4).

Strategy Usage. This analysis does not answer the obvious questions about the reasons for the differences in variability between these two frequencies. Let us dig deeper, this time into the various subgoal episodes. Figure 11.5 shows a timeline for a typical participant. His subgoals for the noisier and lower priority merchant problem are limited to performing target-motion analysis (TMA) and evaluating the signal strength (signal-to-noise ratio, SNR). This is typical for all the participants. There was no need to maneuver the ship in order to localize the merchant.

On the other hand, the hostile submarine presented a more difficult problem. Not only was the sub a threat, but she was also quieter and, in some scenarios, she was maneuvering. One problem with the more realistic simulation is that more information was visible at all times. Thus, strategy usage could not be tracked as closely as in the MouseLab experiment. However, the TMA/DET-DBY/DET-RANGE cycle is identical to one of the sets observed in the earlier study. The com-

[8]Other measures also showed a similarity between these goals, providing corroborating evidence for our observation that the merchant and submarine schema were simply two instances of the same contact schema structure.

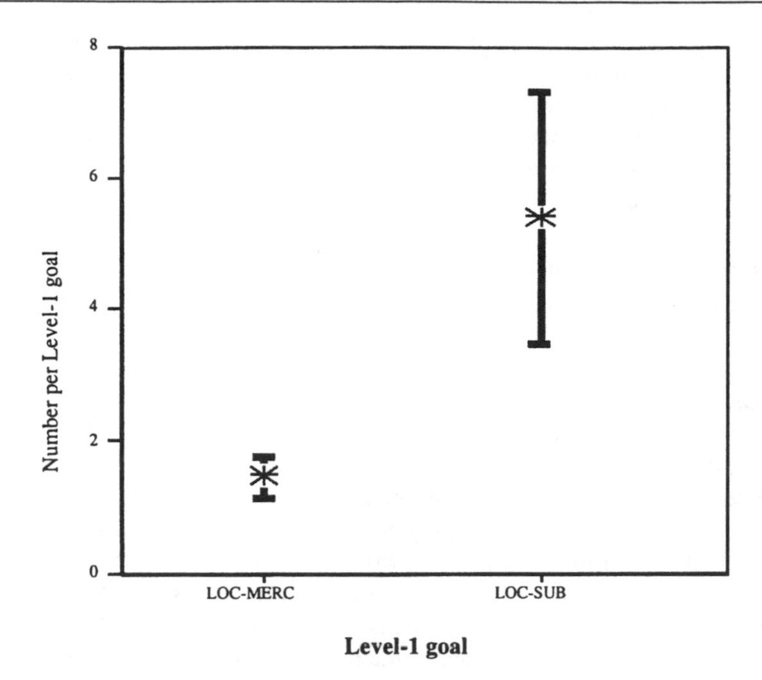

FIG. 11.4. Mean Number of Subgoals per Goal for LOC-MERC and LOC-SUB. (Also shown are the 95% confidence intervals for the standard error of the mean.) *Note.* Derived from *Subgoaling and Subschemas for Submariners: Cognitive Models of Situation Assessment* (p. 52), by S. S. Kirschenbaum, W. D. Gray, and E.B. Ehert, 1997, Newport, RI: Naval Undersea Warfare Center Division Newport.

parison strategy was evident from the iteration between raw (as in EVAL-SNR, DET-DBY, or EVAL-TRACE) and processed (TMA or T-PIC) information seen within LOC-SUB (Fig. 11.5a). Across the full set of protocols, significant support was found in the CSEAL study for each of the expert strategies observed in the MouseLab study. (In the next experimental iteration, Ehret et al., 2000, this research program has retreated from such a robust simulation and should be able to again pinpoint strategies, but that is still work in progress.)

Adaptive Subgoals. Common strategies are only a part of the puzzle. A second experimental question was how does the expert knowledge schema in the head, interact with the dynamic, changing state of the world. Two results resolve this problem. First, the goal stack surprised us by being shallow. During the majority of the problem, it was only two or three deep. In contrast, expert performance for a static problem often consists of a long series of practiced actions, subgoals (Anzai & Simon, 1979). The second key result was the observation that

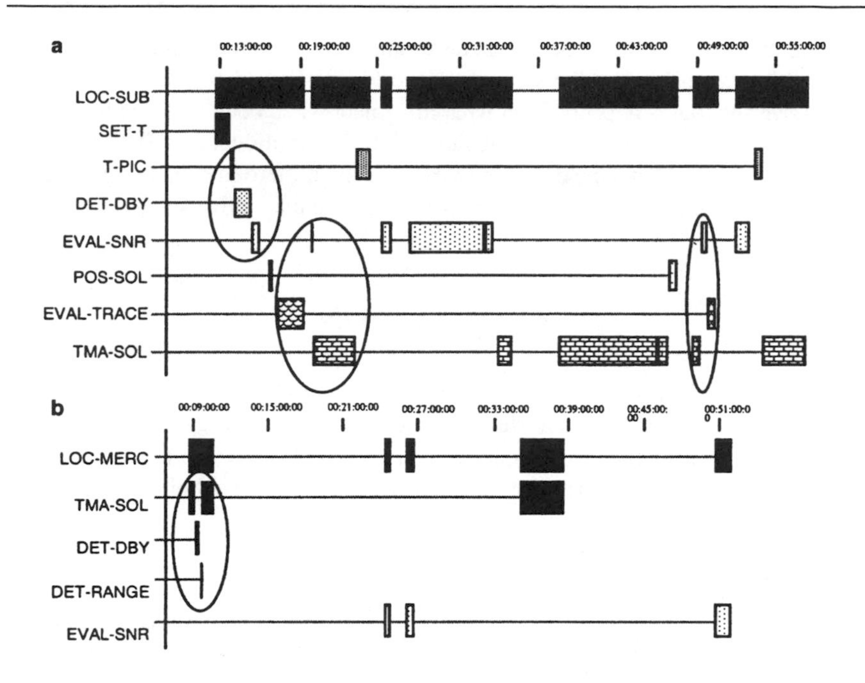

FIG. 11.5. Timelines for Subgoals of (a) LOCATE-SUB and (b) LOCATE-MERC, Goals for a Typical Participant. Circled areas indicate expert strategies. *Note*. Derived from *Subgoaling and Subschemas for Submariners: Cognitive Models of Situation Assessment, Vol. 2*, (p. I–9), by S. S. Kirschenbaum, W. D. Gray, and E. B. Ehert, 1997, Newport RI: Naval Undersea Warfare Center Division Newport.

subgoals for the submarine problem are more variable than those for the merchant problem. The interplay between schema instantiation and information gathering must logically be schema-directed in order for the decision maker to employ his expertise. Schema knowledge plays a more central role in the submarine problem both because solving the submarine problem is a higher priority and because it is the more dynamic problem. Strategy usage in a dynamic situation is highly variable because the strategy selected at any point in time is a result of the current schema instantiation state.

Putting It Together. In summary, the submarine Approach Officer uses his expertise to recognize and instantiate appropriate schema. He uses this schema understanding as the basis for action in an iterative feedback loop. By analyzing these data in greater detail, we have discovered two mechanisms that, together, allow this feedback loop to support decision making that is knowledge driven and adaptive to changes in the problem. This two-part mechanism includes (a) expert, schema-driven information-gathering strategies to control information seeking and (b) a shallow goal structure to allow the decision maker to stay close to and adapt to changes in the state of the world (see Fig. 11.6). Selection of each successive subgoal is dependent on what information had been returned by the previous subgoal (see Gray & Kirschenbaum, 2000, for additional detail). The shallow subgoal structure allows the decision maker to stay close to the changing situation. Thus, we see a schema-directed pattern of shallow and adaptive subgoaling.

CONCLUSION

This analysis is not intended to be an exhaustive presentation of results. Rather, it presents selected examples to show the value obtained by digger deeper into the wealth of data available from process-tracing protocols. At one level, these data could be described in terms of schema-driven, RPD-like behavior. Looking

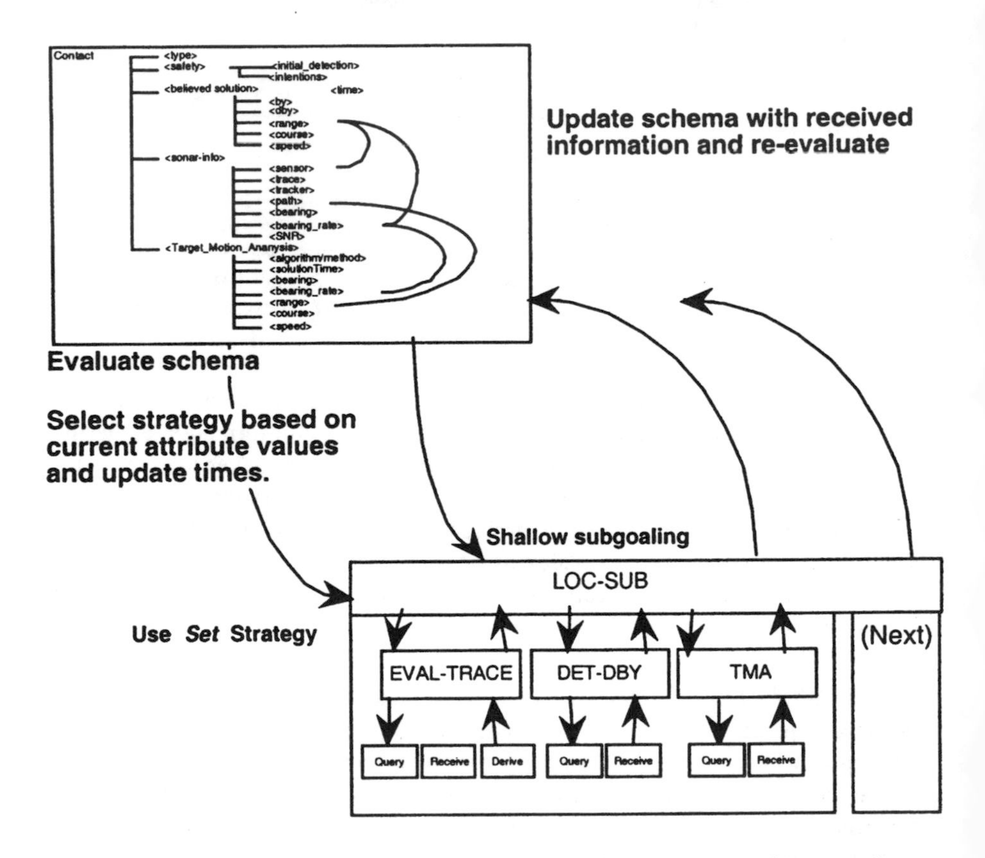

FIG. 11.6. Iterative Feedback Loop. Loop is schema strategy driven and depends on shallow subgoal structure to update the schema.

deeper shows how common information-gathering strategies and shallow subgoaling support the expertise of the submarine decision maker in his dynamic, uncertain, and risky world. We see neither the instant "aha" of recognition primed decision making nor the algorithmic-like decision making that emerges from the analyses of many laboratory tasks. RPD is predicated on access to information. We see a flexible but shallow goal hierarchy dominated by strategic goals to determine (seek) or evaluate specific sets or kinds of information.

This picture is incomplete. One major question remains: What is the control relationship between specific states of knowledge about the current situation (i.e., the current instantiation state of all active schema) and specific strategy selection? Answering this question requires linking the schema state to the selected strategy and then showing how the retrieved information updates the schema. This requires tracing many cycles through the information-gathering steps. One approach is to use the data as the foundation of a computational cognitive model (Gray, Young, & Kirschenbaum, 1997) of the submarine AO. This model, employing the ACT-R (Anderson & Lebiére, 1998) cognitive architecture, will embed the explanation of schema instantiation developed here within a unified theory of cognition. It will also take the explanation to yet another level, the cycle level (see Fig. 11.1). By computational cognitive modeling of the schema instantiation, strategy selection, and information-gathering subgoaling process, we hope to provide a plausible trace of the process. A family of models (at different levels of expertise, etc.) will then allow us to test hypothesized mechanisms against human behavior. However, that is a story for another time. For now, I hope to have shown the value added to a maturing science by digging deeper into the naturalistic decision making phenomena by using the methods and mindset of cognitive science.

ACKNOWLEDGMENTS

The first experiment was supported in part by a grant from the Office of Naval Research, Task No. BR01508x5. The second experiment has been supported by The author's research was also supported by the Office of Naval Research (ONR), Susan Chipman, Scientific Officer (Project No. A10328) and the Naval Undersea Warfare Center's (NUWC) Independent Research (IR) Program, as Project E102077. The IR program is funded by ONR; the NUWC program manager is Stuart C. Dickinson (Code 102). It was also supported by a separate grant to the author's collaborator, Wayne D. Gray, George Mason University, from the Office of Naval Research (ONR), Susan Chipman, Scientific Officer (Grant No. N00014-95-1-0175).

I wish to thank my collaborator on this research, Wayne D. Gray, both for his efforts and insights on the project and for his comments on earlier drafts of this manuscript. Lastly, many thanks are due to Brian D. Ehret, a graduate student at George Mason University during much of the research, for his many hours encoding these data and his many other contributions.

REFERENCES

Anderson, J. R. (1990). *The adaptive character of thought.* Hillsdale, NJ: Lawrence Erlbaum Associates.

Anderson, J. R., & Lebiére, C. (Eds.). (1998). *Atomic components of thought.* Hillsdale, NJ: Lawrence Erlbaum Associates.

Anzai, Y., & Simon, H. A. (1979). The theory of learning by doing. *Psychological Review, 86*(2), 124–142.

Bartlett, F. C. (1995). *Remembering.* New York: Cambridge University Press. (Original work published 1932).

Beach, L. R., & Lipshitz, R. (1993). Why classical decision theory is an inappropriate standard for evaluating and aiding most human decision making. In G, A. Klein, J.. Orasanu, R. Calderwood, & C. E. Zsambok (Eds.), *Decision making in action: Models and methods* (pp. 21–35). Norwood, NJ: Ablex.

Card, S. K., Moran, T. P., & Newell, A. (1983). *The psychology of human-computer interaction.* Hillsdale, NJ: Lawrence Erlbaum Associates.

Ehret, B. D., Gray, W. D., & Kirschenbaum, S. S. (2000). Contending with complexity: Developing and using a scaled world in applied cognitive research. *Human Factors, 42*(1), 8–23.

Einhorn, H. J., & Hogarth, R. M. (1985). Ambiguity and uncertainty in probabilistic inference. *Psychological Review, 92,* 433–461.

Endsley, M. R., & Smith, R. P. (1996). Attention distribution and decision making in tactical air combat. *Human Factors, 38*(2), 232–249.

Ericsson, K. A., & Kinstsch, W. (1995). Long-term working memory. *Psychological Review, 102*(2), 211–245.

Ericsson, K. A., & Simon, H. A. (1984). *Protocol analysis: Verbal reports as data.* Cambridge, MA: MIT Press.

Flin, R., Slaven, G., & Stewart, K. (1996). Emergency decision making in the offshore oil and gas industry. *Human Factors, 38*(2), 262–277.

Gray, W. D., John, B. E., & Atwood, M. E. (1993). Project Ernestine: Validating a GOMS analysis for predicting and explaining real-world performance. *Human-Computer Interaction, 8*(3), 277–309.

Gray, W. D., & Kirschenbaum, S. S. (2000). Analyzing a novel expertise: The unmarked road. In S. F. Chipman, J. M. Schraagen, & V. L. Shalin (Eds.), *Cognitively task analysis* (pp. 275–290). Mahwah, NJ: Lawrence Erlbaum Associates.

Gray, W. D., Young, R. M., & Kirschenbaum, S. S. (1997). Introduction to this special issue on cognitive architectures and human-computer interaction. *Human-Computer Interaction, 12*(4), 301–309.

Howell, W. C. (1993). Engineering psychology in a changing world. *Annual Review of Psychology, 44,* 231–263.

Johnson, E. J., Payne, J. W., Schkade, D. A., & Bettman, J. R. (1986). *Monitoring information processing and decisions: The MouseLab system* (Report). Pittsburgh, PA: Carnegie-Mellon University, Graduate School of Industrial Administration.

Kaempf, G. L., Klein, G. A., Thordsen, M. L., & Wolf, S. (1996). Decision making in complex naval command-and-control environments. *Human Factors, 38*(2), 220–249.

Kirschenbaum, S. S. (1990). *Command decision making: Lessons learned* (NUSC TM No. 902149). Naval Underwater Systems Center.

Kirschenbaum, S. S. (1992). Influence of experience on information-gathering strategies. *Journal of Applied Psychology, 77*, 343–352.

Kirschenbaum, S. S. (1994). *Command decision-making: A simulation study*. (Tech. Rep. 10350). Newport, RI: Naval Underwater Warfare Center Division Newport.

Kirschenbaum, S. S. (in press). Decision making studies in a simulated submarine. In M. McNeese, E. Salas, & M. Endsley (Eds.), *New trends in collaborative activities: Dynamics in complex systems*. Santa Monica, CA: Human Factors and Ergonomics Society.

Kirschenbaum, S. S., Gray, W. D>, & Ehret, B. D. (1997). *Subgoaling and subschemas for submariners: Cognitive models of situation assessment* (Tech. Rep. NUWC-NPT-TR 10764–1). Newport, RI: Naval Undersea Warfare Center Division Newport.

Klein, G. A. (1989). Strategies of decision making. *Military Review*, May, 56–64.

Klein, G. A. (1993). Recognition-primed decision (RPD) model of rapid decision making. In G. A. Klein, J. Orasanu, R. Calderwood, & C. E. Zsambok (Eds.), Decision making in action: Models and methods (pp. 138–147). Norwood, NJ: Ablex.

Marr, D. (1982). *Vision*. San Francisco: Freeman.

Marshall, S. P. (1995). *Schemas in problem solving*. New York: Cambridge University Press.

Newell, A., & Card, S. K. (1985). The prospects for psychological science in human computer interaction. *Human-Computer Interaction, 1*(3), 209–242.

Newel, A., & Simon, H. A. (1972. *Human problem solving*. Englewood Cliffs, NJ: Prentice-Hall.

Orasanu, J. & Connolly, T. (1993). The reinvention of decision making. In G. A. Klein, J. Orasanu, R. Calderwood, & C. E. Zsambok (Eds.), *Decision making in action: Models and methods* (pp. 3–20). Norwood, NJ: Ablex.

Pascual, R., & Henderson, S. (1997). Evidence of naturalistic decision making in military command and control. In C. E. Zsambok & G. Klein (Eds.), Naturalistic decision making (pp. 217–226). Mahwah, NJ: Lawrence Erlbaum Associates.

Payne, J. W., Bettman, J. R., & Johnson, E. J. (1993). *The adaptive decision maker*. New York: Cambridge University Press.

Rosch, E., Mervis, C., Gray, W. D., Johnson, D. M., & Boyes-Braem, P. (1976). Basic objects in natural categories. *Cognitive Psychology, 8*, 382–439.

Roth, E. M., Mumaw, E. M., Vicente, K. J., & Burns, C. M. (1997, September). *Operator monitoring during normal operations: Vigilance or problem solving?* Paper presented at the Annual Conference of the Human Factors and Ergonomics Society, Albuquerque, NM.

Sanderson, P. M., & Fisher, C. (1994). Exploratory sequential data analysis: Foundations. *Human-Computer Interaction, 9*(3 & 4), 251–317.

Tversky, A. (1972). Elimination by aspect: A theory of choice. *Psychological Review, 79*, 281–299.

Urick, R. J. (1983). *Principles of underwater sound* (3rd ed.). New York: McGraw-Hill.

VanLehn, K. (1989). Problem solving and cognitive skill acquisition. In M. I. Posner (Ed.), *Foundations of cognitive science* (pp. 527–579). Cambridge, MA: MIT Press.

12

Cognitive and Contextual Factors in Aviation Accidents: Decision Errors

Judith Orasanu
NASA Ames Research Center
Lynne Martin
National Research Council / NASA Ames Research Center
Jeannie Davison
San Jose State University / NASA Ames Research Center

Naturalistic decision making (NDM) focuses on understanding how people with considerable domain expertise use their knowledge to make decisions (Cannon-Bowers, Salas, & Pruitt, 1996; Klein, Orasanu, Calderwood, & Zsambok, 1993; Zsambok & Klein, 1997). Highly integrated knowledge structures account for the robust and efficient performance of experts (Ericsson & Leman, 1996; Klein, 1993a). However, sometimes people who are considered experts make decisions that turn out badly.

The National Transportation Safety Board (NTSB; 1994b) identified 37 aircraft accidents in which flight crew behavior contributed to the accident. Twenty-five of these involved what the Board considered "tactical decision errors." Given that aircrews are highly experienced and skilled, what are we to make of these cases? Did these bad outcomes reflect inadequate decision-making processes? Were the crews trapped in system or organizational contexts that predisposed them to err? Were the crews simply "unlucky" and overtaken by conditions beyond their control? Or were the putative decision errors simply a function of hindsight bias?

In this chapter we address the concept of "decision error" within an NDM framework, specifically focusing on decisions in the aviation domain. Our effort is guided by the view that errors may be an inevitable consequence of experts behaving like experts, that is, applying their knowledge while performing tasks, and frequently following the principle of cognitive economy and efficiency (March & Simon, 1958). We concur with the positions of Rasmussen (1997), Reason (1997), and Woods, Johannesen, Cook, and Sarter (1994) that we must move beyond trying to pin the blame for accidents on a culprit and must seek instead to understand the systemic causes underlying the outcomes.

The goal of our research effort is to reduce the frequency of negative outcomes, specifically aviation accidents, by a more complete understanding of factors that lead to these consequences. This understanding can then serve as a basis for improving system performance.

Three specific issues will be addressed in this chapter:

1. What is the nature of decision "errors" in aviation?
2. What factors may contribute to negative outcomes in aviation decision making?
3. What could be done to reduce negative outcomes in the future?

Before addressing these issues, however, we examine the broader question of how to define decision "error" in naturalistic environments.

DEFINING DECISION ERRORS: OUTCOME VERSUS PROCESS

Any discussion of decision errors must begin with a definition. However, defining decision errors in naturalistic contexts is fraught with difficulties. Three stand out. First, errors typically are defined as deviations from a criterion of accuracy. However, the "best" decision in a natural work environment such as aviation may not be well defined, as it often is in highly structured laboratory tasks. Second, a loose coupling of decision processes and event outcomes works against using outcomes as reliable indicators of decision quality. Redundancies in the system can "save" a poor decision from serious consequences. Conversely, even the best decision may be overwhelmed by events over which the decision maker has no control, resulting in an undesirable outcome (Woods et al., 1994). A third problem is the danger of hindsight bias. Fischhoff (1975), Hawkins and Hastie (1990), Woods et al. (1994), and others point out a tendency to define errors by their consequences. But the analyst does not know how often exactly the same decision process was used or the same decision was made in similar situations with no negative consequences. Were those prior decisions also "errors"?

These difficulties suggest that a viable definition of decision error must take into account both the nature of the decision process and the event outcome. Lipshitz (1997) offered a definition that we adopt for this paper: Decision errors are "deviations from some standard decision process that increase the likelihood of bad outcomes" (p. 152).

Although outcome alone may not be a good indicator of decision mistakes, intended outcome, or the decision maker's goal, remains important. In naturalistic work contexts, decisions contribute to performance of ongoing tasks. Decision makers strive to achieve their goals through the decisions they make and the actions they take. Decisions do not stand alone as events to be judged independent of the broader task. As Zapf, Maier, Rappensperger, and Irmer (1994) pointed out, "If we want to know whether or not an error has occurred we have to know the user's goal. ...error detection depends on knowledge about the action outcome" (p. 501).

On the other hand, Doherty (1993) called for a normative process model for evaluating the quality of decision making online, independent of outcomes, both to detect erroneous decision processes and to avoid hindsight bias. However, as Lipshitz (1997) noted, there is no normative NDM process model that allows definitive identification of errors independent of outcomes, as in behavioral decision theory. He pointed out that one contribution of the NDM approach has been to develop new methods of analyzing decision processes in complex dynamic contexts, namely cognitive task analysis or expert models. Although not normative in the traditional sense, these models can serve as a basis for defining effective performance, as well as for identifying and characterizing error.

A slightly different perspective was offered by Woods et al. (1994), who suggested that errors be reframed as a lack of system/person fit in which decision errors are the inevitable result of normal human activity in response to novel situations. Errors are potentially adaptive. Successful adaptation involves expending less effort to produce more and better output. However, sometimes it results in overextension of usually efficient skills in new contexts. "In unstable conditions, 'correct' models cannot be specified a priori.... The only way to test one's decision is to judge the response from the environment" (Woods et al., 1994, p. 59).

A systems perspective suggests that a more productive approach to identifying decision errors is to examine the interaction between situational features and individual cognitive factors. As Rasmussen (1997) noted, the decision-making field has moved through two phases, from formal prescriptive models of individual cognition to predictive models for evaluating total systems. Current trends emphasize analyses of "behavior shaping features of the environment," that is, models in terms of "objectives, constraints, options for action, and subjective performance criteria" (p. 73). This latter perspective guides our current effort to understand decision errors in aviation.

WHAT IS THE NATURE OF DECISION "ERRORS" IN AVIATION?

In order to provide a framework for discussing decision errors, we first describe a process model that was developed to characterize decision making in aviation

(Orasanu & Fischer, 1997). Aviation decisions are typically prompted by cues that signal an off-nominal or changed condition that may require adjustment of the planned course of action. The decision process model involves two components: situation assessment (SA) and choosing a course of action (CoA).[1] In aviation, situation assessment involves defining the problem, assessing the level of risk associated with it, and determining the amount of time available for solving it. After the conditions are assessed, a course of action is chosen based on the structure of the options available in the situation. Building on Rasmussen (1985), Orasanu and Fischer (1997) specified three types of response structures: rule-based, choice, and creative.[2] All involve application of knowledge but vary in the degree to which the response is determined by the situation.

How Can Decision Processes Go Wrong?

Under this model, there are two major ways in which error may arise: (a) Pilots may develop an incorrect interpretation of the situation, which leads to an inappropriate decision, or (b) they may establish an accurate picture of the situation, but choose an inappropriate course of action.

Faulty Situation Assessment. Situation assessment errors can be of several types: cues may be misinterpreted, misdiagnosed, or ignored, resulting in an incorrect picture of the problem (Endsley, 1995; Prince, Stout, & Salas, 1995); risk levels may be misassessed (Johnston, 1996; Orasanu, Dismukes, & Fischer, 1993); or the amount of available time may be misjudged (Keinan, 1988; Maule, 1997; Orasanu & Strauch, 1994). Problems may arise when conditions change gradually and pilots do not update their situation models (Woods & Sarter, 1998).

For example, one accident that can be traced to an incorrect assessment of the situation was the decision by the crew of a B-737 to shut down an engine—unfortunately, the wrong one:

> The crew sensed a strong vibration while in cruise flight at 28,000 ft. A burning smell and fumes were present in the passenger cabin, which led the crew to think there was a problem in the right engine (because of the connection between the cabin air conditioning and the right engine). The captain throttled back the right engine and the vibration stopped. However, this was coincidental. In fact, the left engine had thrown a turbine blade and gone into a compressor stall. The captain ordered the right engine shut down and began to return to the airport. He again questioned which engine had the problem, but communication with air traffic control

[1]This model is grounded in Klein's (1993a) Recognition-Primed Decision Theory, but has been tailored to the situational features and typical practices of pilot performance in aviation.

[2]Rasmussen's (1985) typology involved skill-based, rule-based, and knowledge-based decisions.

> and the need to reprogram the flight management computer took precedence, and they never did verify the location of the problem. The faulty engine failed completely as they neared the airport, and they crashed with neither engine running (Air Accidents Investigations Branch, 1990).

The problem was incorrectly defined because the cues (vibration and burning smell) supported the interpretation of a right engine problem. The crew did not verify this interpretation before taking an irrevocable action (irrevocable at that point in the flight).

Faulty Selection of Action. Errors in choosing a course of action may also be of several types. In rule-based decisions (Rasmussen, 1985), the appropriate response may not be retrieved from memory and applied, either because it was not known or because some contextual factor mitigated against it. Conversely, an inappropriate rule may be applied, especially a frequently used one. In choice decisions, options may not be considered (although one hallmark of experts is that they are likely to select the best option first; Calderwood, Klein, & Crandall, 1988). Constraints that determine the adequacy of various options may not be used in evaluating them. Relative strengths of competing goals play an important role in the choice of options.

Creative decisions may be the most difficult because they involve the least support from the environment; candidate solutions must be invented to fit the goals and existing conditions. Any solution that meets one's goal may be considered a success. In all types of decisions, the consequences and uncertainties associated with candidate actions may not be considered, resulting in a course of action error.

An accident in which an inappropriate course of action was chosen in the face of fairly complete information about the nature of a problem occurred near Pinckneyville, IL.

> About two minutes into a night flight in instrument conditions, a Hawker-Siddley commuter aircraft lost its left generator. In error, the first officer isolated the right generator and then was unable to restart it. This meant total loss of ability to generate electrical power which was needed to run all cockpit instruments. Under the best of circumstances batteries might be expected to last for 30 minutes. The captain decided to continue to the destination airport 45 minutes away, rather than diverting. Continued use of non-essential electrical equipment shortened the battery life. A complete electrical failure and subsequent loss of flight instruments critical for IFR flight led the plane to crash. (NTSB, 1985).

This crew's decision to continue as planned, despite the mechanical failure, rather than to land as soon as possible, was fatal.

The decision process model provides a framework for examining factors that may contribute to aviation accidents by their influence on these decision processes.

WHAT FACTORS CONTRIBUTE TO NEGATIVE OUTCOMES IN AVIATION?

Traditional decision research has typically considered the source of decision "error" to be the limited cognitive capacities or inappropriate strategies used by the decision maker (Tversky & Kahneman, 1974). However, these approaches did not consider the role of domain expertise in decision making, nor did they typically consider the impact of contextual features on decision processes. The systems approach recommended by Rasmussen (1997), Reason (1997), and others emphasizes the importance of examining contextual factors in order to understand performance errors in operational situations.

Klein (1993b) offered three factors that contribute to decision errors, taking into account both cognitive aspects and situational ones. These include lack of information, lack of knowledge, and failure to simulate consequences. *Lack of information* primarily reflects situational factors. It may be grounded in poor design of systems or displays or in a malfunction of an information source. Klein's latter two factors are internal cognitive ones. However, whereas *lack of knowledge* is clearly a cognitive element, its source may be external, such as inadequate job-specific training provided by the company. *Failure to simulate consequences*, also a cognitive factor, may reflect habitual decision strategies, situational stressors, or distractions. Elaboration of how these three factors may influence aviation decision errors are considered after a brief discussion of aviation accidents.

NTSB ANALYSES OF AVIATION ACCIDENTS

To explore the influence of cognitive and contextual factors on aviation decision errors, we reexamined a set of errors associated with severe negative consequences, namely, the 37 airline accidents previously analyzed by the NTSB (1994b, Berman, 1995). Our purpose was, first, to determine the relative frequency of various error types; second, to determine what kinds of decision errors were most common; and third, to search for common themes or patterns in the contexts within which they occurred. We used the NTSB data set because all 37 accidents had been reviewed, coded, and classified by the same analysts, providing both consistency and independence.

First, we provide a brief summary of all error types to establish a framework; then we examine decision errors in depth. NTSB (1994b) analysts identified 302 crew errors in the 37 accidents, an average of 7 errors per accident. Crew performance errors were classified into nine types, including tactical decision errors.

Table 12.1 illustrates that procedural errors were most prevalent, occurring in 31 of the 37 accidents and accounting for 24.1 % of the errors. Tactical decision

TABLE 12.1
Distribution of Error Types in NTSB (1994) Accident Data

Error Category	*Percentage of Total Errors*
Primary errors	**(*N* = 302)**
Procedural - PR	24.1
Tactical decision - TD	16.8
Aircraft handling - AH	15.2
Situation awareness - SA	5.9
Systems operation - SO	4.6
Communication - CO	4.3
Resource management - RM	3.6
Navigational - NV	1.9
Secondary errors	
Monitoring and challenging - MC	22.8

errors were the second most frequent type of primary error (occurring in 25 of the 37 accidents). Monitoring and challenging errors were the second most common type of error overall (22.8% of total errors), but they are secondary errors, which means they depend on the prior occurrence of some other type of error, such as an improper procedure or a questionable decision.

Next, we examined the most frequent types of decision errors. The NTSB had coded the 51 tactical decision errors on dimensions that included Phase of Flight and Omission/Commission. To look for patterns, we sorted the tactical decision errors into a matrix defined by those two dimensions. While examining descriptions of the specific decision errors, a theme emerged: Certain decision errors seemed to represent a continuation of the original flight plan in the face of cues that suggested changing the course of action. These included failures to perform a missed approach, failures to go around in questionable weather, or continuing to land during an unstable approach. It appears that crews either misinterpreted existing cues as confirmation that they should continue (a situation assessment error) or made a course-of-action error by deciding that the original plan was still the best (cf. O'Hare & Smitheram). To capture this distinction, we sorted the decision events into what we called Plan Continuation (PC) and Non-Plan Continuation (non-PC) event categories.

As shown in Table 12.2, several observations emerged from this classification. First, decision errors were concentrated in the Approach and Landing phases (28

TABLE 12.2
Distribution of Types of Decision Errors

	Types of Errors			
	Omission		*Commission*	
Phase of Flight	*Plan Continuation*	*NON-Plan Continuation*	*Plan Continuation*	*NON-Plan Continuation*
Gate/Taxi	3	0	1	2
Takeoff	2	1	2	4
Cruise	4	0	0	4
Approach	6	0	5	2
Landing	13	0	2	0
Total	28	1	10	12

Note. The data are from *A Review of Flightcrew-Involved, Major Accidents of U. S. Carriers, 1978 through 1990* (pp. 86–104) by National Transportation Safety Board, 1994, Washington, DC: National Transportation Safety Board.

out of 51 errors). Second, about 75% of the tactical decision errors fell into the plan continuation category (38/51). Moreover, plan continuation events were likely to be associated with errors of omission (28 out of 38 PCE were omissions), while non-PC events were associated disproportionately with errors of commission (12 out of 13). In errors of commission, crews generally performed unusual but unnecessary actions. For example, in the Hawker-Siddley crash, the crew reconnected nonessential electrical equipment (e.g., cabin lights), despite having only battery power. Other examples include using reverse thrust on push-back and rejecting a takeoff above the maximum rejected takeoff speed.

Although it is not possible to determine the causes of these patterns from posthoc analyses, our efforts were drawn to examining factors that might lead crews into plan continuation types of tactical decision errors. Both cognitive and contextual factors that may contribute to tactical decision errors are hypothesized. First, we examine cognitive factors, the more traditional approach. Then, we consider contextual factors that may make decisions difficult and lead to errors, especially plan continuation ones.

Individual Cognitive Factors

Rasmussen (1997) noted that operational decisions are not based on a "rational" analysis of all parameters but are focused on the elements in a situation that allow the decision maker to distinguish between reasonable options, consistent with Simon's (1957) concept of "bounded rationality." Consider that more than one half of the decision errors in the NTSB data base (29 out of 51) involved omissions, that is, failures to do something that should have been done. Crews may have been captured by a familiar schema in these cases, leading them to do what they normally do, that is, to carry on with the usual plan, even though other action may have been called for. Ease of access supports the "wrong but strong" schema-driven approach (Reason, 1990). The case of routine knowledge guiding performance also supports the "cognitive miser" perspective (March & Simon, 1958): people will do as little as possible to get the job done at a satisfactory (though not necessarily optimal) level, a strategy Simon (1957) called "satisficing."

A cognitive economy or a schema dominance explanation, however, fails to account for the remaining errors of commission (22 out of 51). These are cases in which crews took actions that were out of the ordinary, such as attempting to blow snow off their aircraft using the engine exhaust from the aircraft ahead of them (NTSB, 1982). These cases may reflect "buggy" mental models or gaps in knowledge (VanLehn, 1990). Buggy mental models may include strategies that are successful in some cases, so decision makers may have great confidence in their models of the situation.

Lack of relevant knowledge can lead to both misdiagnosis of a problem and to choice of a poor solution. For example, one may suspect lack of knowledge was the basis for a crew based in Florida deciding that it was safe to take off with snow on the aircraft and misleading engine indications on takeoff during a snowstorm (NTSB, 1982). In several aviation accidents, lack of current knowledge seemed to play a role. Although pilots are experienced in general, they may lack currency with specific situations. For example, pilots may be rusty due to having been off duty for medical or military reasons (e.g., National Guard duty), they may be new to the aircraft (recently transitioned), or they may be unfamiliar with the airport, the approach, or the runway.

Failure to simulate outcomes (Klein, 1993b) usually leads to errors in choosing a course of action. Pilots may not anticipate the consequences of an action, especially when the situation is evolving. They may not project potential risk into the future. Failure to evaluate fully the consequence of an option may result from excess workload or time pressure. The behavior of the Hawker-Siddley crew described earlier looks like a case of failure to simulate the consequences of their decision to continue the flight despite having no functioning generators.

Error Inducing Contexts

In addition to individual cognitive factors, many features of the situation may interact with individuals' cognitive processes to induce decision errors. Four factors were abstracted from accident analyses and may contribute to precursor events, leading to errors such as those noted in Table 12.1. The four factors are (1) ambiguity, (2) dynamic risk, (3) organizational and social pressures, and (4) stress. How these four factors might influence cognitive processes is described next.

Ambiguity. Cues that signal a problem are not always clear-cut. Conditions can deteriorate gradually, and the decision maker's situation assessment may not keep pace. Ambiguous cues permit multiple interpretations. If this ambiguity is not recognized, a pilot or crew may be confident in their interpretation of a situation, when in fact they are wrong. People are not likely to question their interpretation of a situation unless there are powerful cues to suggest that their current interpretation is wrong.

In addition to inappropriate situation assessment, ambiguous conditions can influence decision errors in two further ways. First, if the ambiguity is recognized, people may not know clearly what to do, but may realize that several possibilities exist. Second, they may find it difficult to justify a change in plan in the face of ambiguous cues. For decisions that have expensive consequences, such as rejecting a takeoff or diverting, the decision maker may need to feel very confident that the change is warranted. If the situation is ambiguous, a change of plan may be more difficult to justify than if the situation is clear-cut, which may contribute to plan continuation events.

Ambiguity may exist due to an absence of good data or system information (Lipshitz & Strauss, 1997). Poor interface design, especially in modern "glass cockpits," that does not provide adequate diagnostic information or feedback can lead a crew astray (Woods & Sarter, 1998). For example, the crash in which the flight crew shut down the wrong engine resulted in part because the information about which engine had the problem was poorly displayed (AAIB, 1990).

Dynamically Changing Risks. In several accidents, crews were aware of cues that should have signaled a change in course of action, but they appear to have underestimated the level of risk associated with not changing their plans. For example, when approaching Dallas for landing, the first officer of an L-1011 commented on the lightning in the storm lying on their flight path (NTSB, 1986a). Yet, they flew into it and encountered wind shear. We know that risk is important to pilots, because potential risk was the dominant dimension considered by captains from several airlines when making judgments about flight-related decision situations (Fischer & Orasanu, 2000; Fischer, Orasanu, & Wich, 1995). Why then do crews appear to underestimate risk in potentially critical situations?

One possible explanation is that crews lack the relevant experience or are unable to retrieve the knowledge needed to assess risk appropriately in those specific circumstances (cf. Klein, 1993b). Another is grounded in pilots' routine experience: if similar risky situations have been encountered in the past and a particular course of action has succeeded, the crew will expect also to succeed the next time with the same response, for example, landing at airports where weather conditions frequently are poor. Given the uncertainty of outcomes, in many cases they will be correct, but not always. Reason (1990) called this "frequency gambling." Contrary to allegations that decision makers fail to use base rates, this would appear to be an appropriate use of them.

Support for this notion can be found in several sources. Hollenbeck, Ilgen, Phillips, and Hedlund (1994) found that past success influences risk-taking behavior. Baselines become misrepresented over time as a situation becomes familiar and the individual becomes more experienced. Sitkin (1992) argued that if you have only good experiences, you have no baseline by which to determine when the situation is taking a turn for the worse. In sum, the familiarity or regularity of a situation influences the way individuals process the information and the degree of importance they place on it.

Organizational and Social Pressures. An organization's emphasis on productivity may unwittingly set up goal conflicts with safety. As Reason (1990, 1997) and Woods et al. (1994) have documented, organizational decisions about levels of training, maintenance, fuel usage, keeping schedules, and so forth, may set latent pathogens that undermine safety in the face of vocal support for a "safety culture." For example, on-time arrival rates are broadcast to the public. One now-defunct airline went so far as to pay passengers $1 for each minute their flight was late—until a crew flew through, instead of around, a thunderstorm and crashed (Nance, 1986). Companies also emphasize fuel economy and getting passengers to their destinations rather than diverting, perhaps inadvertently sending mixed messages to their pilots concerning safety versus productivity. Mixed messages create conflicting motives, which can affect pilots' risk assessments and the course of action they choose.

The crash of a U.S. Air Force (USAF) plane carrying Secretary of Commerce Ron Brown to a meeting in Dubrovnik appears to be a case in which organizational factors may have led the crew to take risks that in hindsight were too high (USAF, 1997). The aircraft they were flying was not legal for the approach into Dubrovnik (it needed two Automatic Direction Finder radios, but had only one). The crew had not received theater-specific training that would have enabled them to interpret the charts accurately for the instrument approach, and the Command allowed them to use civilian rather than military area charts, which were not authorized. Being a highly visible mission, expectations were strong for the crew to fly it, despite these handicaps. Pilots typically are rewarded for a "can do" attitude. Saying "can't do" or "won't do" may be difficult.

Social factors may also create goal conflicts. Implied expectations among pilots may encourage risky behavior or may induce one to behave as if one is an expert, even when ignorant. This may result in unwillingness to admit that one does not know something and to continue in the face of uncertainties. For example, a runway collision in near zero visibility (due to fog) resulted when one aircraft stopped on an active runway because the crew did not realize where they were (NTSB, 1991). The captain was unfamiliar with the airport and was making his first unsupervised flight after a long period of inactivity. The first officer boasted of his knowledge of the airport but, in fact, gave the captain incorrect information about taxiways. Rather than questioning where they were, the captain went along. Meeting social and organizational goals often appears to outweigh safety goals, especially in ambiguous conditions.

Ambiguous cues, dynamically changing risks, and organizational and social factors may not in themselves be sufficient to cause decision errors. However, when the decision maker's cognitive limits are stressed, these factors may combine to induce errors.

Stress. Ambiguity and goal conflicts imply that several interpretations of the situation are possible and that alternative courses of action should perhaps be examined. Both cases require extra cognitive work relative to straightforward decisions in clear-cut situations. This includes gathering information to clarify the problem or projecting and evaluating the consequences of alternate actions. These cognitive demands may cause stress for the individual, especially when task load is high (Cannon-Bowers & Salas, 1998). Stress typically constrains working memory capacity (Hockey, 1979) and may thus limit the decision maker's ability to entertain multiple hypotheses about a problem or to project the situation into the future, mentally simulating decisions (Wickens, Stokes, Barnett, & Hyman, 1993). Other physiological conditions, such as fatigue, can have similar effects, as suggested by the NTSB in its analyses of accidents in Guam (NTSB, 1999) and Guantanamo Bay (NTSB, 1994a).

Certain phases of flight typically induce higher levels of stress, namely, those in which time is limited, workload is heavy, and there is little room for recovering from an error (Strauch, 1997). These tend to be takeoff and landing situations. Under stress, decision makers often fall back on their most familiar responses (Hockey, 1979). When something goes wrong during takeoff or landing, emerging familiar responses may not be appropriate to the situation. For example, a vibration throughout one aircraft about one minute after takeoff induced the captain to retard power on all four engines (NTSB, 1986b). Reducing power during this critical phase of flight, while close to the ground, was not appropriate because insufficient time was available for recovery. The same action might have been fully appropriate at a higher altitude. Because most abnormal events tend to be quite infrequent, the appropriate responses may not be readily available to the crew. Stress may interfere with recognizing the inappropriateness of certain actions, such as throttling back the engines.

Other potentially dangerous conditions that require decisions may permit more time to diagnose the problem and consider what to do (e.g., fuel leaks or hydraulic, electrical, or communication failures). However, under stress, people often behave as though they are under time pressure, when in fact they are not (Keinan, 1988). Stressful conditions may induce the use of decision strategies or heuristics, which often yield satisfactory results (Klein, 1997). However, when cues are ambiguous and goal conflicts exist, more thorough analysis may be required for a safe decision.

These four contextual factors are not an exhaustive list, and their causal role in specific types of aviation decision errors has not been directly established.

However, they do illustrate the complexity of the influences on a crew's decision processes. These factors may increase the crew's workload, stress, and sense of uncertainty, thereby contributing to decision errors.

HOW CAN DECISION OUTCOMES BE IMPROVED?

If the system complements the decision maker's expertise, especially at times of stress, negative consequences may be avoided. Following are several solutions that are suggested by the contextual factors described earlier. These might help to prevent or catch errors, thereby reducing negative outcomes.

Render Situations Less Ambiguous

Integrated flight displays that present accurate, up-to-date information, especially for unpredictable dynamic factors such as weather and traffic, may render situations less ambiguous. Trend information appears most critical. Information must be relevant as well as available; hence, the information provided should be applicable to the particular phase of flight (Vicente & Rasmussen, 1992).

Support Pilot Risk Assessment in Dynamic Conditions

Assessing the risks pertaining to a flight decision may be difficult because they are context dependent and dynamically changing. Still, some risk estimates could be embedded in aircraft systems. Weather and traffic displays already portray some risk information (e.g., color coding of weather severity, Traffic/Collision Avoidance System warnings). These might include estimates of how long it will take for a condition to degrade to a critical state, for example, fuel consumption or reserve battery life span, how soon weather will improve or a storm hit, or when traffic will dissipate in a target region. Plan continuation may result because the crew does not recognize the extent to which the situation has or is deteriorating.

Reduce Organizational Goal Conflicts

Reducing the pressures on pilots to continue with outmoded plans may require employers to recognize the goal conflicts their pilots often experience. Although it is certainly appropriate for companies to be concerned with productivity, safety might be enhanced if they are unequivocal in their support for pilots who take a safe course of action rather than a riskier one, even if a cost accrues. This is a policy question that can only be informed by scientific knowledge. Changing organizational culture to affect social expectations is perhaps more difficult because it is more insidious.

Increase Likelihood of Outcome Evaluation

Another potential area of support lies in assisting pilots with strategies for evaluating a course of action. Aiding may consist of prompting crews to consider constraints on options prior to acting and to consider the disadvantages of the selected option. Klein (1997) and Cohen, Freeman, and Thompson (1997) included such recommendations in training modules for military decision making. Tools for doing "what if" reasoning and managing multiple hypotheses may also be helpful, encouraging forward thinking. Means, Salas, Crandall, and Jacobs (1993) stressed the importance of making decision makers consider the worst-case scenario and of training them to plan for managing this possibility rather than the best-case scenario.

Support Error Catching

Aviation is a domain in which errors are inevitable (Woods et al., 1994). Therefore, the aviation system needs to support error catching, as well as error prevention. For example, a C-130 Air Force transport was deiced erroneously with washing fluid, adding to the icing problem rather than reducing it. Only the refusal of the ramp attendant to remove the chocks from under the wheels because he saw icicles hanging from the wing prevented the crew from attempting to take off (Park, 1997). Just as equipment is designed with redundancies to prevent a total system failure, crews build on their redundancy to assist error catching. Methods include monitoring each other as well as the situation and communicating to support shared problem models (Helmreich, Wilhelm, Klinect, & Merritt, 1998; Orasanu et al., 1998; Cannon-Bowers, Salas, & Converse, 1993).

SUMMARY

Our reexamination of NTSB accident reports emphasized that certain types of decision problems are specific to the aviation domain. Of particular concern are plan continuation errors. Even though human cognitive capacities are limited, human expertise is the primary source of strength and flexibility that can be used both to prevent and to catch errors. Designing aids that support decision making depends on understanding the context—how specific features of the environment interact with cognitive factors to induce decisions that may have negative consequences. Better understanding of how tools and strategies interact with these features of the domain is essential to increasing aviation safety.

REFERENCES

Air Accidents Investigations Branch (AAIB). (1990). *Report on the accident to Boeing 737-400 G-OBME near Kegworth, Leicestershire on 8 January, 1989* (Aircraft Accident Rep. 4/90). London: HMSO.

Berman, B. (1995). Flightcrew errors and the contexts in which they occurred: 37 major US air carrier accidents. In *Proceedings of the 8th International Symposium on Aviation Psychology* (pp. 1291–1294). Colombus: Ohio State University.

Calderwood, R., Klein, G. A., & Crandall, B. W. (1988). Time pressure, skill, and move quality in chess. *American Journal of Psychology*, *101*, 481–491.

Cannon-Bowers, J. A., & Salas, E. (1998). Individual and team decision making under stress: Theoretical underpinnings. In J. A. Cannon-Bowers & E. Salas (Eds.), *Making decisions under stress* (pp. 17–38). Washington, DC: American Psychological Association.

Cannon-Bowers, J. A., Salas, E., & Converse, S. (1993). Shared mental models in expert team decision making. In N. J. Castellan (Ed.), *Individual and group decision making: Current issues* (pp. 221–246). Hillsdale, NJ: Lawrence Erlbaum Associates.

Cannon-Bowers, J. A., Salas, E., & Pruitt, J. S. (1996). Establishing the boundaries of a paradigm for decision research. *Human Factors, 38,* 193–205.

Cohen, M. S., Freeman, J. T., & Thompson, B. B. (1997). Training the naturalistic decision maker. In C. Zsambok & G. Klein (Eds.), *Naturalistic decision making* (pp. 257–268). Mahwah, NJ: Lawrence Erlbaum Associates.

Doherty, M. E. (1993). A laboratory scientist's view of naturalistic decision making. In G. E. Klein, J. Orasanu, R. Calderwood, & C. E. Zsambok (Eds.), *Decision making in action: Models and methods* (pp. 362–388). Norwood, NJ: Ablex.

Endsley, M. R. (1995). Toward a theory of situation awareness. *Human Factors, 37*, 32–64.

Ericsson, K. A., & Leman, A. C. (1996). Expert and exceptional performance: Evidence of maximal adaptation to task constraints. *Annual Review of Psychology, 47*, 273–305.

Fischer, U., & Orasanu, J. (2000). *Do you see what I see? Effects of crew position on interpretation of flight problems.* NASA Tech. Mem. #2000-[TK]. Moffett Field, CA: National Aeronautics and Space Agency.

Fischer, U., Orasanu, J., & Wich, M. (1995). Expert pilots' perception of problem situations. In R. S. Jensen & L. A. Rakovan (Eds.), *Proceedings of the Eighth International Symposium on Aviation Psychology* (pp. 777–782). Columbus: Ohio State University.

Fischhoff, B. (1975). Hindsight ≠ foresight: The effect of outcome knowledge on judgment under uncertainty. *Journal of Experimental Psychology: Human Perception and Performance, 1*, 288–299.

Hawkins, S. A., & Hastie, R. (1990). Hindsight: Biased judgments of past events. *Psychological Bulletin, 107*, 311–327.

Helmreich, R. L., Wilhelm, J. A., Klinect, J. R., & Merritt, A. C. (1998). *Culture, error and crew resource management* [On-line]. Available University of Texas: http://www.psy.utexas.edu/psy/helmreich/HWKM_ERM.htm

Hockey, G. R. L. (1979). Stress and the cognitive components of skilled performance. In V. Hamilton & D. M. Warbuton (Eds.), *Human stress and cognition: An information-processing approach (PAGE NUMBERS TK).* Chichester, England: Wiley.

Hollenbeck, J., Ilgen, D., Phillips, J., & Hedlund, J. (1994). Decision risk in dynamic two-stage contexts: Beyond the status-quo. *Journal of Applied Psychology, 79*(4), 592–598.

Johnston, N. (1996). Managing risk in flight operations. In B. J. Hayward & A. R. Lowe (Eds.), *Applied aviation psychology: Achievement, change and challenge* (pp. 1–19). Brookfield, VT: Ashgate.

Keinan, G. (1988). Training for dangerous task performance: The effects of expectations and feedback. *Journal of Applied Social Psychology, 18*(4, pt.2), 355–373.

Klein, G. (1993a). A recognition-primed decision (RPD) model of rapid decision making. In G. Klein, J. Orasanu, R. Calderwood, & C. Zsambok (Eds.), *Decision making in action: Models and methods* (pp. 138–147). Norwood, NJ: Ablex.

Klein, G. (1993b). Sources of error in naturalistic decision making. In *Proceedings of the Human Factors and Ergonomics Society 37th Annual Meeting* (pp. 368–371). Santa Monica, CA: Human Factors and Ergonomics Society.

Klein, G. (1997). The current status of the naturalistic decision making framework. In R. Flin, E. Salas, M. Strub, & L. Martin (Eds.), *Decision making under stress: Emerging themes and applications* (pp. 11–28). Aldershot, England: Ashgate.

Klein, G. E., Orasanu, J., Calderwood, R., & Zsambok, C. E. (Eds.). (1993). *Decision making in action: Models and methods*. Norwood, NJ: Ablex.

Lipshitz, R. (1997). Naturalistic decision making perspectives on decision errors. In C. Zsambok & G. Klein (Eds.), *Naturalistic decision making* (pp. 151–162). Mahwah, NJ: Lawrence Erlbaum Associates.

Lipshitz, R., & Strauss, O. (1997). Coping with uncertainty: A naturalistic decision making analysis. *Organizational Behavior and Human Decision Processing, 69,* 149–163.

March, J. G., & Simon, H. A. (1958). *Organizations*. New York: Wiley.

Maule, A. J. (1997). Strategies for adapting to time pressure. In R. Flin, E. Salas, M. Strub, & L. Martin (Eds.), *Decision making under stress: Emerging themes and applications* (pp. 271–279). Aldershot, England: Ashgate.

Means, B., Salas, E., Crandall, B., & Jacobs, T. O. (1993). Training decision makers for the real world. In G. Klein, J. Orasanu, R. Calderwood, & C. Zsambok (Eds.), *Decision making in action: Models and methods* (pp. 51–99). Norwood, NJ: Ablex.

Nance, J. (1986). *Blind trust.* New York: William Morrow.

National Transportation Safety Board (NTSB). (1982). *Aircraft accident report: Air Florida, Inc., Boeing 737-222, N62AF, Collision with 14th Street Bridge, near Washington National Airport, Washington, DC, January 13, 1982* (NTSB-AAR-82/08). Washington, DC: Author.

National Transportation Safety Board (NTSB). (1985). *Aircraft accident report: Air Illinois Hawker Siddley, HS 748-2A, N748LL, near Pinckneyville, Illinois, October 11, 1983* (NTSB-AAR-85/03). Washington, DC: Author.

National Transportation Safety Board (NTSB). (1986a). *Aircraft accident report: Delta Air Lines, Inc., Lockheed L-1011-385-1, N726DA, Dallas/Fort Worth International Airport, Texas, August 2, 1985* (NTSB-AAR-86/05). Washington, DC: Author.

National Transportation Safety Board (NTSB). (1986b). *Aircraft accident report: Galaxy Airlines, Inc., Lockheed Electra-L-188C, N5532, Reno, Nevada, January 21, 1985* (NTSB- AAR-86-01). Washington, DC: Author.

National Transportation Safety Board (NTSB). (1991). *Aircraft accident report: NW Airlines, Inc., Flights 1482 and 299 runway incursion and collision Detroit Metropolitan/Wayne County Airport Romulus, Michigan, December 3, 1990* (NTSB-AAR-91/05). Washington, DC: Author.

National Transportation Safety Board (NTSB). (1994a). *Aircraft accident report: Uncontrolled collision with terrain, American International Airways Flight 808, Douglas DC-8-61, N814CK, US Naval Air Station Guantanamo Bay, Cuba, August 18, 1993* (NTSB-AAR-94/04). Washington, DC: Author.

National Transportation Safety Board (NTSB). (1994b). *A review of flightcrew-involved, major accidents of U.S. air carriers, 1978 through 1990* (NTSB/SS-94/01). Washington, DC: Author.

National Transportation Safety Board (NTSB). (1999). *Aircraft accident report: Controlled flight into terrain, Korean Air Flight 801, Boeing 747-300, HL7468 Nimitz Hill, Guam, August, 6, 1997* (NTSB-AAR-99-02). Washington, DC: Author.

O'Hare, D., & Smitheram, T. (1995). "Presing on" into deteriorating conditions: An application of behavioral decision theory to pilot decision making. *International Journal of Aviation Psychology, 5*(4), 351–370.

Orasanu, J., Dismukes, R. K., & Fischer, U. (1993). Decision errors in the cockpit. In *Proceedings of the Human Factors and Ergonomics Society 37th Annual Meeting* (pp. 363–367). Santa Monica, CA: Human Factors and Ergonomics Society.

Orasanu, J., & Fischer, U. (1997). Finding decisions in natural environments: The view from the cockpit. In C. Zsambok & G. Klein (Eds.), *Naturalistic decision making* (pp. 343–357). Mahwah, NJ: Lawrence Erlbaum Assoicates.

Orasanu, J., Fischer, U., McDonnell, L. K., Davison, J., Haars, K. E., Villeda, E., & VanAken, C. (1998). How do flight crews detect and prevent errors? Findings from a flight simulation study. In *Proceedings of the Human Factors and Ergonomics Society 42nd Annual Meeting* (pp. 191–195). Santa Monica, CA: Human Factors and Ergonomics Society.

Orasanu, J., & Strauch, B. (1994). Temporal factors in aviation decision making. In *Proceedings of the Human Factors and Ergonomics Society 38th Annual Meeting* (pp. 935–939). Santa Monica, CA: Human Factors and Ergonomics Society.

Park, J. T. (1997). De-deicing. *Flying Safety, 53*(October), 30.

Prince, C., Stout, R., & Salas, E. (1995). *Team situation awareness: Preliminary lessons learned: Identification of skills, behavior, design guidelines, and training strategies from interviews and research.* Orlando, FL: Naval Air Warfare Center, Training Systems Division.

Rasmussen, J. (1985). The role of hierarchical knowledge representation in decision making and system management. *IEEE Transactions on Systems, Man and Cybernetics, 15*(2), 234–243.

Rasmussen, J. (1997). Merging paradigms: Decision making, management, and cognitive control. In R. Flin, E. Salas, M. Strub, & L. Martin (Eds.), *Decision making under stress: Emerging themes and applications* (pp. 67–81). Aldershot, England: Ashgate.

Reason, J. (1990). *Human error.* Cambridge, England: Cambridge University Press.

Reason, J. (1997). *Managing the risks of organizational accidents.* Brookfield, VT: Ashgate.

Simon, H. (1957). *Models of man: Social and rational.* New York: Wiley.

Sitkin, S. (1992). Learning through failure: The strategy of small losses. *Research in Organizational Behavior, 14*, 231–266.

Strauch, B. (1997). Automation and decision making—Lessons from the Cali accident. In *Proceedings of the Human Factors and Ergonomics Society 41st Annual Meeting* (pp. 195–199). Santa Monica, CA: Human Factors and Ergonomics Society.

Tversky, A., & Kahneman, D. (1974). Judgment under uncertainty: Heuristic and biases. *Science, 185,* 1123–1124.

United States Air Force (USAF). (1997). *Accident investigation board report. USAF CT-43A (73–1149), 3 April 1996 Dubrovnik, Croatia.* Washington, DC: Author.

VanLehn, K. (1990). *Mind bugs: The origins of procedural conceptions.* Cambridge, MA: MIT Press.

Vicente, K. J., & Rasmussen, J. (1992). Ecological interface design: Theoretical foundations. *IEEE Transactions SMC, 22*(4), 589–607.

Wickens, C. D., Stokes, A., Barnett, B., & Hyman, F. (1993). The effects of stress on pilot juagment in a MIDIS simulator. In O. Svenson & A. J. Maule (Eds.), *Time pressure and stress in human judgment and decision making* (pp. 271–292). Cambridge, England: Cambridge University Press.

Woods, D. D., Johannesen, L. J., Cook, R. I., & Sarter, N. B. (1994). *Behind human error: Cognitive systems, computers, and hindsight.* Wright-Patterson Air Force Base, Ohio: Crew Systems Ergonomics Information Analysis Center.

Woods, D. D., & Sarter, N. B. (1998). *Learning from automation surprises and "going sour" accidents: Progress on human-centered automation* (NASA Rep. NCC 2-592). Moffett Field, CA: NASA Ames Research Center.

Zapf, D., Maier, G. W., Rappensperger, G., & Irmer, C. (1994). Error detection, task characteristics, and some consequences for software design. *Applied Psychology: International Review, 43,* 499–520.

Zsambok, C., & Klein, G. (Eds.). (1997). *Naturalistic decision making.* Mahwah, NJ: Lawrence Erlbaum Associates.

IV

Expertise

13

What Does It Mean When Experts Disagree?

James Shanteau
Kansas State University

Decision researchers often seem perplexed when they observe sizable and consistent disagreements between subjects. This is particularly true when the subjects are experts. As shown by the following quotations, however, disagreements have been viewed historically as necessary:

> "By different methods different men excel." (Churchill, 1764)

> "The history of scholarship is a record of disagreements." (Hughes, 1936)

> "The tough-minded . . . respect differences." (Benedict, 1940)

The position taken here is that disagreement between experts is not a problem, but rather is a normal part of an expert's job.

The purpose of this chapter is to explore why it is natural for experts to disagree with each other. The chapter is organized as follows. First, there is a review of the literature on disagreement between experts with a discussion of a commonly held hypothesis about expertise. Second, the chapter presents 10 structural and functional factors that help explain why experts often disagree. Third, domain differences in the extent to which experts agree are considered. Fourth, an alternate hypothesis is offered that more closely corresponds to how experts view their tasks. Finally, the chapter concludes with implications for future research directions.

BACKGROUND

Since the start of systematic analyses of decision making in the 1950s, investigators have expressed surprise and dismay at the extent to which domain experts disagree. For example, if a researcher asks two financial experts to assess an investment, the expectation of most investigators is that they should make the same recommendation. If the experts arrive at different recommendations, then the researcher wonders whether the experts are as skilled as they claim.

In a seminal paper, Einhorn (1974) argued that *consensus* or between-expert reliability is a necessary condition for expertise. He reported, however, significant differences in diagnoses by three expert medical pathologists. The average between-expert correlation (r) was .55 (where .0 is chance and 1.0 is perfect).

Similar evidence was presented in a study of four expert livestock judges evaluating overall breeding quality of swine (Phelps, 1977). Despite a high level of internal consistency (average r = .96), the consensus was much lower, r = .50. Apparently, livestock experts have internally consistent strategies, but do not agree with each other about what those strategies should be.

Comparable results have been reported by researchers for other domains. For example, Hoffman, Slovic, and Rorer (1968) and Goldberg and Werts (1966) reported consensus values of less than .40 for judgments by professional stockbrokers and clinical psychologists, respectively.

Several studies of financial experts (auditors) have reported somewhat higher between-expert consensus. For instance, Kida (1980) asked 27 audit partners to evaluate 40 financial profiles based on five accounting ratios; the average between-expert correlation was .76. Ashton (1974) had 63 professional auditors evaluate the strength of internal controls for 32 cases; the mean r between auditors was .70. Finally, Libby, Artman, and Willingham (1985) observed an average correlation of .68 between 12 auditors making control reliance judgments.

Several studies have explored whether agreement increases with experience. Ettenson, Shanteau, and Krogstad (1987) compared 11 accounting students with 10 (mid-level) audit seniors and 10 audit partners; the mean between-judge correlations increased from .66 to .76 to .83, respectively. Messier (1983) reported similar results—audit partners with more than 15 years experience had greater consensus than partners with less experience. In contrast, Hamilton and Wright (1982) found no difference in consensus for internal control assessments by three groups of auditors varying in experience; the average correlation for all three groups was .72.

These results suggest three conclusions. First, experts in a variety of nonfinancial domains often disagree; the consensus correlations range from .40 to .55. Second, the agreement between financial experts is higher, with correlations ranging from .68 to .83. Third, there is suggestive evidence that increased experience may lead to greater consensus. Still, most researchers conclude that because of the sizable disagreement between experts, there is reason to be concerned about the competence of experts. In the next section, I explore the hypothesis behind this concern.

EXPERTS-SHOULD-CONVERGE HYPOTHESIS

The unimpressive consensus correlations for nonfinancial experts led many researchers to question the abilities of experts in general. Following Einhorn's logic, these investigators assume that agreement is a necessary condition for expertise. The lack of agreement, therefore, suggests that "experts are no damn good" (Gettys, personal communication, 1980). This analysis and interpretation of reliability data are derived from an implicit hypothesis about experts. The hypothesis, labeled *Experts-Should-Converge* (ESC), is based on the following five arguments:

1. For most tasks performed by experts, there is assumed to be a "gold standard" or unique "ground truth." If this truth is easy to access, people can get it directly. For expert tasks, however, the truth is outside the realm of common knowledge or direct sensory experience of most people. Thus, unique correct answers exist, at least in theory, but they are difficult to obtain.
2. Because of their special skills and experience, experts should be able to tell the rest of us about this "ground truth." That is, experts can access what others cannot access.
3. Because, by definition, there can be only one "ground truth," all experts should give us the single correct answer. The special abilities of experts should allow them to obtain the same truth.
4. If experts disagree, then someone is wrong—they cannot all be correct. Some or all (or none) of them must not really be experts. Thus, disagreements are a reflection of incompetence.
5. Because nonexperts do not know which of the so-called "experts" are correct, the only safe course of action is to distrust all of them. That is, disagreement between experts implies that one should be suspicious of their claimed special abilities.

This hypothesis, of course, is not a formal chain of logic. But, it is implicit in the way that most researchers reason about results reporting disagreements between domain experts.

CONTRIBUTING FACTORS

The ESC hypothesis is supported by two disconnected lines of thought: one that affects theoretical reasoning, and the other, empirical research. First, research on decision making has been linked historically to theories from economics (e.g., expected utility theory) and statistics (e.g., Bayes theorem; see Edwards, 1983). When applied to well-specified problems, such theories lead to unique solutions. Given a specified set of antecedent conditions, these formalisms provide point estimates of the optimal decision strategy. It is interesting that a great deal of laboratory work has been devoted to showing the inadequacy of such theories (Yates, 1990). Nevertheless, economics and statistics continue to provide a point of comparison in much of decision research.

This reasoning has also been applied to analyses of domain experts. Although most experts work on problems that are not well specified, researchers nonetheless assume that unique solutions exist—just as they do for simple cases. Some investigators have questioned the relevance of economic/statistical theories as a basis for making real-world decisions (Klein, Orasanu, Calderwood, & Zsambok, 1993). Other scholars have questioned the narrow assumptions required and have argued instead that these theories are not descriptive of real situations (Gigerenzer, 1993). Although economic and statistical theories often cannot be applied to the domains where experts work, it is nonetheless commonly assumed that point solutions exist, at least in the abstract.

The second line of thought shows that nearly a century of experimental psychology (particularly in the United States) has been focused on analysis of the "Generalized Normal Human Adult Mind" (*GNHAM*; Edwards, 1983). According to Boring (1929), this emphasis dates to the beginnings of experimental psychology, particularly Wilhelm Wundt and Edward Titchener: "Titchener's interest lay in the generalized, normal, human adult mind that had also been Wundt's main concern" (p. 90). Later writers reflected this theme: "Psychology may gather its materials from many sources, but its aim is to understand the generalized human mind" (Heidbreder, 1933, pp. 125–126).

In fact, there is no evidence that Wundt ever used this term (Shanteau, 1999). References to the concept, however, are common in Titchener's writing, for example, "psychology is concerned with the normal, human, adult mind" (1916, p. 2). Modern historians of psychology now believe that Boring created a "myth of origin" (Samelson, 1974) to provide a justification for Titchener's (who was his mentor) place in psychology. As noted by Hebb (1972, p. 291), although Boring's history is "commonly considered the standard work, and beautifully clear in its exposition, this book is thoroughly misleading in its emphasis (on the relationship between) . . . Wundt and Titchener."

According to the GNHAM view, the goal of behavioral research is to investigate commonalties among humans, not differences between them. That is, the focus of psychology belongs on the generalized mind, not on individual minds. Because the

Titchener–Boring view dominated (at least in the United States), research was directed to the search for "universal truths" of behavior. As a result, experimental psychology developed neither the paradigms nor the theories to deal with outlier behavior. And expertise, by definition, is outlier behavior. (For a further discussion of the influence of GNHAM on decision research, see Shanteau, 1999.)

Research in decision making, therefore, has been influenced by two research streams that assumed (a) that decision problems should have unique correct answers and (b) that differences between individuals are not important to empirical investigations. The persistence of observed differences between experts provided evidence inconsistent with these views. Thus, disagreements between experts led to the conclusion that something must be wrong, that is, expertise is a sham.

In the next two sections, I explore 10 factors behind the premise that researchers should not be surprised when experts disagree. This is followed by an alternate hypothesis along with some supporting data. The chapter then concludes with a discussion of implications and conclusions.

STRUCTURAL FACTORS

Analysis of the context in which most experts work provides five structural factors behind the explanation of why experts may disagree. These factors reflect the situational constraints under which most experts work.

1. In the contexts where experts work, the ground truth is a fiction. Single-point optimal solutions do not exist. Despite the tremendous analytic ability of master players and the incredible computation speed of computer programs, such as *Deep Blue*, for instance, the game of chess still does not yield optimal solutions. If this is true for a well-structured game such as chess, how can it be possible to find a correct answer in an ill-structured setting? The reason society needs experts in the first place is that they offer answers that we could not obtain any other way.
2. A distinction can be made between the different levels of decisions made by experts. Using terminology from medicine, it is possible to distinguish between three levels: The first is *diagnosis* (What is it?) based on categorization and/or classification; the second is *prognosis* (What is the likely outcome?) based on forecasting future scenarios; and the third is *treatment* (What to do about it?) involving selection of a course of action. There are thousands of diagnoses and hundreds of prognoses, but relatively few treatments. It should not be surprising, therefore, to find that experts might disagree at one level (diagnosis), but agree at another (treatment).

3. Despite the assumption behind the ESC hypothesis, experts are seldom asked to make single-outcome decisions. The concept of a *point prediction* is largely a fiction created for the convenience of the researcher and is not descriptive of the tasks that experts do. As Golde (1969, p. 213) noted, although "an expert does sometimes make decisions, his [her] role is usually much more of an advisor . . . [they] let me know the kinds of decisions or actions that I must take." In other words, the job of the expert is to clarify alternatives and describe possible outcomes for clients.
4. As Klein (1993) emphasized, experts generally work in dynamic situations with frequent updating. Thus, the problems faced by experts are unpredictable, with evolving constraints. In such situations there are rarely any best or correct answers. Therefore, whereas the ESC model assumes a stationary target, the reality faced by experts is generally more like a moving target.
5. A long-term perspective reveals that experts work in realms in which the basic science is still evolving. For instance, the rapid changes in medicine mean that the current "best answers" are soon obsolete. Why should researchers expect experts to agree on a single correct answer, say for the treatment of Acquired Immune Deficiency Syndrome, when new knowledge will likely provide better solutions tomorrow?

FUNCTIONAL FACTORS

An analysis of the strategies used by experts to make decisions reveals at least five functional or process factors behind why experts may disagree. These factors have to do with how experts think about the decisions and judgments that make.

1. Most experts operate as if they have flat loss functions for deviations from optimality. They see small deviations as having minor consequences. In comparison, researchers often operate as if they have steep loss functions. That is, they view any deviation from optimality, no matter how slight, as having large consequences. A similar argument can be made for how disagreements between experts are viewed.
2. Whereas most researchers view an error as any deviation between behavior and the correct answer, experts have a different definition of error. As noted, experts are usually more concerned about avoiding big mistakes, whereas researchers are looking for perfection. Thus, the same outcome could well be called an *error* by the researcher and a *success* by the expert. In other words, experts may see agreement where investigators see disagreement.

3. In many (most?) settings, experts expect to disagree with each other. In a discussion among any two academics, for instance, researchers know that they invariably will find something to argue about. Even when they agree on 99% of the issues, they will quickly find the last 1% and disagree about that. Similarly, experts in most any field bypass items of agreement to focus instead on disagreements. Thus, experts view disagreements as a normal part of their job.
4. Disagreements are often the route by which experts increase understanding of their field. By seeking out areas of disagreement, experts examine the limits of their own knowledge and stretch their range of competency. Therefore, experts see disagreements as a key step in increasing their grasp of their field.
5. After a domain has advanced to the point where all issues are resolved, there are few disagreements among experts because there is nothing to argue about. When a field has developed to that degree, however, the answers are known and agreed upon. Thus, total agreement among experts is an indication that there is no longer much of a role for experts to play in that domain.

DOMAIN DIFFERENCES

Experts in different domains perform different tasks. Yet, decision researchers persist in treating all experts alike, so that the term *expert* is used generically. For instance, Kahneman (1991) concluded "there is much evidence that experts are not immune to the cognitive illusions that affect other people" (p. 143). Yet nearly all researchers are aware that at least some experts, for example, weather forecasters, show little sign of biases or "cognitive illusions." Thus, despite the generalizations drawn about experts, there are many exceptions to the rule.

In an effort to account for these domain differences, I constructed Table 13.1 to differentiate between those domains where experts do well and those where experts do not. The table is based on a continuum from high to low performance (see Shanteau 1992a, 1992b for earlier versions of this table). In the left column are those domains where experts make aided decisions using Decision Support Systems (DSS) or other computerized tools, for example, in weather forecasting. The next column contains domains where experts make skilled but largely unaided decisions, for example, livestock judges. The third column lists domains where experts show limited competence, for example, clinical psychologists. The behavior of experts in the last column is close to random, for example, stockbrokers.

It should be noted that assignment of domains within the table was based on a review of the literature, that is, the assessment of competence is drawn from

TABLE 13.1
Progression of Domains from High to Low Performance

Highest Levels of Performance		*Lowest Levels of Performance*	
Aided Decisions	*Competent*	*Restricted*	*Random*
Weather forecasters	Chess masters	Clinical psychologists	Polygraphers
Astronomers	Livestock judges	Parole officers	Managers
Test pilots	Grain inspectors	Psychiatrists	Stock forecasters
Insurance analysts	Photo interpreters	Student admissions	Parole officers
Physicists	Soil judges	Intelligence analysts	Court judges
Nurses	Nurses	Nurses	
Physicians	Physicians	Physicians	
Auditors	Auditors	Auditors	

Note. The space in the table separates domains (top) where the competence of experts can be classified into one category from domains (bottom) where the evidence of competence is varied. For example, various studies have reported that the behavior of auditors ranges from moderately to extremely competent.

researchers who study each domain. Also three domains—nurses, physicians, and auditors—appear in several columns. That is because the literature in these fields provides mixed evidence of the competence of experts.

There are many ways to describe the differences in this table (see Shanteau, 1992a, 1992b). For present purposes, it makes most sense to note that domains to the left side possess more stable (*static*) properties. That is, the stimuli and the problem "hold still" for experts to evaluate. The domains to the right side, however, involve more changeable (*dynamic*) properties. Thus, the stimuli and problem are less stable, harder to specify, and more like "moving targets." It makes sense, therefore, that expert agreement will be higher on the left side and lower on the right side.

To test this idea, Table 13.2 gives reliability values from studies of domain experts in the four categories of Table 13.1. Two domains are listed under each category, with the between-expert agreement (consensus) given as average correlations. As can be seen, the average consensus r value for weather forecasters is .95, whereas average values for livestock judges, clinical psychologists, and stock forecasters are .50, .40, and .32, respectively. Comparable results appear for other domains on the second line. The trend supports the prediction outlined earlier—better structured domains lead to high consensus and less structured domains to less consensus.

For comparison, the average within-expert reliability (consistency) correlations for these same domains are listed in Table 13.3. The trends are similar, with better structured domains leading to higher internal consistency. As expected, the consistency values (except for pathologists) are higher than the corresponding

TABLE 13.2
Reliability (Consensus) Values for Experts

Highest Levels of Performance		*Lowest Levels of Performance*	
Aided Decisions	*Competent*	*Restricted*	*Random*
Weather forecasters[a]	Livestock judges[b]	Clinical psychologists[c]	Stockbrokers[d]
$r = .95$	$r = .50$	$r = .40$	$r = .32$
Auditors[e]	Grain inspectors[f]	Pathologists[g]	Polygraphers[h]
$r = .76$	$r = .60$	$r = .55$	$r = .33$

Note. The values cited in this table were drawn from the following studies (from left to right): Stewart, Roebber & Bosart (1997)[a]; Phelps & Shanteau (1978)[b]; Goldberg & Werts (1966)[c]; Slovic (1969)[d]; Kida (1980)[e]; Trumbo, Adams, Milner, & Schipper (1962)[f]; Einhorn (1974)[g]; and Lykken (1979)[h].

TABLE 13.3
Reliability (Internal Consistency) Values

Highest Levels of Performance		*Lowest Levels of Performance*	
Aided Decisions	*Competent*	*Restricted*	*Random*
Weather forecasters[a]	Livestock judges[b]	Clinical psychologists[c]	Stockbrokers[d]
$r = .98$	$r = .96$	$r = .44$	$r = <.40$
Auditors[e]	Grain inspectors[f]	Pathologists[g]	Polygraphers[h]
$r = .90$	$r = .62$	$r = .50$	$r = .91$

Note. The values cited in this table were drawn from the following studies (from left to right): Stewart, Roebber & Bosart (1997)[a]; Phelps & Shanteau (1978)[b]; Goldberg & Werts (1966)[c]; Slovic (1969)[d]; Kida (1980)[e]; Trumbo, Adams, Milner & Schipper (1962)[f]; Einhorn (1974)[g]; and Raskin & Podlesny (1979)[h].

consensus values in Table 13.2. In two domains (livestock judges and polygraphers), there are notable discrepancies between the consensus and consistency correlations; the reasons for these differences are not clear.

DECISION RESEARCHERS' VIEW OF EXPERTS

Compared to other fields of inquiry, decision researchers have taken an idiosyncratic view of the abilities of experts. Investigators in artificial intelligence, expert system design, cognitive science, systems analysis, and computer science have all concluded that experts are superior decision makers. That is why knowledge engineers build computer simulations around what experts know.

Furthermore, most domain-specific researchers (such as those in medicine and weather forecasting) view experts as possessing unique information essential for making good decisions. In short, investigators in these fields see human expertise as something to be emulated.

In contrast, decision researchers, especially in the United States, have concluded that experts are flawed and prone to making simple errors (for example, Kahneman, 1991). Moreover, experts and novices are viewed as sharing the same shortcomings. For instance, Tversky (quoted in Gardner, 1985, p. 360) stated, "whenever there is a simple error that most laymen fall for, there is always a slightly more sophisticated version of the same problem that experts fall for."

Investigators have overlooked the fact that in most real-world problems, unique solutions do not exist. Instead, there are multiple solution paths. For instance, in medicine there may be many ways to treat an illness—when one approach does not work, a physician seeks another. It should not be surprising, therefore, to find experts disagreeing about which is the appropriate course of action to take. Other inquiry systems, therefore, accept multiple points of view across experts as inevitable. In contrast, decision researchers with their simplified, single-answer view of the world find a multiple-solution perspective difficult to understand.

MULTIPLE-SOLUTION HYPOTHESIS

The position taken here is that previous researchers have unknowingly adopted an ESC view of expertise. According to ESC, disagreement between experts is a sign that something is wrong. This leads to the conclusion that experts are not as skilled or as competent as they claim to be.

In this section, I propose an alternate hypothesis that is closer to the view of how experts see themselves. There are five arguments behind this *Multiple Solution Model* (MSM).

1. The primary job of an expert is not to make decisions but to help clients reach a broadly defined target state. For example, the goal of the client may be to design a better investment portfolio or to find a better loan strategy. These goals do not involve single answers, but instead require something more elaborate from the expert, such as a strategic plan.
2. To reach the client's goal state requires dealing with multiple, constantly changing, dynamic factors. As noted by Klein et al (1993), the situations faced by experts are different and more complex than the simplified settings studied in research laboratories. Thus, experts work on problems that are considerably more complex than those studied in laboratory settings.

3. Using their knowledge and experience, the expert recognizes patterns and finds consistencies in a dynamic problem space. The expert's job is to clarify the issues for the client. In other words, the challenge for an expert is "to make sense out of chaos."
4. Based on their experience and insights into the nature of problems, experts try to help clients clarify their thinking. For instance, an expert will often identify several alternate paths to the desired goal states. The expert's role is to lay out the options and the consequences in a clear and comprehensible fashion for the client.
5. In the end, it is the client, not the expert, who actually makes most decisions. The expert offers insights and observations, but the client makes and implements the final choice(s). Thus, the final responsibility for the decision rests on the client, not the expert.

The view is summarized nicely by the management consultant Golde (1969): "We seem to expect too much and the wrong things of our experts" (p. 213). As expressed by MSM, experts generally act more like knowledgeable consultants. Rarely do they function as the all-knowing single-answer decision makers envisioned by most researchers. Instead, experts help clients by giving them the insights and information needed to make their own decisions.

When experts disagree, therefore, it is because they see different paths to the client's goal state. In turn, savvy clients may seek out the views of various experts precisely because they want different perspectives on their problems. Thus, disagreements between experts are not only expected, but also may actually be useful.

CONCLUSIONS

By relying on comparisons to economics/statistical analysis and by incorporating GNAHM assumptions into their research designs, decision investigators unknowingly have adopted a distorted view of what experts do. The remainder of the chapter looks further at the implications arising from these two assumptions.

Economic/Statistical Thinking

By drawing a parallel to economic/statistical theory, decision researchers have adopted a single-correct-answer approach to assessing expertise. When an expert (or anyone else) gives an answer different from the correct answer, he or she is said to have a *bias* (Tversky & Kahneman, 1974). And, when two or more experts give different answers, the claim of experts to special competence is questioned (Einhorn, 1974).

The position here is that researchers have relied on an incorrect view of how experts function. For instance, the environment in which experts work is much different from that incorporated into traditional laboratory research. The complex, changeable environment that experts operate in is considerably more complicated than the small, stable context constructed by investigators. In reality, problems rarely are simple enough to lead to a single correct answer. Instead, there are multiple answers (or at least multiple routes to answers). If so, it should not be surprising to find that experts often take different approaches (as envisioned by MSM) to making recommendations.

The underlying problem is that researchers misunderstand what experts do and what is expected of them. Investigators seem to think that experts see the world as they do—with simplifying assumptions and normative solutions. However, experts generally have a different world view—with many complexities and contingencies, but few optimal solutions. In addition, experts have a flexible approach to adaptation (for example, hedge clipping) and are better at managing uncertainty.

From the MSM perspective, disagreements between experts are to be expected. Although researchers view disagreements as evidence of incompetence, experts see disagreements as more-or-less inevitable and even as a useful, part of the job.

GNAHM Research Paradigm

By adopting the GNAHM approach to research, most investigators have largely ignored individual differences. The average is emphasized, variability is not. More important, researchers almost never look at the distribution of responses over subjects. Consequently, exceptions are overlooked and statements made about results that imply that all people follow the same pattern of behavior.

Aside from risk preferences (risk adverse vs. risk seeking), there are few individual-subject variables that have received much attention in the decision-making literature. Even risk preferences are more talked about than studied. Yet, there is an emerging research stream showing consistent individual differences in decision-making strategies. For instance, Yates, Lee, and Bush (1997) found sizable and stable cross-cultural differences between subjects coming from different backgrounds. Such investigations, however, have yet to arouse mainstream interest.

As a field, decision research has focused on the behavior of student subjects. Rather than using tools to discover what makes experts unique, researchers rely on paradigms derived from studies of nonexpert behavior. Consequently, distinct decision-making paradigms for investigating domain experts do not exist. Instead, the methods to study experts are borrowed from standard procedures. Because these borrowed paradigms are often ill-suited to investigate expertise, it should not be surprising to find that our understanding of experts has not advanced at the same rate as our understanding of nonexpert subjects.

That is not to say that research methods developed from a GNAHM perspective cannot be adapted to study expert behavior. The problem is not the methodology, per se. Rather, it is how researchers use the methodology—any technique can be misused (Birnbaum, 1973).

The dangers of reliance on GNAHM thinking for researchers have been recognized for some time (Edwards, 1983). However, the warnings have not been heard. Edwards (p. 509) offered two messages: "One is that psychologists have failed to heed the urging of Egon Brunswik (1955) that generalizations from laboratory tasks should consider the degree to which the task . . . resembles or represents the context to which the generalizations are made. The other message is that experts can in fact do a remarkably good job of assessing and working with probabilities" (for example, in weather forecasting).

FINAL COMMENTS

To gain insights into expertise, it is necessary to understand what it is that domain experts do. In my experience, decision researchers rarely take the opportunity to gain an in-depth understanding of how experts think. If they did, they would learn that single-solution approaches to defining correct and incorrect answers do not work. They would also learn that the GNAHM perspective is a trap that limits understanding of individual differences generally and expertise specifically.

Therefore, disagreement between experts should not be viewed as a source of concern about the competence of experts. It is time for researchers to rethink their view that lack of agreement leads to supposed incompetence of experts. Persistence in such beliefs says more about the biases of investigators than it does about experts. If disagreement between experts is not a focal issue, then what should be? Let me suggest three useful goals for future research on expertise.

First, various investigators have concluded that the superiority of domain experts depends on their ability to distinguish between relevant and irrelevant information (Ettenson et al., 1987; Jacavone & Dostal, 1992; Mosier, 1997; Schwartz & Griffin, 1986; Shanteau, 1992b). One goal in future research should be to learn how experts make these discriminations and to find ways to enhance the process.

A second goal should be to understand the kinds of intellectual and physical tools used by experts to enhance their judgments. Experts seldom, if ever, make unaided judgments of the sort emphasized in laboratory research. In fact, researchers make use of the very tools denied their subjects. "The experimenters themselves, using tools and expertise, are able to perform (laboratory) tasks rather well" (Edwards, 1983, p. 511). The type of tools used by experts needs to be better understood.

The final goal should be to develop insights into domain differences. As argued by Edwards (1983, p. 512), "we have no choice but to develop a taxonomy of intellectual tasks themselves. Only with the aid of such a taxonomy can we think with reasonable sophistication about how to identify among the myriad types of experts and the myriad types of tasks . . . just exact what kinds of people and tasks deserve our attention." The analyses in Tables 13.1, 13.2, and 13.3 offer one perspective on such a taxonomy.

Research on such goals will help researchers expand their understanding of expertise. In contrast, concern about the supposed incompetence of experts based on disagreement offers little opportunity for enhancing understanding of expertise. As argued here, the future of research on experts lies in other directions. Specifically, researchers should focus their efforts on analyses of relevance/irrelevance, tool usage, and domain differences. These are directions that future researchers should explore.

ACKNOWLEDGMENTS

Preparation of this manuscript was supported, in part, by National Science Foundation Grant DMI 96-12126 and by support from the Institute for Social and Behavioral Research at Kansas State University. The author wishes to thank Ward Edwards, Julia Pounds, and Rob Ranyard for their helpful comments on earlier versions of the manuscript.

REFERENCES

Ashton, R. H. (1974). An experimental study of internal control judgments. *Journal of Accounting Research, 12*, 143–157.

Bartlett, J. (1992). In J. Kaplan (Ed.), *Bartlett's familiar quotations* (16th ed.). Boston, MA: Little, Brown and Company.

Birnbaum, M. (1973). The devil rides again: Correlation as an index of fit. *Psychological Bulletin, 79*, 239–242.

Boring, E. G. (1927). *A history of experimental psychology*. New York: Appleton-Century-Crofts.

Brunswik, E. (1955). Representative design and probabilistic theory in a functional psychology. *Psychological Review, 62*, 193–217.

Edwards, W. (1983). Human cognitive capacities, representativeness, and ground rules for research. In P. Humphreys, O. Svenson, & A. Vari (Eds.), *Analyzing and aiding decision processes*. Budapest, Hungary: Akademiai Kiado.

Einhorn, J. (1974). Expert judgment: Some necessary conditions and an example. *Journal of Applied Psychology, 59*, 562–571.

Ettenson, R., Shanteau, J., & Krogstad, J. (1987). Expert judgment: Is more information better? *Psychological Reports, 60*, 227–238.

Goldberg, L. R., & Werts, C. E. (1966). The reliability of clinicians' judgments: A multitrait-multimethod approach. *Journal of Clinical Psychology, 30*, 199–206.

Gardner, H. (1985). *The mind's new science: The history of the cognitive revolution.* New York: Basic Books.

Gigerenzer, G. (1993). The superego, the ego, and the id in statistical reasoning. In G. Keren & C. Lewis (Eds.), *A handbook for data analysis in the behavioral sciences: Methodological issues* (pp. 311–339). Hillsdale, NJ: Lawrence Erlbaum Associates.

Golde, R. A. (1969). *Can you be sure of your experts?* New York: Award Books.

Hamilton, R. E., & Wright, W. F. (1982). Internal control judgments and effects of experience: Replications and extensions. *Journal of Accounting Research, 20*, 756–765.

Hebb, D. O. (1972). *Textbook of psychology* (3rd ed.). Philadelphia: W. B. Saunders.

Heidbreder, E. (1933). *Seven psychologies*. New York: Appleton-Century.

Hoffman, P., Slovic, P., & Rorer, L. (1968). An analysis of variance model for the assessment of configural cue utilization in clinical judgment. *Psychological Bulletin, 69*, 338–349.

Jacavone, J., & Dostal, M. (1992). A descriptive study of nursing judgment in assessment and management of cardiac pain. *Advances in Nursing Science, 15*, 54–63.

Kahneman, D. (1991). Judgment and decision making: A personal view. *Psychological Science, 2*, 142–145.

Kida, T. (1980). An investigation into auditors' continuity and related qualification judgments. *Journal of Accounting Research, 8*, 506–523.

Klein, G. (1993). Sources of error in naturalistic decision making tasks. In *Proceedings of the Human Factors and Ergonomics Society*, pp. 368–371.

Klein, G. A., Orasanu, J., Calderwood, R., & Zsambok, C. E. (1993). *Decision making in action: Models and methods*. Norwood, NJ: Ablex.

Libby, R., Artman, J. T., & Willingham, J. J. (1985). Process susceptibility, control risk, and audit planning. *The Accounting Review, 60*, 212–230.

Lykken, D. T. (1979). The detection of deception. *Psychological Bulletin, 80*, 47–53.

Messier, W. F. (1983). The effect of experience and firm type on materiality/disclosure judgments. *Journal of Accounting Research, 21*, 611–618.

Mosier, K. L. (1997). Myths of expert decision making and automated decision aids. In C. Zsambok & G. Klein (Eds.), *Naturalistic decision making*. Hillsdale, NJ: Lawrence Erlbaum Associates.

Phelps, R. H. (1977). *Expert livestock judgment: A descriptive analysis of the development of expertise*. Unpublished doctoral dissertation, Kansas State University, Manhattan.

Phelps, R. H., & Shanteau, J. (1978). Livestock judges: How much information can an expert use? *Organizational Behavior and Human Performance, 21*, 209–219.

Raskin, D. C., & Podlesny, J. A. (1979). Truth and deception: A reply to Lykken. *Psychological Bulletin, 86*, 54–59.

Samelson, F. (1974). History, origin myth and ideology: "Discovery" of social psychology. *Journal for the Theory of Social Behavior, 4*, 217–231.

Schwartz, S., & Griffin, T. (1986). *Medical thinking: The psychology of medical judgment and decision making*. NY: Springer-Verlag.

Shanteau, J. (1992a). Competence in experts: The role of task characteristics. *Organizational Behavior and Human Decision Processes, 53*, 252–266.

Shanteau, J. (1992b). How much information does an expert use? Is it relevant? *Acta Psychologica, 81*, 75–86.

Shanteau, J. (1999). Decision making by experts: The GNAHM effect. In J. Shanteau, B. Mellers, & D. Schum, (Eds.), *Decision science and technology: Reflections on the contributions of Ward Edwards* (pp. 105–130). Norwell, MA: Kluwer Academic.

Slovic, P. (1969). Analyzing the expert judge: A descriptive study of a stockbroker's decision processes. *Journal of Applied Psychology, 53*, 255–263.

Stewart, T. R., Roebber, P. J., & Bosart, L. F. (1997). The importance of the task in analyzing expert judgment. *Organizational Behavior and Human Decision Processes, 69*, 205–219.

Titchener, E. B. (1916). *A beginner's psychology*. New York: Macmillan.

Trumbo, D., Adams, C., Milner, M., & Schipper, L. (1962). Reliability and accuracy in the inspection of hard red winter wheat. *Cereal Science Today, 7*, 62–71.

Tversky, A., & Kahneman, D. (1974). Judgment under uncertainty: Heuristics and biases. *Science, 185*, 1124–1131.

Yates, J. F. (1990). *Judgment and decision making*. Englewood Cliffs, NJ: Prentice-Hall.

Yates, J. F., Lee, J. W., & Bush, J. G. (1997). General knowledge overconfidence: Cross-national variations, response style, and "reality." *Organizational Behavior and Human Decision Processes, 70*, 87–94.

14

Representing Expertise

Laura G. Militello
Klein Associates, Inc.

This chapter proposes a scheme for categorizing Cognitive Task Analysis (CTA) methods according to the types of representations they offer. CTA can be defined as "a set of methods to elicit, explain, and represent the mental processes involved in performing a task" (Klein & Militello, in press). The representation or the output of the CTA is the view of expertise in the context of a specific task available to designers and developers of training, decision support systems, expert systems, and other applications. Clearly, the type of representation influences the conclusions reached and the eventual design of the end product.

Since 1985, the Naturalistic Decision Making (NDM) community has seen considerable growth in the number of CTA methods. This has led to increasing interest in finding ways to categorize existing CTA methods in order to understand which are most beneficial under which circumstances. Several categorization schemes have been proposed (Cooke, 1994; Hutton et al., 1997; Seamster, Redding, & Kaempf, 1997), rating CTA methods on multiple dimensions. The primary focus of these categorization schemes has been on the mechanics of conducting the CTA including the knowledge elicitation portion of the methods, resources required, amount of training required to conduct the methods, and so forth. In contrast, this chapter considers only one dimension, focusing specifically on the types of expertise elicited and represented using various CTA methods.

Because the ability to elicit and represent expertise is a critical measure of the utility of a CTA method for designing a usable system or successful training, I would suggest that this dimension warrants in-depth coverage. As the NDM community begins to evaluate CTA methods, define boundary conditions, and attempt to link specific methods to specific types of tasks or applications, focusing on the knowledge representation portion of the CTA is key to understanding what each method offers. The importance of documenting and representing expertise has been an impetus for the development of many CTA methods, but, surprisingly, has not been used as a means of comparison across the methods. Given the many fields and traditions from which these methods have arisen, it seems important to understand what different CTA methods offer for representing expertise.

FRAMEWORK

The framework described in this chapter is intended to be a starting point for discussion. The need for metrics or categories to aid in understanding the strengths and limitations of various CTA methods becomes increasingly important as more techniques become available. This framework focuses specifically on the view of expertise that is elicited and represented. The framework includes four categories of methods: expertise in context, conceptual links, operation sequences, and simulations of expert performance. Each is described in turn.

Expertise in Context

This set of CTA methods focuses on expertise in the context of a specific incident. By focusing on a specific incident, the investigator is able to obtain concrete, detailed information about perceptual cues and patterns of cues, as well as the implications of specific cues and cue sets. These techniques also allow the investigator to observe and ask questions about the decision-making process, judgments that are made, and problem-solving strategies at a level of depth that is not possible without an incident from which to work.

One important benefit of this approach is that one is able to examine expertise over the course of an incident, capturing expertise in a dynamic setting. The methods allow a view of the expert's changing perspective, understanding of the situation, and strategies as an incident unfolds. This view of the situation from the expert's perspective in the changing context of an incident provides insight into the expert's use of information, strategies, metacognition, and heuristics.

However, the focus on specific incidents and retaining detailed contextual elements may preclude the representation of abstracted mental models. It seems to be difficult to transition from detail to abstraction using these methods. This may be related to the dynamic nature of these representations and the static nature of mental model representations.

Examples of incident-based representations include the use of annotated text in which the Subject Matter Expert's (SME) explanation of the significance of and implications of specific cues, patterns, and events are interjected into the narrative of the incident (Fig. 14.1). A second example is the use of a timeline in which the SME's assessment is linked to specific events. The actions taken based on this assessment, as well as other options available and insights into expert/novice differences are also captured (Fig. 14.2).

Knowledge elicitation methods used to build these types of representations include the Critical Decision Method (Hoffman, Crandall, & Shadbolt, 1998; Klein, Calderwood, & MacGregor, 1989) and various simulation techniques[1] (Bell & Hardiman, 1989; Clarke, 1987; Hall, Gott, & Pokorny, 1995; Militello & Hutton, 1998). Critical Decision Method interviews are structured around real, lived experiences. The Subject Matter Expert (SME) is asked to recount a challenging incident and the interviewer uses focused probes to better understand the SME's decision making, judgments, and use of context in the incident. Simulation techniques incorporate observations and interview techniques. In these interviews the investigator supplies the incident in the form of a simulation or scenario. The investigator is able to both observe the SME working through the scenario and to interview the SME using cognitive probes.

Conceptual Links

These methods focus on representing a mental model of the task, including specific concepts and their relationships. These methods allow the investigator to explore goal hierarchies, spatial relationships, and causal networks. In order to adequately represent these relationships, a certain level of context is required. However, the focus here on representing a mental model requires a sufficient level of abstraction, which limits the amount of specific contextual information that can be included.

These methods focus on understanding how the expert organizes information. The representations emphasize organizing structures, completeness, and consistency. The representation is developed and refined during knowledge elicitation sessions. This focus on the representation during knowledge elicitation encourages the SME to stay in the present situation. The SME is asked to represent his or her understanding of a task, goal, or concept at a static point in time. No attempt is made to ask the SME to recall an incident from the past or to relate his or her interpretation of events or information at different points of time in different contexts.

---

[1]Knowledge elicitation techniques utilizing scenarios and simulations as a means to elicit expert knowledge should be distinguished from knowledge representation techniques that attempt to simulate expert performance. The latter are discussed in the Simulations of Expert Performance section.

CASE ACCOUNT	ELICITED KNOWLEDGE
The newborn in this incident was a large baby[1] who was on the unit because he was a drug withdrawal baby[2]. As I was about to restart the IV, the baby went into a seizure[4] and then into complete cardiac arrest[5]. Of course at that point he shut down completely so we couldn't get a line in.	

A 2-3 day old, full term baby.

Typically, drug withdrawal babies are jittery, scream constantly with a high-pitched cry, and are extremely irritable. They have difficulty sleeping, and often sleep for only 2-4 hours a day. This can continue for 3 to 4 weeks. A seizure[2a] is an extreme withdrawal symptom.

However, seizures are common in premature babies. Experienced NICU nurses are likely to have witnessed "hundreds." This was not a typical "premie" seizure.

The baby became rigid and paled out[4a].

A "pale" baby becomes gray/white. The loss of color happens first on the hands and feet.

During cardiac arrest, the veins clamp down. They constrict to the point that an IV needle cannot be inserted.

FIG. 14.1. Annotated text depicting SME's explanation of the significance of and implications of specific cues, patterns, and events in the incident narrative from an interview with a neonatal intensive care unit nurse.

Timeline of Events

TIME:	12:00:00	12:00:10	12:00:40	12:00:50	12:01:20	
EVENT:	Air tracker identifies 7 aircraft	ESM received from aircraft	PCF jumped in one aircraft	Missile signal received	Continuing missile signal	Aircraft break off
SA:	This is an unusual event	These are Soviet craft	Aircraft is preparing for missile lockon	Don't take provocative actions. "He's not going to fire"	Aircraft are less likely to fire	Incident ends
ACTIONS:		Cycled through all 7 different frequencies – used audio. Armed chaff launchers. Went from AECM standby to ON.	Told watch supervisor about missile	No actions taken		
OPTIONS:		1. Illuminate 2. Use ECM 3. Fire chaff 4. Get escorts from carrier		1. Illuminate 2. Use ECM (SLQ-32) 3. Fire chaff		
EXPERT/NOVICE DIFFERENCES:		Audio confusing for different freqs. Could have armed wrong launcher. Could have paid too much attention to chaff.	A novice would have been more likely to take one of the actions listed above. This would have been a mistake.			

FIG. 14.2. Timeline linking SME's assessment to specific events from interview with an electronic warfare technician.

The use of the representation in the knowledge elicitation session makes gaps and conflicts visible so that they can be addressed. This type of representation does not allow for contradictory or ambiguous states. The representation makes gaps and conflicts visible so that can be resolved by the SME. The use of mapping/graphing procedures allows the flexibility to represent concepts and relationships that are not linear. The resulting representation has an organizing structure, is complete, and resolves all inconsistencies.

It is clear organizing structure and lack of inconsistencies that makes these types of representations particularly useful in the development of artificial intelligence tools and expert systems. The clear links between tasks, concepts, and goals make it possible to abstract rules needed to drive the transition from a representation of expertise to software. The level of abstraction presented in these types of representation also lends itself to training development and can be used to organize courses or curriculum.

One limitation of these representations is that this level of abstraction does not allow for changing environments, unusual situations, or new information that may become available to the expert as a situation unfolds. Investigators are attempting to capture a view of the expert's understanding of a task, concept, or goal at a specific point in time. The focus here is on a view of the organization of concepts inside the expert's head. In contrast the *expertise in context* representations attempt to represent the ways in which experts understand, interpret, and act on information as it becomes available in the context of an unfolding incident.

Conceptual graph analysis (Gordon & Gill, 1992) and concept mapping (Gowin & Novak, 1984; McNeese et al., 1990) are two well-known techniques used to capture and represent mental models and conceptual links. Both techniques use a knowledge representation as a focus for the interview session. The representation is built, revised, and fleshed out based on interviews with SMEs. Conceptual graph analysis includes a set of specific probes aimed at eliciting goals, actions, events, rationales, implications, consequences, and properties. Goals, events, concepts, and states are specified via textual descriptors, circled, and represented as nodes in the graph. Directional arrows are drawn connecting nodes and defining the relationship between the nodes. The resulting representation (Fig. 14.3) consists of nodes and arcs depicting causal relationships, goal hierarchies, and taxonomic structures. Concept mapping advocates less structure both in the knowledge elicitation sessions and in the resulting representation. Although the resulting representation bears a strong resemblance to a Conceptual graph, the content of each node is not categorized (i.e., goal vs. event vs. concept vs. state) and relationships between nodes may or may not be defined (Fig. 14.4).

Operation Sequences

These representations capture goal hierarchies, timelines, function sequences, and interactions. Tasks are broken into component operations and represented in

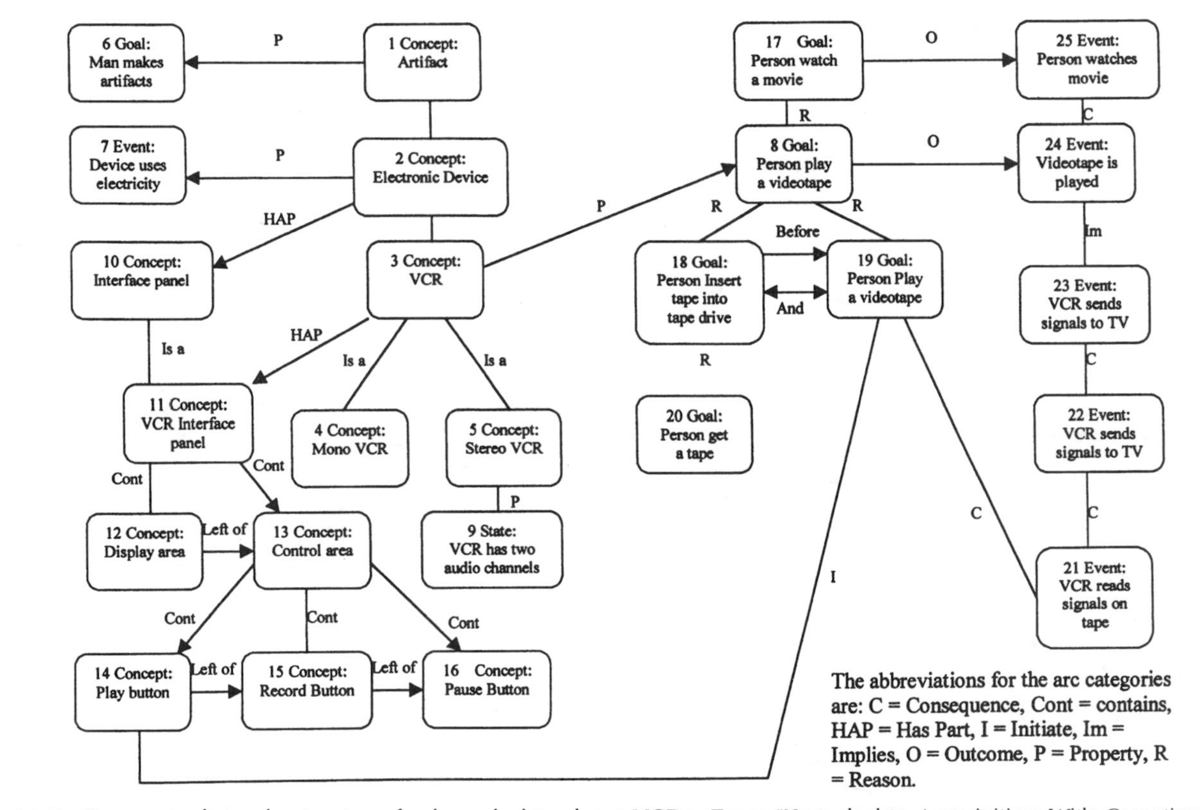

FIG. 14.3. Conceptual graph structure for knowledge about VCRs. From "Knowledge Acquisition With Question Probes and Conceptual Graph Structures," by H. W. Gordon and R. T. Gill, 1992, in T. Laurer, E. Peacock, and A. Graesser (Eds.), *Questions and Information Systems*. Hillsdale, NJ: Lawrence Erlbaum Associates. Copyright 1992 by Lawrence Erlbaum Associates. Reprinted with permission.

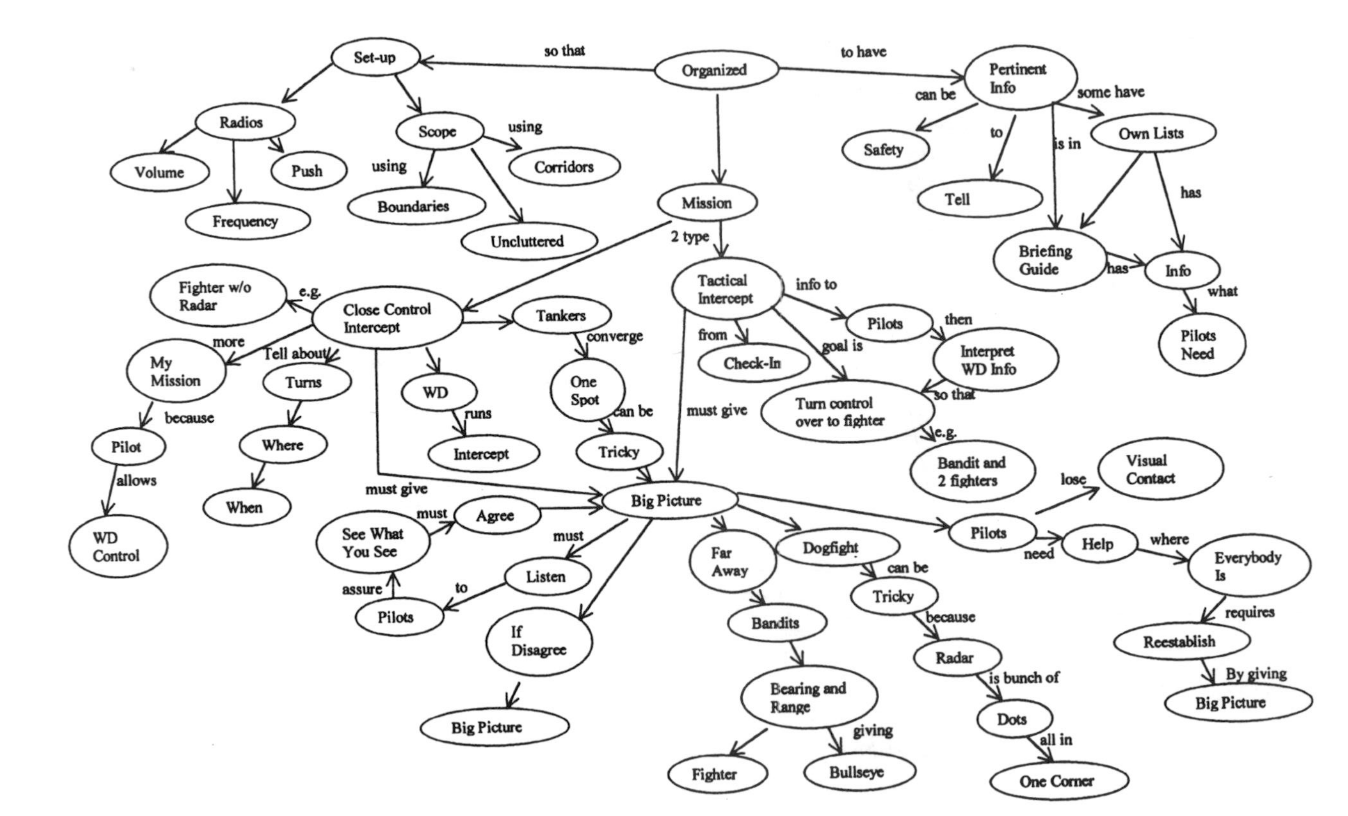

FIG. 14.4. Concept map of weapons director task. From *Designing for Performance: A Cognitive Systems Engineering Approach to Modifying an AWACS Human Computer Interface* (p. 15), by D. W. Klinger, S. J. Andriole, L. G. Militello, L. Adelman, & G. Klein, Tech Rep. No. AL/CR-TR-1993-0093, 1993, Fairborn, OH: Klein Associates Inc. Copyright 1993 by Klein Associates. Reprinted with permission.

hierarchical sequences. The component parts of the task are presented sequentially according to the order in which they are implemented in order to complete the task. Furthermore, the component parts can be broken down into increasingly detailed components, from higher level operations to tasks required to complete the operations to subtasks required to complete the tasks, and so on. Resulting representations tend to be generic, incorporating little context.

These representations tend to be at a high level of abstraction and are, therefore, highly generalizable. They have been used extensively to drive training and to provide a framework for understanding the many people and machines that contribute to the accomplishment of complex processes, particularly in manufacturing environments and process control tasks.

Although this flexibility is appealing, the lack of context limits the utility of these representations beyond serving as a larger, organizing structure. Although it is possible to use these methods to compare expert and novice goal hierarchies, action sequences, timelines, and so forth, few examples exist.

Operation sequence techniques include hierarchical task analysis, timeline analysis, work flow analysis, and many others. Several examples of these types of representations are found in Kirwan & Ainsworth (1992). Figure 14.5 depicts a hierarchical task analysis, which breaks down the steps required to operate an overhead projector. Figure 14.6 depicts a timeline analysis, specifying time critical tasks and predictable pauses between steps. The representation provides a clear view of synchronization and timing issues. This type of representation is most useful for settings in which there is a fixed range of tasks and in which operators have clearly defined roles. These techniques may identify decision points, important judgments, goals, and other cognitive aspects of the task and the point at which these elements become important in accomplishing the task. However, little or no information is elicited or represented regarding how the expert approaches and handles these challenges. Cognitive requirements of the task are depicted as goals, tasks, and operations within the context of the sequential hierarchy. Knowledge elicitation methods consist of observation techniques and interview techniques that aid experts in decomposing the task.

Simulations of Expert Performance

These methods use computer models to represent expertise within a specific task. Some methods build models based on probability assessment and value weights established by SMEs in addition to measures of reaction time in actual or simulated events, questionnaires, and interviews. Other methods such as COGNET[2]

[2]The primary focus of COGNET has been to elicit and represent the underlying information processing mechanisms, the acquired experience and expertise an SME brings to the tasks, and the context in which the task takes place. Although the COGNET simulation software is a recent advancement, we focus on it here as an example of a simulation of expert performance.

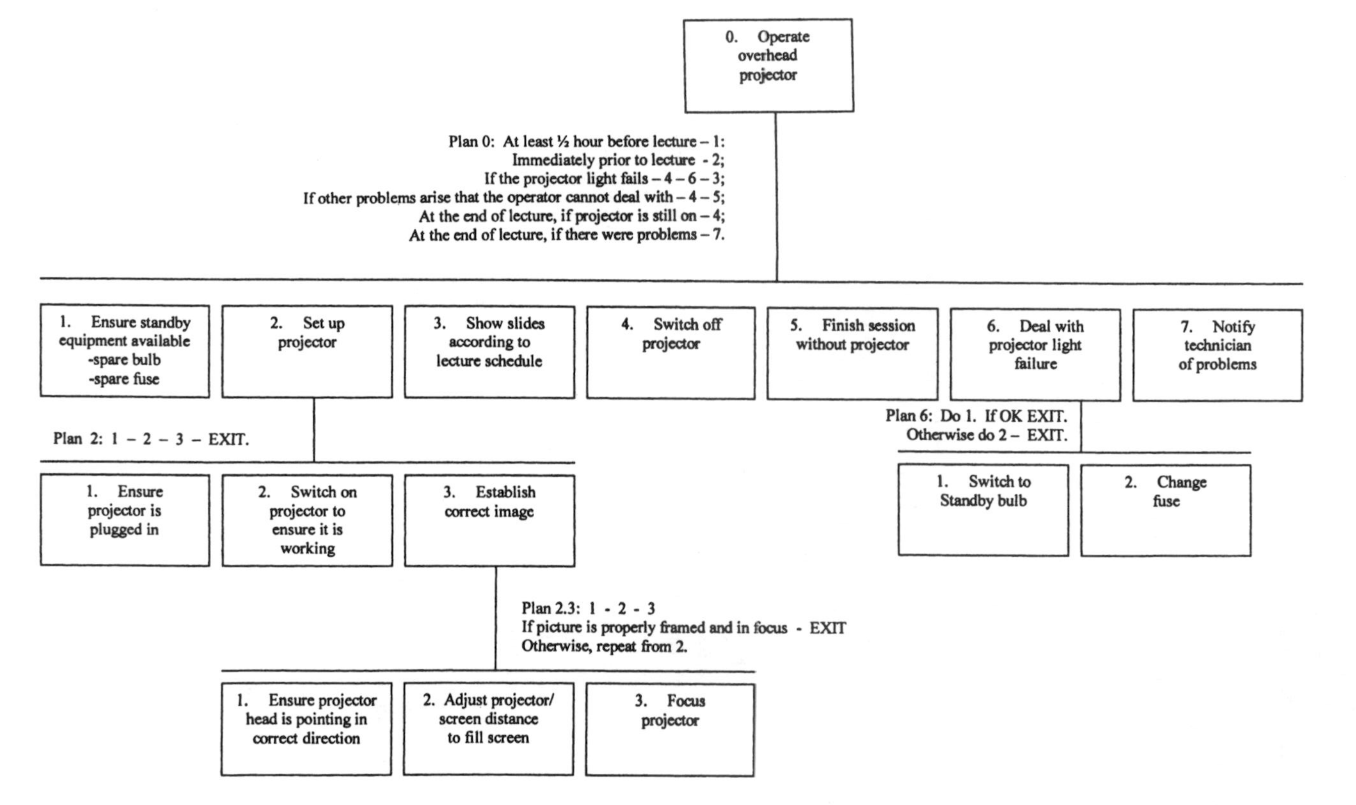

FIG. 14.5. Hierarchical task analysis for operating an overhead projector. From *A Guide to Task Analysis* (p. 106), by B. Kirwan and L. K. Ainsworth (Eds.), 1992, London, England: Taylor and Francis. Copyright 1992 by Taylor and Francis. This figure was originally developed by Dr. Andrew Shepherd. Reprinted with permission.

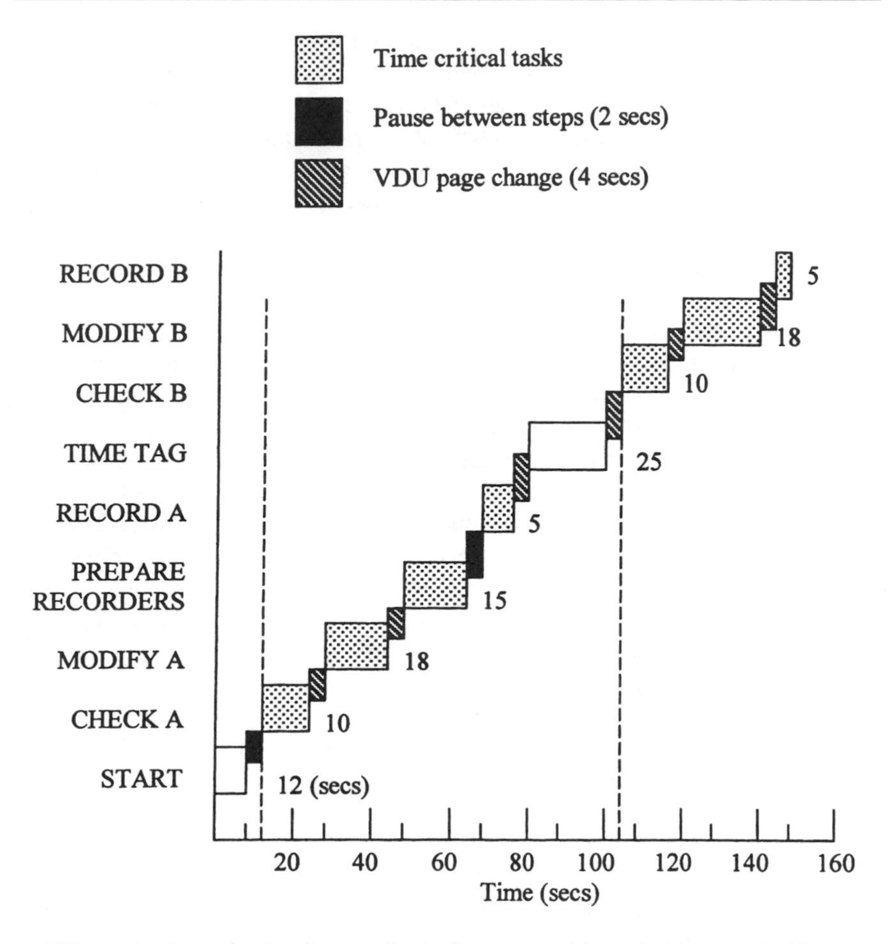

FIG. 14.6. Sample timeline analysis for a complex task. From *A Guide to Task Analysis* (p. 137), by B. Kirwan and L. K. Ainsworth, (Eds.), London, England: Taylor and Francis. Copyright 1992 by Taylor and Francis. Reprinted with permission.

(Zachary et al., 1998), developed to represent cognitively complex tasks, rely on interviews with SMEs immediately following performance in a simulated incident to build the model of expert performance. These methods allow investigators to represent human performance in the context of a specific task in a software format. The representation is a computer model that simulates expert performance.[3]

Simulations of expert performance present a thorough model of the task. Goals, subtasks, sequencing, resources used, and restrictions are all captured. The

[3]Simulations as a tool in knowledge elicitation are discussed in the Expertise in Context section of this chapter.

model includes a description of the information the expert attends to, the procedures or actions the expert invokes, and how the expert prioritizes actions. The model relies on observable elements of expertise (i.e., reaction times, eye scan measures, and keystroke analysis) augmented by the expert's explanation of the recorded observable event. Based on this model, predictions and simulations are generated depicting the functioning of individual operators and the system as a whole.

Although these methods are very powerful in terms of representing action sequences, including the cognitive challenges that might drive the actions, the complexity of the expert's cognition is not represented. For example, the resulting simulation might depict the individual steps, sequencing, and information used to solve a problem. The simulation will not reveal the expert's interpretation of that information, metacognitive strategies, or the use of heuristics.

It is important to point out that some of these software simulation-based representations rely on CTA data that do articulate these cognitive processes. The cognitive processes may even be visible in interim representations such as the COGNET blackboard structures described later in this section. However, the simulation software does not make these aspects of cognitive performance visible in a simulation run.[4] Other methods such as MicroSaint have recently attempted to provide flexible representations that make visible instantiations of specific decision strategies such as RPD (Archer, Warwick, & Oster, 2000; Warwick et al., 2000, in prep).

Examples of computer simulation techniques include COGNET (Zachary, et al., 1998) and MicroSaint (Micro Analysis and Design, 1996). COGNET is theoretically driven, building on information processing theory. Many of these types of techniques rely on observations including reaction time measures such as keystroke analysis, combined with interviews and questionnaires. MicroSaint focuses primarily on actions and action sequences combined with human performance data to design a workspace.

Although it is not possible to include a sample computer simulation as part of this chapter, we can examine the intermediate representations used to build a COGNET[5] (Fig. 14.7). These include a blackboard structure, depicting a mental model of the task, combined with a task list. For each of the elements included in the blackboard, blackboard panels are created containing more specific information such as critical attributes of the element. These representations drive the computer simulation.

[4]This is a debatable issue in the NDM community. Further discussion of this issue will be important to future efforts.

[5]These are simplified examples used merely to illustrate a point. COGNET is generally used to represent complex tasks that require much more in-depth blackboard structures and task lists.

Blackboard Structure

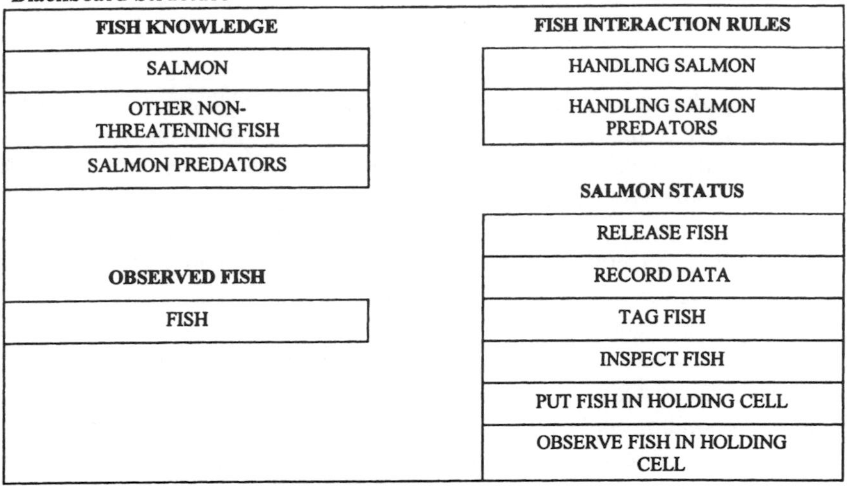

FISH KNOWLEDGE	FISH INTERACTION RULES
SALMON	HANDLING SALMON
OTHER NON-THREATENING FISH	HANDLING SALMON PREDATORS
SALMON PREDATORS	
	SALMON STATUS
	RELEASE FISH
OBSERVED FISH	RECORD DATA
FISH	TAG FISH
	INSPECT FISH
	PUT FISH IN HOLDING CELL
	OBSERVE FISH IN HOLDING CELL

Task List

- Attend to new fish
- Protect Salmon from Predators
- Record fish information

Blackboard Panels

Fish Knowledge
- Salmon
  - *Attributes*: shape, color, pattern, maximum and minimum size, maximum and minimum weight, maximum and minimum length, species, spawning, and sex
- Other Non-threatening Fish
  - *Attributes*: shape, color, pattern, maximum and minimum size, maximum and minimum weight, maximum and minimum length, species
- Salmon Predators
  - *Attributes*: shape, color, pattern, maximum and minimum size, maximum and minimum weight, maximum and minimum length, species, relative danger rating

Observed Fish
- Fish
  - *Attributes*: shape, color, pattern, ID, weight, length, sex, tag number, tagged status, species, attended status, time in holding tank

FIG. 14.7. COGNET sample representations for fish sorting task. From *Integrated Cognitive/Behavioral Task Analysis in Real-Time Domains*, by W. W. Zachary & J. M. Ryder, 1997, in workshop presented at the 41st Annual Meeting of the Human Factors and Ergonomics Society, Albuquerque, NM. Copyright 1992 by CHI Systems. Reprinted with permission.

APPLICATIONS OF ALTERNATIVE REPRESENTATIONS

The next step after defining each category is to discuss which representations are most useful for different types of applications. Clearly, in most cases a combination of methods would be optimal. However, NDM researchers are generally working on applied problems, which are constrained by time, money, SME availability, and so forth. Furthermore, few researchers have the training and experience needed to conduct CTA methods resulting in representations from all four categories. Therefore, observations made here regarding links between knowledge representation and application assume one does not have time to exhaustively research the task from all angles, using multiple methods, and an unlimited supply of SMEs. Instead, this discussion focuses on the strengths of each representation for building applications.

The first category, expertise in context, is best suited for applications for which it is important to understand the contextual elements that influence the task as well as the decision-making processes of the operator or user. These representations provide insight into how the operator uses contextual information to assess a situation, maintain situation awareness, and monitor his or her own performance.

Methods in the expertise in context category can be used to drive training. These methods have been used to make visible the critical decisions and judgments students should be exposed to, as well as making the experienced decision makers' processes and strategies available to students (Kaempf, Wolf, Thordsen, and Klein, 1992; Cohen, Freeman, Wolf, and Militello, 1995). Furthermore, these representations have been used to develop training that provides students information about how to interpret specific cues and patterns of cues (Crandall & Gamblion, 1991).

Expertise in context representations have also been used to provide the content and frame for decision support systems, including a system to help engineers create bids for machined parts (Klinger, 1994) and a system to aid targeteers in assessing the effectiveness of specific munitions for particular targets (Miller & Lim, 1993). A third type of application for which these representations have been found to be useful is to guide interface design and redesign. Successful efforts include the redesign of the Tactical Action Officer interface in the command information center of an AEGIS cruiser (Kaempf, Wolf, Thordsen, & Klein, 1996) and the design of screens to aid Landing Signal Officers in the difficult task of assessing whether an aircraft is in position for a safe landing on an aircraft carrier (Thordsen, 1998).

One limitation of expertise in context representations is that it is difficult to move from a detailed, incident-based representation to a more abstract mental model. These methods do not attempt to depict thoroughly the expert's structure for organizing a set of concepts and relationships. Although the expert's understanding of specific concepts and relationships may be represented within an incident, no attempt is made to link concepts together into a larger, explicit organizing structure.

In contrast, if the application requires an expert's mental model of a task, function, or piece of equipment, the conceptual links representations are the most useful. These methods can be used to provide a frame or an advance organizer for training, expert system, or decision support development. The representation is a format that is relatively easy to translate into computer code. The use of a unified and consistent representation of knowledge without gaps or contradictions is key if the intent is to try to mimic or predict expert performance in a computer system.

Conceptual graph analysis has been used to design training in physics and engineering mechanics and to develop expert systems in forestry and education (Gordon & Gill, 1992). Concept mapping was used in conjunction with the Critical Decision method to redesign the weapons director station on the Airborne Warning and Control Systems aircraft (Klinger & Gomes, 1993).

One limitation of these types of representations is that, when used exclusively, it is difficult to incorporate unexpected events, elements, and cues. Conceptual links representations tend to present a snapshot (or a series of snapshots at best), making it difficult to predict performance in a dynamic environment.

The third category, sequential hierarchies, is most useful if the application requires a detailed task description depicting the sequence of events and actions that must occur for the task to be completed. These methods have been used extensively to design training, and in some cases, to drive system development (Meister, 1989).

The lack of context included in these representations is a limitation for representing cognitively complex tasks with these types of methods. The representation does not differentiate straightforward elements with predominantly behavioral components from cognitively complex elements. This can lead to difficulty in extracting the cognitive elements from the hierarchical representation. These representations do not provide a means for representing specific perceptual discriminations, an expert's interpretation of information within a particular context, or metacognitive strategies used by experts in making decisions, judgments, and so forth.

The primary strength of the fourth category, simulations of expert performance, is that the representation is a dynamic model. Based on a mental model and a task list, one can actually play an incident forward in time. A researcher can view the operator's performance as certain environmental variables are changed.

Simulations of expert performance have been used to predict workload to drive task assignment and crew utilization issues (i.e., Kuperman & Perez, 1988). MicroSaint has been used extensively in manufacturing environments to design workspaces (Micro Analysis & Design, 1996). Efforts have been made to adapt MicroSaint methods for use with more cognitive tasks. COGNET was developed specifically to model cognitive tasks and has been applied to model the Anti-Air Warfare Coordinator position in the command information center of an AEGIS cruiser (Zachary et al., 1998). COGNET applications include an improved interface for oncologists retrieving information from PDQ and CANCERLIT databases (Zachary et al., 1996).

The primary limitation of these representations, as with any dynamic simulation, is that the variables the simulation can handle have to be identified a priori. Thus the simulation can represent a reasonable set of circumstances, but it is impossible to anticipate all the variables that might influence performance. The resulting representation, although dynamic, represents only that subset of variables envisioned by the researchers.

DISCUSSION

Based on this categorization scheme, two dimensions within expertise can be abstracted. The first relates to the level of contextual detail elicited and retained in the representation. The methods in the category expertise in context retain a contextual richness that is not captured with other methods. In contrast, the methods in the conceptual links and simulations of expertise categories sacrifice contextual detail for a level of abstraction that allows a much more visible organizing structure and a representation of an expert's mental model. Sequential hierarchy methods retain the least amount of context.

The second dimension concerns the view of expertise that is available. Methods in the conceptual links category have been developed with understanding and representing mental models as an objective. Simulations of expertise are often built based on mental models. These methods seek to depict an expert's mental frame at a static point in time, so that predictions can be made and simulations can be run based on that framework. The expertise in context methods instead seek to capture the expert's cognitive processes as an incident unfolds. The representation is an actual record of the expert's decision making, assessments, judgments, and so forth in a lived or simulated incident. This view depicts more dynamic aspects of expertise (as opposed to the simulations of expertise category, which depicts dynamic predictions of expert performance based on mental models). Task decomposition methods provide an outline of the behavioral and cognitive aspects of the task in which experts operate.

This chapter presents a proposed framework for categorizing CTA methods according to the ways in which expertise is elicited and represented. Given the importance of understanding expert cognition in designing effective training and developing useful systems, closer examination of the utility of CTA methods in addressing aspects of expertise is warranted. As more CTA methods become available, it will be increasingly important to understand the strengths and limitations of each and to link them to potential applications. Developing and articulating relevant dimensions for rating CTA methods will be a key part of this process.

REFERENCES

Archer, S., Warwick, W., Oster, A. (2000). Current efforts to model human decision making in a military environment. *Proceedings of Advanced Simulation Technologies Conference* (ASTC), Washington, DC (Apr 16–20). (pp. 151–155).

Bell, J., & Hardiman, R. J. (1989). The third role—The naturalistic knowledge engineer. In D. Diaper (Ed.), *Knowledge elicitation: Principles, techniques, and applications* (pp. 49–85). New York: Wiley.

Clarke, B. (1987). Knowledge acquisition for real-time knowledge-based systems. *Proceedings of the First European Workshop on Knowledge Acquisition for Knowledge Based Systems.*

Cohen, M. S., Freeman, J., Wolf, S., & Militello, L. (1995). Training metacognitive skills in naval combat decision making (Tech. Rep. for Contract No. N61339-92-C-0092). Arlington, VA: Cognitive Technologies.

Cooke, N. J. (1994). Varieties of knowledge elicitation techniques. *International Journal of Human-Computer Studies, 41*, 801–849.

Crandall B., & Gamblian V. (1991). *Guide to early sepsis assessment in the NICU* [Instruction manual prepared for the Ohio Department of Development under the Ohio SBIR Bridge Grant program]. Fairborn, OH: Klein Associates.

Gordon, H. W., & Gill, R. T. (1992). Knowledge acquisition with question probes and conceptual graph structures. In T. Laurer, E. Peacock, & A. Graesser (Eds.), *Questions and information systems* (pp. 29–46). Hillsdale, NJ: Lawrence Erlbaum Associates.

Gowin, D. B., & Novak, J. D. (1984). *Learning to learn.* New York: Cambridge University Press.

Hall, E. M., Gott, S. P., & Pokorny, R. A. (1995). *A procedural guide to cognitive task analysis: The PARI method.* (AL/HRM Tech. Rep. AL/HR-TR-1995–0108). Brooks AFB, TX: USAF Armstrong Laboratory.

Hoffman, R. R., Crandall, B. W., & Shadbolt, N. R. (1998). A case study in cognitive task analysis methodology: The critical decision method for the elicitation of expert knowledge. *Journal of Human Factors, 40*(2), 254–276.

Hutton, R. J. B., Pliske, R. M., Klein, G., Klinger, D. W., Militello, L. G. M., & Miller, T. E. (1997). *Cognitive engineering implications for information dominance* (Contract No. F41624-97-C-6014 for Armstrong Laboratory, Wright Patterson Air Force Base, OH). Fairborn, OH: Klein Associates.

Kaempf, G.L., Wolf, S., Thordsen, M.L., & Klein G. (1992). *Decision making in the AEGIS combat information center* (Contract N66001-90-C-6023 for the Naval command, Control and Ocean Surveillance Center, San Diego, CA). Fairborn, OH: Klein Associates.

Kaempf, G. L., Klein, G., Thordsen, M.L., & Wolf, S. (1996). Decision making in complex command-and-control environments. *Human Factors Special Issue, 38*, 220–231.

Kirwan, B., & Ainsworth, L. K. (1992). *A guide to task analysis.* London: Taylor and Francis Ltd.

Klein, G. A., Calderwood, R., & MacGregor, D. (1989). Critical decision method for eliciting knowledge. *IEEE Transactions on Systems, Man, and Cybernetics, 19*(3), 462–472.

Klein, G., & Militello, L.G. (in press). Some guidelines for conducting a cognitive task analysis. In E. Salas (Ed.), *Human/Technology Interaction in Complex Systems, IO.* Stamford, CT: JAI, Press.

Klinger, D.W. (1994). A decision centered design approach to case-based reasoning: Helping engineers prepare bids and solve problems. In P. T. Kidd & W. Karwowski (Eds.), *Advances in Agile Manufacturing* (pp. 393–396). Manchester, England: IOS Press.

Klinger, D. W., Andriole, S. J., Militello, L. G., Adelman, L., Klein, G., Gomes, M. E. (1993). *Designing for performance: A cognitive systems engineering approach to modifying an AWACS human-computer interface* (Contract AL/CF-TR-1993-0093 for Department of the Air Force, Armstrong Laboratory, Air Force Materiel Command, Wright-Patterson AFB, OH). Fairborn, OH: Klein Associates Inc.

Klinger, D. W., & Gomes, M. G. (1993). A cognitive systems engineering applications for interface design. *Proceedings of the Human Factors and Ergonomics Society 1993 Annual Meeting*, 16–20. Santa Monica, CA: Human Factors and Ergonomics Society.

Kuperman, G. G., & Perez, W. A. (1988). A framed-based mission decomposition model. *Proceedings of the Human Factors and Ergonomics Society 32nd Annual Meeting* (pp. 135–139). Santa Monica, CA: Human Factors and Ergonomics Society.

McNeese, M. D., Zaff, B. S., Peio, K. J., Snyder, D. E., Duncan, J. C., & McFarren, M. R. (1990). *An advanced knowledge and design acquisition methodology: Application for the pilot's associate.* (Final Rep. AAMRL-TR-90-060). Wright-Patterson Air Force Base, OH: Armstrong Aerospace Medical Research Laboratory.

Meister, D. (1989). *Conceptual aspects of human factors.* Baltimore, MD: The Johns Hopkins University Press.

Micro Analysis and Design. (1996). *Micro Saint User's Manual for Windows, Version 2.0.* Boulder, CO: Micro Analysis & Design Simulation Software.

Militello, L. G., & Hutton, R. J. B. (1998). Applied Cognitive Task Analysis (ACTA): A practitioner's toolkit for understanding cognitive task demands. *Ergonomics Special Issue: Task Analysis, 41*(11), 1618–1641.

Miller, T. E., & Lim, L. S. (1993). *Using knowledge engineering in the development of an expert system to assist targeteers in assessing battle damage and making weapons decisions for hardened-structure targets* (Contract DACA39-92-C-0050 for the U.S. Army Engineers CEWES-CT, Vicksburg, MS). Fairborn, OH: Klein Associates.

Seamster, T. L., Redding, R. E., & Kaempf, G. L. (1997). *Applied cognitive task analysis in aviation.* Aldershot, England: Avebury Aviation.

Thordsen, M. (1998). Display design for Navy landing signal officers: Supporting decision making under extreme time pressure. In E. Hoadley & I. Benbasat (Eds.), *Proceedings of the Fourth Americas Conference on Information Systems* (pp. 255–256). Madison, WI: Omnipress.

Warwick, W., Hutton, R., McDermott, P., & McIlwaine, S. (2000, in preparation). Improving Computer-Generated Forces through Modeling Recognitional Decision Making Strategies. Boulder, CO: Micro Analysis and Design.

Zachary, W. W., Ryder, J. M., & Hicinbothom, J. H. (1998). Cognitive task analysis and modeling of complex decision making. In J. A. Cannon-Bowers & E. Salas (Eds.), *Decision making under stress: Implications for training and simulation* (pp. 315–344). Washington, DC: American Psychology Association.

Zachary, W. W., & Ryder, J. M. (1997). Integrated Cognitive/Behavioral Task Analysis in Real-Time Domains. Workshop presented at the 41st Annual Meeting of the Human Factors and Ergonomics Society, Alburquerque, NM.

Zachary, W.W., Szczepkowski, M.A., Le Mentec, J.C., Schwartz, S., Miller, D. & Bookman, M.A. (1996). An interface agent for retrieval of patient-specific cancer information. *Proceedings of the Human Factors and Ergonomics Society 40th* Annual Meeting (pp. 742–746). Santa Monica, CA: Human Factors and Ergonomics Society.

15

Searching for Evidence: Knowledge and Search Strategies Used by Forensic Scientists

Jan Maarten Schraagen
TNO Human Factors
Henk Leijenhorst
Forensic Science Laboratory

The Forensic Science Laboratory of The Netherlands is suffering from a growing backlog due to an increasing number of cases, which results in long delivery times of research reports within a number of departments. The project, "Strategies for Searching Trace Evidence," was started by the Forensic Science Laboratory in order to increase the effectiveness and efficiency of search strategies, with the ultimate aim of achieving shorter delivery times.

The Forensic Science Laboratory has requested the assistance of TNO Human Factors in carrying out the Search Strategies project. TNO's contribution consisted of analyzing and, where necessary, formalizing and optimizing the search strategies used by experienced analysts.

SEARCH STRATEGIES

It was expected initially that by formalizing the strategy of experienced staff, this strategy could be incorporated more easily in procedures and would be more transferable to less experienced staff. The assumption behind this initial expectation was that the distinguishing feature between experts and novices in the area of visual search was the experts' use of better search strategies. In this view, better search strategies are characterized by a more systematic scan pattern, resulting in fewer misses. However, a quick scan of the literature showed that this assumption could not be maintained. In a study of inspection of car tires for defects, Noro (1984) showed that, contrary to what was expected, the more experienced inspector missed more defects than the less experienced inspector, due to excessive speed of visual and tactile search.

It is well-known that when people search the environment for a particular target, and they make both eye and head movements, top-down strategies control the visual scanpath (see, e.g., Lévy-Schoen, 1981). This line of reasoning is supported by the literature on visual inspection strategies. An early study by Schoonard, Gould, and Miller (1973) on visual scan patterns on defective microchips showed that better inspectors had fewer eye fixations per microchip and were highly selective in where they searched on the microchip. These results point to the importance of top-down search strategies. A later study by Abernethy (1990) on visual search patterns of novice and expert squash players showed that experts were able to use cues of the opponent's arm and racquet before the ball was hit, whereas beginners could only use these cues after the ball was hit. By using these cues very early on, experts were able to predict the ball's trajectory, whereas novices were not, thus enabling the experts to anticipate the ball's final position. The search patterns and the type of cues used were the same for both groups, but they differed in the information extracted from the search process. Apparently, experts had developed extensive knowledge of the relationship between cues and the outcome of strokes. In fact, the ability to predict what is going to happen based on knowledge of a class of situations or an explanatory "story" seems to be a more general aspect of perceptual-cognitive expertise (see Klein & Hoffman, 1993, for a review). Pennington and Hastie (1993) developed a theory of explanation-based decision making in which the "story" plays the role of an intermediate summary representation of the available evidence. The story is matched with a set of alternatives, and the alternative that best matches the story is chosen.

The theoretical drivers behind the present study emphasize the cognitive constructs that are developed by forensic scientists rather than their visual scanning patterns used. Based on the available literature, we hypothesize that forensic scientists also develop a story that acts as a causal model to explain the presence or absence of traces on the exhibits they are confronted with. This story directs the forensic scientist in his or her search process, in that it determines the relative importance of the exhibits, where to search on the exhibit, when to stop, and what traces to preserve. The story also justifies why particular traces are not being preserved. We also hypothesize that the ability to develop a story is what distinguishes experts from novices.

BEHAVIOR-SHAPING CONSTRAINTS

Search strategies used by analysts serve a particular purpose. This paragraph identifies the behavior-shaping constraints put on the analyst by the organization of his or her work. This constraint-based approach should be contrasted with an instruction-based approach: the former identifies the limits on goal-directed behavior whereas the latter tries to predict that behavior (cf., Vicente, 1999). With constraint-based approaches, guidance only about the goal state and the constraints on action are provided; guidance about how the task should be accomplished, as in instruction-based approaches, is not. Constraint-based approaches to task analysis provide more worker discretion than instruction-based approaches and, hence, more opportunities for learning. The Forensic Science Laboratory implicitly adopted an instruction-based approach to task analysis at the start of the project. That is, the Laboratory believed that a temporally ordered sequence of actions could be described that are required to complete the task. The advantage of such an approach is that more guidance is provided to workers, with less chance of human errors. However, as discussed by Vicente (1999), the instruction-based approach is only feasible for closed systems that are completely isolated from their environments. Systems that are substantially open to unpredictable disturbances are less amenable to an instruction-based form of task analysis. Because there are unpredictable external disturbances acting on the system, it will not be possible to accurately pre-identify the different flow sequences or timelines that lead to the satisfaction of the goal. The important question to resolve at the outset, therefore, becomes whether the forensic scientist is dealing with an open or a closed system.

The Forensic Science Laboratory searches for the true circumstances surrounding a criminal offence. Orders to carry out such research are given mostly by the technical detectives of the regional police forces. The police hand over so-called *exhibits* to the Laboratory, for instance: pieces of clothing, weapons and ammunition, and samples of handwriting. The police want to know whether particular traces can be found on these exhibits and whether these traces can be linked to a suspect or a victim (often after subsequent DNA profiling). When such a linkage can be established, this is often referred to as the *criminological triangle*, as three elements are involved: exhibit, suspect, and victim. For instance, a trace of blood of a suspect found on a victim's clothes may be used as incriminating evidence in a courtroom.

The system a forensic scientist is dealing with has, at first sight, all the characteristics of a closed system: the exhibits on which traces have to be searched are not subject to external disturbances. However, the forensic scientist works within a system larger than just the exhibits. This can be illustrated by the cases in which either no traces or lots of traces are found. In the first case, it often occurs that with further information on the circumstances surrounding the criminal offense, traces are found after a second round of inspection on particular parts of clothing, for instance. In the second case, a decision has to be made about which traces to preserve for further (DNA) investigation. As DNA profiling is expensive in both time and money, the forensic scientists are acutely aware of the importance of not overloading the DNA department with traces that could all have the same origin. Instead, they need to be highly selective and preserve only the most promising traces, while ignoring the less promising ones. These examples show that the work of a forensic scientist is more than just searching exhibits for evidence. If this was the case, one could teach a novice a standard procedure for searching exhibits, and this would be sufficient for successfully carrying out the job. However, this instruction-based approach is not feasible, because a lot of decisions have to be made concerning where to search, when to stop, and what traces to preserve. These decisions require knowledge of the circumstances surrounding the criminal offense. As the forensic scientist receives only little information from the police regarding these circumstances, he or she needs to make a lot of inferences. The constraints put on the forensic scientist are, therefore, those that impact on the ability to develop a story and are not so much on the exhibits themselves. When asked, experienced forensic scientists will confirm that a case gets more difficult as less information is available concerning the circumstances of the criminal offense. Using several knowledge elicitation techniques originally put forward by Hoffman (1987), we tried to constrain the forensic scientists in the information and time available. These techniques are discussed in more detail next.

METHOD

Participants

Participants were two experienced forensic scientists and one intern, all working at the Department of Serology. This Department mostly deals with sexual offenses and murder cases and, therefore, searches for blood, semen, vaginal excretion, and saliva. The experienced participants had been working at this Department for 4 and 6 years, respectively; the intern, for 6 months.

Case Description

The case was constructed for the purposes of the present research. It was a rape case with concomitant violence. Participants were given a short description stating that the male suspect raped a female victim in the back of her car,which was parked at a parking lot of a sports center. The victim vehemently resisted and was stabbed with a knife, as a result of which she was wounded. The suspect also was wounded. After photo recognition, the suspect was apprehended. A knife was found in his possession. Because the victim had thoroughly cleansed herself after the rape, no traces of semen could be recovered on the victim herself.

Eleven traces of blood and semen were placed on 10 exhibits. The victim's exhibits were a coat, jeans, t-shirt, underwear, and debris of the fingernails; the suspect's exhibits were: sweater, jeans, shoe, and knife. Finally, a plaid blanket (a cloth with a tartan pattern) was recovered from the back seat of the victim's car. The most important exhibits, containing foreign traces, were the victim's jeans (containing blood and semen) and her underwear (semen); the suspect's sweater (containing blood) and knife (blood); and the plaid blanket (containing blood and semen).

Apparatus

The participants were videotaped during their work. The videotapes were analyzed using The Observer, an exploratory sequential data analysis tool (marketed by Noldus Information Technology, Inc.). The Observer enables real-time coding of videotapes according to a predefined coding scheme (see later). Subsequently, various data analysis procedures may be applied for uncovering regularities in the data.

Tasks

Participants were requested to carry out the following tasks, in order:

1. Forming expectations in a limited-information task.

   Participants read the short description of the criminal offense sentence by sentence. After each sentence, they were asked to verbalize all thoughts and expectations regarding the events that occurred and the typical exhibits that could be expected. This task was included to elicit expectations that would be missed if participants read all 10 sentences without intermediate prompting.
2. Ordering of exhibits.

   Participants were shown the list of exhibits and were asked to rank order them in terms of the likelihood of discovering traces. The exhibits themselves were not shown. This task was included to elicit knowledge of typical exhibits given a sexual offense, such as the one presented.
3. Screening of exhibits within a limited time.

   Participants were given 10 minutes in total to inspect the exhibits and to determine what traces, if any were found, to preserve. This task was included to test the hypothesis that if top-down strategies were the driving force behind the forensic scientist's search behavior, little effect of time pressure would be found because the forensic scientist would know immediately where to search.

4. Exhaustive search (familiar task).

   After the screening task, participants could proceed with their regular search procedures and take as much time as they needed. Participants were asked to think aloud, and probe questions were asked when attention shifted to different parts of the exhibits. These probe questions were limited to the actual search process.

5. Retrospective probed recall.

   The day after the search process, the video recordings were discussed with the participants in order to clarify any remaining obscurities. General organizational issues were also raised.

Procedure

Participants were notified in advance that as soon as a suitable case arrived, they would be observed dealing with this case. They were under the impression they were dealing with an actual case. On the first day, Tasks 1 to 4 were carried out. On the second day, the retrospective probed recall task was carried out. Participants were fully debriefed at the end and requested not to talk with their colleagues about the case.

Coding Scheme

A coding scheme was developed based on a video recording of a pilot session. The following categories were used: object (i.e, the exhibit), person (suspect, victim), behavior (e.g., administration, testing, searching by eye, by microscope, etc.), object modifier (parts of exhibits, e.g., the pockets of a pair of jeans, the sleeves of a sweater, etc.), and test modifier (test of blood or semen). Every observable behavior was time-stamped and coded with a unique combination of categories. For instance, the combination: "sweater-suspect-test-blood" would indicate that a participant was at that moment carrying out a blood test on the suspect's sweater. As soon as the participant changed activities, the current combination of categories was ended and a new combination was started.

RESULTS

Forming Expectations

The recordings were literally transcribed. Unique idea units were identified and classified in two categories: expectations regarding the criminal offense, and expectations regarding the exhibits. An example of the former would be: "She

TABLE 15.1
Total Number of Expectations Regarding the Criminal Offense and the Exhibits for Each Participant

Participant	*Expectations offense*	*Expectations exhibits*
Expert A	30	36
Expert B	13	28
Intern	6	11

may be raped on the backseat of the car, as it is cool outside this time of year." An example of the latter would be: "I would expect blood on the knife, certainly with multiple stabs." Table 15.1 lists the number of expectations in each category for the three participants.

Table 15.1 shows that both experts clearly and unexpectedly differed in the number of expectations regarding the criminal offense. This difference was less pronounced for the expectations regarding exhibits. As predicted, the intern generated far fewer expectations in both categories than the experts.

Ordering of Exhibits

Participants generally rank ordered the exhibits in the same way. The most important exhibit, according to all three participants, was the victim's panties, followed by the suspect's sweater and knife. These three exhibits did indeed contain foreign blood or semen. Some other exhibits that were considered likely to contain traces (e.g., the victim's coat or the suspect's jeans) in fact did not contain traces at all (coat) or merely the person's own blood (jeans). Generally, then, the story behind the fake case matched the participants' expectations; however, in some cases, there were some mismatches, which might point to a deficiency on the part of the researchers fabricating the case.

Screening of Exhibits Within a Limited Time

As the results of the screening can only be meaningfully interpreted in comparison with the results obtained after unlimited search, Table 15.2 shows the actual number and percentage of traces found for both limited and unlimited time to search (in case of unlimited time, the actual time taken is identified in brackets).

Table 15.2 shows that the experts, after screening, found more than 50% of the traces they would find after a more extensive search, whereas the intern found only 25% of the traces in the first 10 minutes of screening. This suggests that the experts are able to use top-down strategies in quickly selecting the most promising parts to search for traces. An alternative explanation may be that experts have developed better pattern-recognition capabilities than novices, enabling them to more quickly recognize a blood stain for what it is.

Exhaustive Search

The success rate for the familiar task in which participants were given unlimited time to search is presented in Table 15.2. The most striking result, undoubtedly, is the poor performance by Expert B. One possible explanation for this result is a difference in search times by the three participants. Focusing only on the search times on the exhibits themselves, with the exclusion of administration, testing, and looking under the microscope, analysis of the videotapes showed that Expert A spent 61 minutes, Expert B, 30 minutes, and the intern, 33 minutes searching. The three participants did not differ in the relative effectiveness, expressed as the number of traces found per minute, with which they searched: Expert A found 0.11 traces per minute, Expert B 0.10 traces per minute, and the intern 0.12 traces per minute. This suggests there is a positive linear relationship between search time and probability of discovering a trace: the longer one searches, the more traces are found.

A longer search may be caused by several factors, for instance: every part of an exhibit may have been searched for a longer time; every part may have been searched for an equally long time, but it may have been searched more frequently; or the maximum search times for particular parts may have been extremely long, whereas other parts may have been searched for a much shorter time. In order to attribute the longer search time of Expert A to one of these causes, a detailed analysis was carried out on the search strategies for each exhibit. The results of this analysis showed that the longer search time was caused by several factors at the same time, dependent also on the precise exhibit. The most frequent cause for the longer search time used by Expert A turned out to be his strategy of searching the same part of an exhibit with different means. For instance, the pair of jeans of both victim and suspect was searched with both the naked, unaided, eye and a

TABLE 15.2
Number of Traces (and Percentages) Found
by the Three Participants by Time Available

Participant	*Limited time (10 min)*	*Unlimited time*
Expert A	4 (36%)	7 (64%) (61 min)
Expert B	2 (18%)	3 (27%) (30 min)
Intern	1 (9%)	4 (36%) (33 min)

straylight by Expert A, whereas the other participants used either the eye or the straylight. It also turned out that Expert A searched longer on most parts, given one particular means of searching and given that all participants searched that part at all. This implies that his search was more thorough than that of the other participants. A third factor explaining the longer search time was Expert A's strategy of returning to a previously investigated part, adding to his total search time.

The possibility of a relationship between Expert A's strategy and his success in finding traces was investigated by examining the exhibits where A was successful and the other participants were unsuccessful in finding traces. The exhibits concerned were the t-shirt, the panties, and the plaid blanket. In the case of the t-shirt, a direct link could be established between Expert A's strategy and his success: Expert A turned out to be the only one investigating the particular spot where a stain was put. This may be a reflection of his thoroughness. In the case of the panties, Expert A and the intern both found the semen traces, while Expert B did not. This was probably caused by longer search times using the microscope on the part of Expert A and the intern. In the case of the plaid blanket, Expert A searched for a shorter time than the intern but was more successful. This may be explained by the intern's lack of a systematic procedure for searching the large plaid blanket: he folded the blanket in different sections but got confused about which section he had already searched. Expert B did not search with focused light. In conclusion, Expert A was more successful because of different reasons: he covered the whole exhibit, he searched longer, and he was more systematic (basically the same reason as the first one). There is no conclusive evidence that the use of different means for searching and the higher frequency with which the same parts were investigated led to a higher proportion of traces detected.

Apart from the individual differences between the experts, a second point in Table 15.2 that warrants attention is the relation between time invested in exhaustive search and the number of extra traces found. In other words, does every additional minute of search time beyond the initial 10-minute screening time yield a proportional number of additional traces, or is there a diminishing return on invested time? A measure of the proportion of additional time divided by the proportion of additional traces was defined. This measure is 1.0 if every unit of additional search time leads to the same unit of additional traces found. In case of Expert A, this measure is: (61/10)/(7/4) = 3.49. For Expert B, the measure turns out to be 2.0, and for the intern, 0.82. This means that Expert A was relatively effective in his initial screening and that he had to spend 3.49 units of search time beyond screening for each unit of additional traces. On the other hand, the intern only had to spend 0.82 units of search time beyond screening for each unit of additional traces. Both experts, therefore, showed diminishing returns on invested time beyond the initial 10-minute screening. This result has important implications for increasing the effectiveness of the search process.

Retrospective Probed Recall

All tasks were recorded on videotape and these recordings were discussed with the participants. In particular, we took a detailed look at how the search process was conducted on each exhibit. It became clear that the visual scanpaths of all participants were roughly similar in the sense that all went over each exhibit in a systematic fashion. For instance, a sweater was started at the end of the left sleeve, the eyes moved upwards toward the shoulder and then downwards along the chest, then upwards again toward the right shoulder, and finally downwards along the right sleeve. After this, the sweater was turned around and the back was inspected. The experts also inspected the inside of the sweater, because they thought it was strange that so few traces had been found so far. The intern, on the other hand, did not inspect the inside as he thought the sweater would not have been worn inside-out. There were also qualitative differences among the participants. In the case of the sweater, for instance, Expert A used his fingers to spread the material so as to make it smoother. He also took a detailed look at first, and a more global look from a distance later on, because he had been unsuccessful so far and, in his experience, it had helped before to take a different angle at the same material. As mentioned, he also used more means for searching the sweater (the binocular, which was not used by the others), he returned to previously investigated parts often, and he searched each part much more thoroughly. So, although all participants went over the exhibits in roughly the same way, there was a sufficient number of differences to allow us to conclude that the way they interpreted the information was different. It was this interpretation, in the light of their expectations, that drove their search process.

During the retrospective probed recall, it also became apparent that both experts expressed serious doubts about the credibility of the story. The intern, on the other hand, accepted the story at face value. The experts remarked that the description of the criminal offense provided by the police would lead one to expect many more traces than they had found. They thought they had not received the clothes that were actually worn by the victim at the time of the rape or that those clothes had been cleaned, or that they were dealing with a false rape claim by the victim (on average, in 4% of the sexual offense cases presented to the Department of Serology, claims, mostly by teenage women, that they have been raped turn out to be false and are later withdrawn).

DISCUSSION

The goal of the present research was to increase the effectiveness and efficiency of search strategies, by formalizing the strategies used by experienced forensic scientists and incorporating these strategies in procedures. At first sight, this requires an instruction-based approach to task analysis. This approach assumes that a temporally ordered sequence of actions can be described that are required for an effective and efficient search procedure. The results showed that large individual differences in search strategies exist, even among experts. Moreover, there is a trade-off between effectiveness and efficiency: the most effective expert was the least efficient one. Therefore, it is difficult to decide whose strategy to formalize and to incorporate in a procedure, if both effectiveness and efficiency should be served at the same time.

A different angle on the same issue is provided by a constraint-based approach. This approach leads to the specification of a number of constraints and leaves it up to the forensic scientist to meet those constraints in any way he or she chooses. In this case, the constraints are provided by the information the forensic scientist receives from the police, the time available for searching, and the number of traces to be preserved for further DNA profiling. Increasing the effectiveness and efficiency of search strategies can only be accomplished when all constraints are met. The effectiveness of the search process can be increased when the forensic scientists receive more specific information regarding the criminal offense than they now do. The efficiency can be increased by adopting a multiphased search process: first, start with screening; second, resort to more exhaustive search strategies if screening does not yield enough traces. For this kind of multiphased search process to work, it is necessary to specify in advance, separately for each case or for each class of cases, the minimum number of traces that should be preserved. Experts now use rules of thumb to determine when to stop (e.g., two traces for each exhibit, and two exhibits for each offense). The viability of these rules of thumb should be investigated. If it turns out to be possible to state explicitly a norm for each case in terms of the required number of traces to preserve, it becomes possible to start with a highly efficient screening process that stops as soon as the required number of traces is found. Our results show that screening may be a highly efficient strategy for experts, and that a longer search process leads to a diminishing return on invested time. Two important caveats are in order here: first, in practice one does not know the origin of a particular trace, and hence its value in the courtroom, until DNA profiling has confirmed the origin. This makes it essential to preserve the best traces possible, in terms of criminalistic value. This, in turn, requires the ability of the forensic scientist to construct a highly plausible story explaining the origin of the traces. Again, detailed information provided by the police is essential in this regard. A second caveat concerns the use of screening by less experienced forensic scientists. Our study shows that the intern profited less from screening and more from extensive search than the experts. Further research should determine whether screening can be taught to novices in the absence of the ability to construct a story.

REFERENCES

Abernethy, B. (1990). Expertise, visual search, and information pick-up in squash. *Perception, 19,* 63–77.

Hoffman, R. R. (1987). The problem of extracting the knowledge of experts from the perspective of experimental psychology. *AI Magazine* (Summer 1987), 53–67.

Klein, G. A., & Hoffman, R. R. (1993). Seeing the invisible: Perceptual-cognitive aspects of expertise. In M. Rabinowitz (Ed.), *Cognitive science foundations of instruction* (pp. 203–226). Hillsdale, NJ: Lawrence Erlbaum Associates.

Lévy-Schoen, A. (1981). Flexible and/or rigid control of oculomotor scanning behavior. In D. F. Fisher, R. A. Monty, & J. W. Senders (Eds.), *Eyemovements: Cognition and visual perception* (pp. 299–314). Amsterdam: North-Holland.

Noro, K. (1984). Analysis of visual and tactile search in industrial inspection. *Ergonomics, 27,* 733–743.

Pennington, N. & Hastie, R. (1993). A theory of explanation-based decision making. In G.A. Klein, J. Orasanu, R. Calderwood, & C. E. Zsambok (Eds.), *Decision making in action: Models and methods* (pp. 188–201). Norwood, NJ: Ablex.

Schoonard, J. W., Gould, J. D., & Miller, L. A. (1973). Studies of visual inspection. *Ergonomics, 16,* 365–379.

Vicente, K. J. (1999). *Cognitive work analysis: Toward safe, productive, and healthy computer-based work.* Mahwah, NJ: Lawrence Erlbaum Associates.

16

Using and Gaining Experience in Professional Software Development

Sabine Sonnentag
University of Amsterdam

This chapter addresses the question of how expert software professionals accomplish their tasks and how they meet the complex requirements typical for their work situation. By adopting the expertise research framework (Ericsson & Smith, 1991), high and moderate performers are compared. Two empirical studies provided evidence that high and moderate performers differ in their approach to task accomplishment: High performers put more emphasis on problem comprehension, preparatory activities, local planning, and feedback seeking. Moreover, high performers engage more in communication activities. It is interesting that high and moderate performers do not differ with respect to years of professional experience. This chapter suggests that aspects of experience other than just many years in the field might be more important for high performance.

WORK REQUIREMENTS IN PROFESSIONAL SOFTWARE DEVELOPMENT

Professional software development is a knowledge-intensive work domain in which professionals have to integrate knowledge from various disciplines and domains (Clegg, Waterson, & Axtell, 1996; Curtis, Krasner, & Iscoe, 1988). Overall, the cognitive requirements are high (Glass, Vessey, & Conger, 1992).

Moreover, due to the high innovation rate within this domain with continuously changing working methods, procedures and tools, learning requirements are high as well (Brodbeck, 1994). Additionally, many software professionals have to meet high cooperation and communication requirements (Curtis et al., 1988; Walz, Elam, & Curtis, 1993) and have to accomplish their tasks under moderate to high time pressure (Fujigaki & Mori, 1997). Taken together, professional software development is characterized by high multiple requirements and takes place in a complex context.

Naturalistic Decision Making (NDM) research studies how individuals and groups act in dynamic and uncertain situations in which they are confronted with ill-structured problems and time pressure that have to be dealt with in a larger organizational context (Orasanu & Connolly, 1993; Zsambok, 1997). Professional software development in which many of the features relevant within NDM research are present is a good field for the study of NDM. Moreover, researchers described studies of expertise and experience as belonging to the core of NDM research (Cannon-Bowers, Salas, & Pruitt, 1996; Zsambok, 1997). Thus, the investigation of expert software professionals and the way they make use of their experience can add to knowledge about high performance in the field of NDM.

PAST RESEARCH ON EXPERTISE IN SOFTWARE DEVELOPMENT AND PROGRAMMING

On a general level, psychological research on expertise seeks to "understand and account for what distinguishes outstanding individuals in a domain from less outstanding individuals in that domain, as well as from people in general" (Ericsson & Smith, 1991, p. 2). For answering the question of high performance in professional software development on an empirical basis, one needs studies that compare higher and lower performers working professionally in this domain. High and moderate performers' process of task accomplishment should be studied with realistic tasks incorporating crucial features of real world work tasks.

Table 16.1 provides an overview over past quasi-experimental expertise research on software development and programming in which experts were compared with nonexperts. In a literature search, 57 studies—published between 1981 and 1997—that examined expertise in this domain were located.[1] In most studies, expertise was operationalized by years of experience, that is, individuals with (relatively) many years of experience were regarded to be experts and individuals with a small number of years of experience were regarded to be nonexperts. In

[1]A complete list of all studies is available from the author on request.

TABLE 16.1
Quasi-Experimental Studies on Expertise in
Software Development and Programming (1981 to 1997)

	Percentage of studies
Operationalization of expertise	
Years/months of experience	84%
High performance	12%
No detailed information provided	4%
Samples studied	
Students as nonexperts	68%
Professionals as nonexperts	14%
No detailed information provided	18%

Note. Sample size is 57.

only 12% of the studies was expertise operationalized as "high performance"—the operationalization that Ericsson and Smith's (1991) definition would suggest. In nearly all studies, experts were professionals working in the field of software development and programming. However, in only 14% of the studies were the nonexperts professionals as well. In more than half of the studies, the nonexperts were students with very little experience in the domain. The tasks to be accomplished by the study participants were simple and could be finished in a short period of time.

Thus, past expertise research concentrated more on experience than on high performance and compared students with professionals. Comparisons within the group of professionals were rare. Of course, comparisons between students and professionals offer important insights about how professional competencies develop (cf., Ackerman & Humphreys, 1990). However, generalizations from students who just started to learn a programming language to software professionals are limited. It is one of the basic findings of expertise research that students and professionals differ in their problem-solving processes (Jeffries, Turner, Polson, & Atwood, 1981). Furthermore, comparisons between students and professionals provide no information on performance differences within the groups of professionals.

A series of studies were conducted to compare high and moderate performers working as software professionals. In the remainder of this chapter, I summarize major findings from two of these studies. Because experience was a crucial variable in past expertise research, I discuss the relationship between experience and high performance within this professional field in greater detail.

STUDY 1: FIELD STUDY

Sample and Method

A first study was conducted as a field study in 29 professional software development projects from 19 German and Swiss organizations (Sonnentag, 1995). The projects produced a variety of software products ranging from administrative software for small and large companies for telephone and communication purposes to applications within banks, insurance companies, and traffic institutions to process control applications. One hundred and eighty persons participated in an interview. Among the participants, 62.1% were systems analysts and programmers, 25.6% were team or subteam leaders, 12.4% were user representatives or held other positions. On average, participants had a mean professional experience in software development of 5.7 years.

Within individual interview sessions, study participants were asked to nominate a person in their team they perceived as an "excellent software professional." In subsequent comparisons, software professionals who were nominated by at least two of their fellow team members were regarded as high performers. Software professionals not nominated at all were considered to be moderate performers. Those who were nominated once were excluded from subsequent comparisons.

Study participants were asked to describe the features that made the nominated person an excellent performer. Answers to this open-ended interview question were later categorized (interrater agreement: 78%). Furthermore, daily work activities of software professionals were assessed during the interview (for a detailed description of this procedure cf., Brodbeck, 1994).

Later in the interview procedure, participants described their process of accomplishing a typical work task. Based on protocols of the task accomplishment process, raters coded preparatory and feedback-seeking activities (mean interrater agreement: 98.7%). In total, data on the task accomplishment process were available from 107 participants. However, reported tasks covered a wide range of activities and were not easily comparable. Analysis, therefore, concentrated on design tasks ($n = 35$). The comparison finally included data from 5 high performers (nominated by at least two coworkers) and 17 moderate performers (not nominated by any coworker). High and moderate performers' tasks did not differ in perceived difficulty.

Results and Discussion

During the interview, study participants provided detailed descriptions of excellent performers. Most often, participants described excellent performers as highly knowledgeable in their domain (68.6%). Additionally, they observed high social competence in excellent performers (54.1%). Social competence included communication and cooperation skills, team leading competencies, and being a "good colleague" with whom it was easy to go along. Furthermore, study participants identified a specific working style to be typical for excellent performers (49.1%). Most prominent characteristics for this working style were a consequent use of software design methodologies and systematic strategies, team orientation, and a high degree of independence. Additionally, excellent performers' high cognitive skills were mentioned by 37.1% of the interviewed participants. User orientation and motivation were seldom described as typical features of excellent performers.

A comparison between high and moderate performers' daily work activities showed that high performers spent more time on specific communication activities. They were significantly more involved in consultations within the team, that is, informal exchange of information and support of others when problems occurred. Additionally, high performers participated more in review meetings. There were no significant differences between high and moderate performers in more narrowly defined software development activities such as design, coding, or testing programs.

Additionally, how software professionals proceeded when accomplishing a typical work task was analyzed. Table 16.2 shows high and moderate performers' involvement in preparatory and feedback-seeking activities while working on a software design task. None of the moderate performers reported having used a manual, other literature, or existing programs for preparatory purposes, whereas

TABLE 16.2
Percentage of High and Moderate Performers
Showing Preparatory and Feedback-Seeking Activities

	High performers	*Moderate performers*
Preparatory activities		
Use of manuals	6.3	0.0
Use of literature	40.0	0.0
Use of existing programs	14.3	0.0
Discussion with team leaders	20.0	0.0
Discussion with coworkers	50.0	12.5
Discussion with customers	0.0	20.0
Feedback seeking activities		
Feedback from team leaders	60.0	0.0
Feedback from coworkers	75.0	47.1
Feedback from customers	0.0	43.8

high performers used these resources to a moderate degree. For preparatory purposes, high performers referred more to literature than moderate performers (Fisher's exact test; $p = .04$). Furthermore, high performers discussed more with team leaders and coworkers in the preparatory stage of task accomplishment (Fisher's exact test; $p = .063$). With respect to feedback-seeking activities, a clear picture emerged: high performers mainly asked team leaders and coworkers for feedback (Fisher's exact test; $p = .006$ for feedback from team leaders), whereas moderate performers heavily relied on feedback from customers.

This first study showed that excellent performers were perceived to be both highly knowledgeable and to possess high communication and cooperation skills. The analysis of high and moderate performers' daily work activities pointed in the same direction. High performers participated more in cooperation activities, such as consultations within the team and in review meetings. These results suggest that social skills are highly relevant within domains such as professional software development (for a similar finding c.f., Riedl, Weitzenfeld, Freeman, Klein, & Musa, 1991).

When interviewing software professionals about their ways of accomplishing a task, specific preparatory and feedback-seeking activities (e.g., consulting the literature, discussing with team leaders and coworkers, and approaching team leaders and coworkers for feedback) were more prominent in high performers. These findings indicate that high performers prepare more intensively for task accomplishment. One can assume that, through more intensive preparatory activities, high performers learn much about the task and the domain. Moreover, high and moderate performers seek feedback from different persons, with high performers turning to those from whom they might receive the more relevant and faster feedback. As a consequence, high performers can use this feedback immediately for adapting their task accomplishment process. Feedback from customers may include important information as well. However, it often comes very late in the software development process and, therefore, will be less helpful for immediate improvements.

Because this analysis was based on retrospective self-report data, it can not be ruled out that high and moderate performers were very similar with respect to their preparatory and feedback-seeking activities but reported them differently. To arrive at a more unequivocal interpretation, the second study assessed concurrent thinking-aloud data.

STUDY 2: THINKING-ALOUD STUDY

Sample and Method

A second study focused on cognitive processes during task accomplishment (Sonnentag, 1998). In total, 40 software professionals participated in the study. Participants' mean professional experience in software development was 6.9 years.

Study participants were asked to work on the Lift Control Problem, a complex and relatively ill-defined software design task (Guindon, 1990). Verbal protocol data based on thinking-aloud processes were gathered and later categorized segment by segment (interrater agreement: 77.5%). The category system included the following main categories: problem comprehension by analyzing requirements (i.e., reading the requirements given in the task description and reflecting on them), problem comprehension by scenarios (i.e., reflecting on typical problems and scenarios within the lift domain), planning ahead (i.e., reflecting on how to proceed, deciding on the future course of action, thinking about what to do first and what to postpone), local planning (i.e., thinking about the next step without extensively reflecting on it), solution development (i.e., designing the outline of the software system), feedback processing (i.e., evaluating the designed solution), and task-irrelevant cognitions (i.e., statements having nothing to do with the lift control task).

High and moderate performers were identified by two methods: (a) a peer nomination procedure similar to that used in the first study and (b) performance on the Lift Control Problem. In further analysis, only participants who were nominated by at least two of their peers to be excellent performers and who showed a design performance above the median (12 high performers) were compared with participants who were not nominated at all and who showed a design performance below the median (12 moderate performers).

Results and Discussion

Analyses revealed clear differences in the task accomplishment process between high and moderate performers. High performers verbalized more local plans, that is, they structured their working process by stating explicitly what they intended to do next. Furthermore, they were more engaged in feedback processing. They more often evaluated their design solution and started early with these evaluations. Compared to high performers, moderate performers' verbal protocols contained more task-irrelevant cognitions. Such task-irrelevant cognitions included statements about participants' emotional state at the present moment, statements about participants' everyday work situation, statements about participants' assumptions about their competencies and knowledge, statements indicating "leaving the field" (i.e., statements concerning issues from completely different contexts, such as the kind of technology used in their work organizations or leisure time activities), and statements about the experimental setting. The most prominent difference between high and moderate performers' task-irrelevant cognitions referred to "leaving the field" statements, with moderate performers verbalizing more of these cognitions.

Furthermore, moderate performers spent more time for problem comprehension by analyzing the requirements, particularly late in the process. This suggests that moderate performers were less able to build an adequate problem representation early in the process. Detailed analysis of transition frequencies between two types of problem comprehension activities (problem comprehension by analyzing requirements and problem comprehension by scenarios) indicated that high performers used a specific strategy for arriving quickly at an adequate problem representation. They immediately linked the analysis of task requirements to problem domain scenarios, that is, they "translated" the abstract problem formulations into concrete examples and simulations. For example, high performers sequentially read single sentences of the task description and then immediately visualized the extracted information by sketching a graphical representation. Such an approach to problem comprehension can be characterized as a *relational strategy* (cf. Pennington, 1987; Shaft & Vessey, 1998, for findings which point in a similar direction). In contrast, moderate performers read the complete task description without decomposing it and without checking what each sentence might tell them about the problem to be solved.

This second study confirmed some of the results of the first study. For example, high performers were found to be more engaged in feedback seeking. Additionally, the second study showed that high and moderate performers' work processes differ with respect to local planning, specific problem comprehension strategies, and the amount of task-irrelevant cognitions. All in all, high performers worked very task- and solution-oriented and adopted a relational strategy.

HOW DOES EXPERIENCE MATTER?

Past expertise research has mainly referred to length of experience as the crucial variable which explains performance differences between students and professionals (cf. Table 16.1). This section addresses the question of whether and how experience matters for expertise within professionals. I discuss specific findings from the two presented studies that shed light on the role of experience for high performers' task accomplishment process.

Length of Experience

In the first study, high and moderate performers were compared with respect to their professional experience. Analysis showed that these two groups did not differ significantly with respect to length of experience. Average professional experience in software development was $M = 6.6$ years for high performers in system analysis and programming jobs, $M = 7.7$ years for high performers in team leading jobs, $M = 4.0$ years for moderate performers in system analysis and programming jobs, and $M = 8.1$ years in team leading jobs. In all groups there was a substantial variability with respect to years of experience. Thus, data from this first study showed that high performers have not worked more years within professional software development than did moderate performers.

In the second study, high and moderate performers were also compared with respect to length of professional experience. Again, high and moderate performers did not differ. High performers had even a somehow shorter experience (M = 6.6 years for high performers; M = 7.8 years for moderate performers); however, this difference did not reach significance. Additionally, it was analyzed whether professionals with many years of experience differed in their way of task accomplishment from less experienced persons. No such differences were found. These findings indicate that, within professional software development, length of experience does not really matter for high performance. Just spending more years in the field does not contribute to performance improvement.

One reason for this could be the obsolescence phenomenon that is typical for the software development area. Within this field, innovations continually emerge and existing knowledge becomes obsolete very quickly. This implies that professionals with many years of experience do not have a great advantage over relatively inexperienced colleagues who became familiar with the most recent innovations during their studies or professional training. Nevertheless, there might be some advantages of many years of experience. For example, experienced software professional might be better at handling more political and conflictual project management issues, which are known to be crucial within professional software development (Symon, 1998). Therefore, it might be that in this domain many years of professional experience contribute to team performance, but not to individual performance.

Variety of Experience

In the first study, substantial differences with respect to variety of experience were found. During their careers, high-performing systems analysts and programmers had worked in 7.4 different projects, compared to 3.5 projects for systems analysts and programmers performing at an average level. Similarly, high performers had worked with more programming languages in about the same period of time. Taken together, it seems that a broad and varied experience is more typical for high performers than is a long experience. However, because of the cross-sectional nature of that study, causality is still unclear. First, it might be that a broad and varied experience helps in developing high performance. Second, it might be that high performers have a greater chance of acquiring a broader experience, for example, by advancing more quickly from one project to another.

Making Use of One's Experience

If one considers the other findings of the two studies, it becomes obvious that high and moderate performers differ in the way and extent they use their experience. High performers use and apply their experience more explicitly in the working process. This became most obvious in the second study in which software professionals worked on the Lift Control Task. High performers more often related task requirements to scenarios of the lift domain. Scenarios of the lift domain, in this context, are mainly based on application domain knowledge, that is, experiences with lifts in one's everyday life. It is not plausible to assume that high and moderate performers differ in their experiences with lifts. Therefore, their application domain knowledge about lifts should be rather similar. However, when working on the design task, high performers applied their experiences about lifts to a higher extent in order to build an appropriate problem representation. For example, based on their everyday experience, they verbalized their knowledge about typical lift operating procedures ("Yes, I expect this when I am in a lift: that I go in one direction and that I do not go up and down while being in the lift" cf., Sonnentag, 1996). Thus, high performers made more use of their experience.

Furthermore, high performers use their experience by sharing it with others, for example, during their participation in review meetings and consultations. This might imply that their experience remains "vivid" and can be referred to relatively easily in the future. Being highly involved in cooperation activities might have an additional positive effect on the development of expertise. Research on the self-explanation effect has shown that explaining facts and procedures to oneself has a positive impact on future performance (Chi, de Leeuw, Chiu, & LaVancher, 1994). The same might be true when explaining something to other people. This would imply that high performers, who are often asked for an explanation (cf. also Sonnentag, de Gilder, & Winkelman, 1999; Turley & Bieman, 1995) might gain something for themselves from these activities.

Gaining Experience From the Working Process

Both studies showed that high performers work in such a way that they gain more from the working process—thus, that they accumulate more "experience." For example, while preparing for task accomplishment, high performers refer more to the literature. While working toward task accomplishment, they use this resource in order to learn more about a topic or a specific problem. One can assume that this knowledge will not be useful only in the short run, that is, for this specific task, but that high performers will also benefit in the long run from their tendency to consult the literature.

High performers seek more feedback, particularly feedback which is available early in the task accomplishment process. Such feedback provides information about errors, inconsistencies, and other suboptimalities in one's working process or product. It is, therefore, helpful in finishing the specific task at a high quality level. Moreover, feedback can point to learning necessities and can stimulate further learning processes. Moderate performers, however, who would need feedback most, search for feedback to a lesser degree and, therefore, miss the opportunity to learn and gain experience from their working processes.

CONCLUSION

Both studies showed that high and moderate performers differ in the way they accomplish their tasks and how they spend their working time. High performers meet the high cognitive requirements typical for professional software development by using software design methodologies, by early building an adequate problem representation, by structuring their working process with local planning, and by quickly evaluating their solutions. They refer to literature and mainly team leaders for getting the necessary information and feedback.

It became obvious that high and moderate performers do not differ in length of experience. Thus, it is not warranted to equate length of experience with expertise in a professional domain. However, this does not imply that experience is irrelevant for expertise. Aspects of experience other than length are more important. High performers were found to have a broader and more varied experience, made more use of their experience, and extracted more experiences out of the working process. Although the causal processes are still unclear, it might be that breadth and variety of experience as well as use of experience plays a highly relevant role in developing expertise.

Taken together, the studies made clear that it is not experience per se which differentiates high from moderate performers, but the way a person actively refers to his or her experience and applies it in the process of task accomplishment.

REFERENCES

Ackerman, P. L., & Humphreys, L. G. (1990). Individual differences theory in industrial and organizational psychology. In M. D. Dunnette & L. M. Hough (Eds.), *Handbook of industrial and organizational psychology* (Vol. 1, pp. 223–282). Palo Alto, CA: Consulting Psychologists Press.

Brodbeck, F. C. (1994). Software-Entwicklung: Ein Tätigkeitsspektrum mit vielfältigen Kommunikations- und Lernanforderungen [Software development: Activities with varied communication and learning requirements]. In F. C. Brodbeck & M. Frese (Eds.), *Produktivität und Qualität in Software-Projekten. Psychologische Analyse und Optimierung von Arbeitsprozessen in der Software-Entwicklung* [Productivity and quality of software projects. Psychological analysis and optimization of work processes in software development] (pp. 13–34). München, Germany: Oldenbourg.

Cannon-Bowers, J. A., Salas, E., & Pruitt, J. S. (1996). Establishing the boundarires of a paradigm for decision-making reserach. *Human Factors, 38*, 193–205.

Chi, M. T. H., de Leeuw, N., Chiu, M.-H., & LaVancher, C. (1994). Eliciting self-explanations improves understanding. *Cognitive Science, 18*, 439–477.

Clegg, C. W., Waterson, P. E., & Axtell, C. (1996). Software development: Knowledge-intensive work organization. *Behaviour and Information Technology, 15*, 237–249.

Curtis, B., Krasner, H., & Iscoe, N. (1988). A field study of the software design process for large systems. *Communications of the ACM, 31*, 1268–1287.

Ericsson, K. A., & Smith, J. (1991). Prospects and limits of the empirical study of expertise: An introduction. In K. A. Ericsson & J. Smith (Eds.), *Toward a general theory of expertise: Prospects and limits* (pp. 1–38). Cambridge, England: Cambridge University Press.

Fujigaki, Y., & Mori, K. (1997). Longitudinal study of work stress among information system professionals. *International Journal of Human-Computer Interaction, 9*, 369–381.

Glass, R. L., Vessey, I., & Conger, S. A. (1992). Software tasks: Intellectual or clerical? *Information & Management, 23*, 183–191.

Guindon, R. (1990). Designing the design process: Exploiting opportunistic thoughts. *Human-Computer Interaction, 5*, 305–344.

Jeffries, R., Turner, A. A., Polson, P. G., & Atwood, M. E. (1981). The processes involved in designing software. In J. R. Anderson (Ed.), *Cognitive skills and their acquisition* (pp. 255–283). Hillsdale, NJ: Lawrence Erlbaum Associates.

Orasanu, J., & Connolly, T. (1993). The reinvention of decision making. In G. Klein, J. Orasanu, R. Calderwood, & E. Zsambok (Eds.), *Decision making in action: Models and methods* (pp. 3–20). Norwood, NJ: Ablex.

Pennington, N. (1987). Comprehension strategies in programming. In G. M. Olson, S. Sheppard, & E. Soloway (Eds.), *Empirical studies of programmers: Second workshop* (pp. 100–113). Norwood, NJ: Ablex.

Riedl, T. R., Weitzenfeld, J. S., Freeman, J. T., Klein, G. A., & Musa, J. (1991). What we have learned about software engineering expertise. *Proceedings of the Fifth Software Engineering Institute Conference on Software Engineering Education* (pp. 261–270). New York: Springer.

Shaft, T. M., & Vessey, I. (1998). The relevance of application domain knowledge: Characterizing the computer program comprehension process. *Journal of Management Information Systems, 15*, 51–78.

Sonnentag, S. (1995). Excellent software professionals: Experience, work activities, and perceptions by peers. *Behaviour & Information Technology, 14*, 289–299.

Sonnentag, S. (1996). Planning and knowledge about strategies: Their relationship to work characteristics in software design. *Behavior & Information Technology, 15*, 213–225.

Sonnentag, S. (1998). Expertise in professional software design: A process study. *Journal of Applied Psychology, 83*, 703–715.

Sonnentag, S., de Gilder, D., & Winkelman, E. (1999). *Excellent performers' involvement in helping and feedback behavior*. Technical report. Amsterdam: University of Amsterdam.

Symon, G. (1998). The work of IT system developers in context: An organizational case study. *Human-Computer Interaction, 13*, 37–71.

Turley, R. T., & Bieman, J. M. (1995). Competencies of exceptional and nonexeptional software engineers. *Journal of Systems and Software, 28*, 19–38.

Walz, D. B., Elam, J. J., & Curtis, B. (1993). Inside a software design team: Knowledge acquisition, sharing, and integration. *Communications of the ACM, 36*, 63–77.

Zsambok, C. E. (1997). Naturalistic decision making: Where are we now? In C. Zsambok & G. A. Klein (Eds.), *Naturalistic decision making* (pp. 3–16). Mahwah, NJ: Lawrence Erlbaum Associates.

17

Expertise in Laparoscopic Surgery: Anticipation and Affordances

Cynthia O. Dominguez
Headquarters, Air Force Material Command

THE SURGICAL CONTEXT

The seemingly simple Hippocratic injunction "First, do no harm" has never been all that simple. Hospitals are filled with varieties of knives and poisons. Every time a medication is prescribed, there is potential for an unintended side effect. In surgery, collateral damage is inherent. External tissues must be cut to allow internal access so that a diseased organ may be removed, or some other manipulation may be performed to return the patient to better health. Harm must be done in order to attain the greater good. Since 1980, however, medical technology and techniques have been developed that have significantly reduced the damage to healthy tissue required in surgery; as a whole, these advances are known as minimally invasive surgery. The "First, do no harm" principle seemingly has been maximized by the reduction in incision size common to these procedures. Patients are out of the hospital and back to normal activity within days of a gallbladder removal, because they do not need to wait for incisions in abdominal muscles to heal.

However, as with all complex systems, such marvelous advances in technology have their costs. The new surgeon/patient interface consists of fiber-optic cameras, TV monitors, tiny incisions, and long-stemmed instruments. This interface adds a barrier between the surgeon and the work environment so that essential perceptual information is more difficult to ascertain and the motor skills required are more technically demanding (Cuschieri, 1995). When a surgical procedure is especially challenging, involving a patient with unusual anatomy or acute inflammation, risk of injury to a nearby structure can be increased by the perceptual barriers involved in minimally invasive surgery. The surgeon has to decide, either before operating or over the course of the surgery, whether the risk of injury outweighs the benefit of minimizing tissue damage from the incision. This is the conversion decision: whether a surgeon will convert from a minimally invasive procedure (laparoscopic procedure, when in the abdominal area) to a more traditional, open-incision procedure (called an open procedure here).

In this chapter, I present results from a field study on decision-making expertise in the domain of laparoscopic surgery, examining the conversion decision just described. In an open procedure, surgeons use large incisions and can directly look into and manipulate the field of operation. Converting essentially widens the field of view and permits direct tactile feedback; it expands the field of safe travel available to a surgeon (Gibson & Crooks, 1938; Dominguez, 1997). Typically, a gallbladder removal procedure begins laparoscopically, and the decision to convert is made when a complication or unintentional injury is predicted, suspected, or confirmed.

The remainder of this introduction is devoted to explaining the goals and context of this research. I then outline background research and theory, detail the methods and approaches used in this field study, and present findings in two areas. The first area deals with prediction and anticipation, and the second discusses perceptual expertise, with an emphasis on affordances.

Research Goals

This research had three goals. The first goal was to examine the decision to open within the larger context of surgery. The second goal was to understand staff-resident differences, that is, how each group differs in approaching the decision to open as well as how surgeons with different levels of experience overcome perceptual handicaps inherent in laparoscopy. The third goal was to take an exploratory look at surgeons' verbal protocols. As this work progressed, surgeons' verbalizations and my measures of them pointed to the importance of metacognition. The post hoc goal that resulted was to understand how metacognition interacts with expertise in laparoscopic surgery. This last topic is examined elsewhere (Dominguez, 1997); in this chapter, I focus on staff-resident differences related to anticipation and affordances.

Context: Theoretical and Medical

As all researchers bring to their work a certain frame of reference, so I began this study with the desire to understand how perception and action are coupled in worlds where technology has inserted a barrier between the environment and ways of acting on this environment. What does this mean, the coupling of perception and action? It reflects a circularity of behavior. Often we clearly see how information we perceive causes us to act, such as when a red light causes braking in a car. What is less apparent, but nevertheless basic to human behavior is how we "tweak" our environment to get information, such as when we lightly touch the brakes to get a better appreciation for iciness on the road. Action is used to gain perceptual information, just as perceptual information is used to determine action.

In surgery, visual perception and tactile perception are tightly coupled with action. The introduction of laparoscopic gallbladder removal techniques (known as laparoscopic cholecystectomy) in 1989 has caused general surgeons to enter the realm of domains where perception-action coupling is interrupted. Other such domains include modern nuclear power plants, commercial flight decks, and remotely piloted vehicles. Nowadays, pilots do not send a crewmember down into the bowels of a bomber to examine the hardware when a bomb fails to release; the state of the weapon must be determined by available indicators. Action-driven perceptual information becomes more difficult to get. When it is not possible to directly tweak one's environment to better understand its state because a layer of automation or a bank of indicators is one's only interface to the system, it is important to understand how humans compensate for these barriers. This understanding is necessary for creating better interfaces, tools, and training programs.

In surgery, it is possible that laparoscopic techniques cause surgeons as they watch the magnified television image of the operating area to pay greater attention to how tissues react when deflected. Participants described numerous navigation rules they use to orient themselves in the remote world seen on the video screen, such as "always go from known to unknown areas." One critical element of safe operation is identification with certainty of the patient's structures. In gallbladder surgery, only two structures need to be identified, clipped, and severed before the gallbladder can be removed from the patient's abdomen. If another nearby structure, such as the common bile duct, is accidentally clipped and severed, the outcome can be catastrophic. How surgeons gain perceptual information to identify structures with certainty in an uncertain environment is a main theme of this research.

The next two sections provide brief reviews of relevant literature relating to prediction and perceptual expertise.

Background: Prediction and Anticipation

Prediction refers to declaring in advance; *anticipation* refers to being prepared and looking forward to something. These terms are used as such to denote slightly different but similar concepts. In a field study looking at anesthesiology, Xiao (1994) observed anesthesiologists over a long period of time, forming hypotheses without formal data collection. He then targeted a series of operations during which he interviewed the anesthesiologist before the surgery, observed and recorded think-aloud data during the procedure, and conducted follow-up interviews to elaborate on the events that had taken place. Xiao proposed that behavior of anesthesiologists does not resemble an information-processing model; instead, the behavior is anticipatory and preparatory. Anesthesiologists prepare their physical and mental workspaces for anticipated events, they control the patient's status in a feedforward manner, and they off-load anticipated activities during slow periods. Xiao employed Rasmussen's decision ladder (Rasmussen, Pejtersen, & Goodstein, 1994) to represent how anesthesiologists use planning to reduce response complexity in future situations.

Similar anticipation of high workload periods was found in Amalberti and Deblon's (1992) study of fighter aircraft process control. In pre-mission planning, which generally took more time than the mission itself, the pilots devoted considerable time to analyzing each leg of the route for possible threats. Those pilots classified as experts differed from the less experienced pilots in planning strategy, both in number and type of waypoints they chose and in the number of potential incidents predicted for each flight leg (experts predicted fewer incidents, organized in a more hierarchical manner). During the actual missions, Amalberti and Deblon found that pilots devoted over 90% of their reasoning time during free-time periods to anticipation. Pilots developed a "tree" of possible events during mission planning, and events in the tree became more or less salient over the course of the flight. For salient possible events, the pilot would mentally simulate a response to see if it would work given the constraints of the situation; if not, the pilot would try to change the current parameters so that the desired response would work. "The implication is that most of a pilots' expertise lies in avoiding situations where they have no solution or no chance of applying known solutions (i.e., situations similar to ones generated by totally unexpected events)." (p. 655) When unexpected problems did occur, the pilots generally responded with poor solutions. Thus, it seems as if anticipation was used to engineer a field of safe travel (Gibson & Crooks, 1938; Dominguez, 1997), defined as a field for which the pilots are ready to handle already predicted problems and threats.

In other research on how pilots and other complex systems operators assess situations, Endsley (1995) identified three levels of situation awareness, namely perception, comprehension, and projection of future status. An operator might perceive local information and comprehend its meaning, but to function effectively (and survive) in a rapidly changing flight situation, it is necessary to anticipate and prepare for upcoming high-workload periods. This third level, projection, describes what Amalberti and Deblon (1992) documented in a general way. It is captured also in the conventional wisdom of the pilot's adage, "always stay ahead of the airplane." Staying ahead means thinking ahead. If a busy period such as approach or landing is imminent, thinking ahead means getting required tasks (i.e., checklists) for that period done early so that there will be more time to handle unanticipated problems should they arise.

Background: Perceptual Expertise

> Novices see only what is there; experts can see what is not there. With experience, a person gains the ability to visualize how a situation developed and to imagine how it is going to turn out. . . . Our emphasis is not on rules, or strategies, or size of knowledge base per se, but on the perceptual and cognitive qualities of experience—experts do not seem to perceive the same world that other people do. (Klein & Hoffman, 1993, p. 203)

This review discusses research and theory examining how people perceptually recognize and cognitively understand information in their environment, and what they are able to do with that information. The preceding section included relevant material; prediction of consequences and risks involved in a course of action are vital to this process. As a result, this section builds on the previous section, rather than presenting a separate "component" of expertise.

Expertise has only recently been a topic of study in cognitive psychology. In research which later inspired a rich literature on the nature of expertise in problem solving, deGroot (1946/1965) and later Chase and Simon (1973) did ground-breaking studies on expertise in chess players. This research suggested that grand masters develop, over time, a tremendous vocabulary, or repertoire, of familiar patterns that may exist on a board. This is the domain-specific knowledge of chess. Over time, experts do not merely know more, they come to know things in a different way. Chess experts do not just accumulate factual knowledge; they learn to identify a configuration of pieces in terms of its meaning and for appropriate moves that are associated with that configuration. This is in essence a *categorization*, recognizing a pattern in terms of its implications for action. Categorization of perceived events may even be tacit, meaning that the expert is unaware of this process, but, nevertheless, the categories are associated with appropriate scripts for action (Means, Salas, Crandall, & Jacobs, 1993).

A concept that encapsulates recognizing patterns of information with regard to their implications for action is that of affordances (Gibson, 1979/1986). *Affordances* are functional relationships between the environment and a particular actor. We evaluate whether something will afford doing what we want to do with (or to) it. We see affordances everywhere, if we look for them. Sometimes, meaningful potential for our intended actions is immediately perceivable, and sometimes trial and error is needed to know whether, for instance, a truck affords a 4-year-old climbing into it. We could also relate Amalberti and Deblon's pilot anticipation just described to this concept; pilots watched for approaching situations that do not afford applying known solutions. Gibson (1979/1986) proposed that people become attuned to invariants in the natural environment that directly specify affordances for them; in other words, people learn to directly perceive affordances. Learning to perceive the affordances of tissues and anatomical structures, and to understand the implications of appearances and configurations of these structures, would seem to be an important part of expertise in surgery.

A final representation of expertise that adds to these concepts of knowledge-based categorization and judging typicality is that of Rasmussen's (1983) skills-rules-knowledge-based cognitive control. This representation goes beyond the fact that experts have superior knowledge bases; it addresses how skill and knowledge are alternately employed in situations that are familiar and unfamiliar to the practitioner. Performance changes fluidly between the three levels of cognitive control, depending on how well the requirements of the current task match with the resources of the individual (Olsen & Rasmussen, 1989). Knowledge-based control is always present in the background to provide oversight about whether the appropriate goals are in focus or to catch errors (Sanderson & Harwood, 1988). This oversight is the metacognitive component of the skills-rules-knowledge model.

METHODS

Videotape Stimulus

The author and an assistant interviewed 20 surgeons as they watched and commented on videotapes of challenging gallbladder removal cases. The videotapes were selected from a library of taped cases by a surgeon who collaborated on this research. This analysis focuses on one case that all 20 surgeons saw. The videotaped case allowed us to establish a close mapping between the domain of surgery and the research situation. Deciding whether to open is a process extended in time. It depends upon a developing composite of details that are difficult to capture with pencil and paper descriptions used in some research paradigms. Although this approach could capture neither interactions between surgeons nor the operating room environment, it veridically captured the visual view of the patient's operating space available during a gallbladder removal.

Participants

Ten of the surgeons interviewed were surgeons in training, senior residents in either their fourth or fifth year of residency. Because the surgeons in this resident group had 8 or 9 total years of medical training, we considered them to be a "journeyman" group rather than a "novice" group. The "expert" group consisted of 10 surgeons who were on the staff at local hospitals; surgeons in this group had between 2 and 28 years since completing residency. The median number of years since residency for the staff group was 5.

Procedure

My colleagues and I interviewed surgeons in libraries or conference rooms of the hospitals where they worked. The interviews involved a combination of structured and unstructured elicitation. In the structured portions of each interview, at three predetermined points chosen by a collaborating surgeon, we asked several predefined questions, such as "What do you think is going on here?" "What are your concerns?" and "What errors might be made in this situation?" Surgeons were asked to role-play as the supervising surgeon in this case and were encouraged to think aloud to provide continuous commentary on events seen on the videotape. Rating scales were also used, but they are not relevant to the findings presented here. All of the interviews were audiotaped and transcribed.

The Case

The case itself involved an 80-year-old woman with an acutely infected gallbladder. The first decision point of the interview occurred before any video was shown; questions probed whether this case should be started as a laparoscopic or an open procedure. Two of the 20 surgeons indicated they would not begin this case laparoscopically; all of the others would "at least take a look" with the laparoscope. This decision revealed a paradox in surgeons' reasoning. Some surgeons reasoned that this case should be done *open* because the patient was old; her aged lungs and heart would be less likely to withstand the pressure from insufflation[1] needed in a laparoscopic case, and she did not need to return to work quickly. Other surgeons reasoned that this case should be done *laparoscopically* because the patient was old; she was more likely to develop pneumonia after an open procedure, and the bed rest needed to recover from the large incisions may decrease the likelihood of this elderly patient ever returning to full functioning. The conflicting opinions and rationales are difficult to untangle, which is precisely why this case is a good one to study in the tradition of examining critical incidents (Klein, Calderwood, & MacGregor, 1989) to understand expertise and decision making (Dominguez, 1997).

Measurement

My colleagues and I developed two sets of measures (dependent variables): those tied to specific videotape events, and those noted at any time they appeared in a transcript. Variables were developed initially and applied in a process where two readers evaluated each transcript, highlighting sections of text where a variable,

[1]Insufflation creates an airspace for operating in the abdominal cavity; carbon dioxide gas is used at a pressure that is monitored by an insufflation machine.

such as 'prediction' or 'if/then rule' applied. Each variable was well-defined for the reader. The two readers then compared results in a meeting in which they justified each variable and agreed on a final analysis for that transcript. As we applied the set of variables, readers suggested new aspects of variables and entirely new variables in order to more completely and descriptively capture the verbal protocols. The "perceptual expertise" variable was conceived of in a descriptive manner to capture what we were seeing in the transcripts. When all 20 transcripts had been initially analyzed and a final measurement scheme was in hand, one reader reread each transcript to assure consistent application of the variables. There were 23 variables in all, for each of which we developed a range of responses identified with particular codes. For this chapter, I limit the discussion to variables capturing prediction/anticipation and perceptual expertise.

My colleagues and I noted predictions made at any time during the interviews. These predictions were culled and lifted into a spreadsheet and examined for trends across specific segments of the case. *Perceptual expertise* statements were defined as follows: When the surgeon provided a substantial description of a cue they saw or felt or might see or feel hypothetically, and either (a) made predictions based on that cue, (b) drew inferences about the patient's disease, or (c) recommended an action based on that information.

PREDICTION AND ANTICIPATION FINDINGS

Quantitative Differences in Predictions

I would expect that surgeons with more experience would be more likely to be able to make predictions. In this research, there was overlap between the prediction and perceptual expertise statements. My colleagues and I applied the perceptual expertise variable when the surgeon cited a specific piece of information that led to the prediction. The prediction variable alone was invoked anytime the surgeon made a prediction, whether or not they provided the underlying reasoning or information used for prediction.

Twenty-nine of the 99 *perceptual expertise* variable statements involved prediction. Twenty of these statements were made by staff surgeons, and nine were made by residents. Adding all of the prediction statements together for staff and residents, we found residents made an average of 4.4 prediction statements per interview, whereas staff surgeons made an average of 7.2. A one-way ANOVA on the average predictions between staff and resident revealed them to be statistically different, $F\ (1, 19) = 13.8, p < .001$. From these data we can conclude that the staff either had greater knowledge that allowed them to make more predictions, or they were more comfortable or accustomed to verbalizing these predictions than the residents were.

Qualitative Nature of the Predictions

Given that staff surgeons made more predictions than residents, what was the nature of these statements? Two distinct types of predictions emerged from examination of the predictions. First, surgeons predicted they would have difficulty in dissecting and identifying structures (which are essential tasks), primarily because the gallbladder and surrounding areas were swollen and inflamed in this patient. Second, they predicted that this patient would have a higher risk of a negative outcome, such as an injury to nearby structures or tearing of the gallbladder wall, which increases risk of abdominal contamination or a postoperative infection from spilled bile. These two types of predictions are associated: the first, inflammation leading to difficult identification, is essentially a lowered ability to assess the situation and understand the configuration of the anatomy. This degraded situation assessment leads to the second type of prediction, higher risk of negative outcome.

Prediction statements in the first category were typically found early in the interview. There were several predictions made about the difficulty or lack of difficulty that would be encountered in identifying structures, finding tissue planes, and generally seeing what is going on. When a gallbladder is acutely inflamed, it becomes more difficult to tell what tissue belongs to a duct, artery, or gallbladder and what is just surrounding tissue; all of the tissue becomes reddish-pink and swollen, so color and texture cues are lost. One staff surgeon, in explaining why he would be dissecting closer to the gallbladder than the video showed, said:

> Because you don't know where structures are, because there's so much inflammation around there, it's kind of blind guessing where the important things are going to be, and they are digging through inflamed tissue.

Another comment that captured the difficulty of assessing the situation in a case like this came from a resident, also during the initial dissection:

> I would be very worried, I have no idea where the cystic duct and artery are, let alone the common bile duct. I would be kind of worried about that, because I can't tell where the gallbladder ends. Grabbing the common bile duct can injure it just by grabbing it alone.

Making these predictions seems to be a way that surgeons can prepare their minds to encounter problems and decide ahead of time how to deal with them, as combat pilots do during mission planning (Amalberti & Deblon, 1992).

In summary, surgeons had a wide range of opinions as to how to deal with the difficulty of identifying structures and the associated risk of injury. The two main types of prediction seen in this research correspond to the two kinds of feedforward control that Xiao (1994) identified in anesthesiology. The first, predicting difficulty in dissection and identification, involves what Xiao (1994) termed "preparing the mental workspace," preparing mentally for dealing with the predicted situation. The second, anticipating injury, have to do with control actions that are taken to prevent undesired outcomes in a feedforward manner. These actions include treating the gallbladder as gently as possible to avoid tearing it; cleaning up thoroughly after any spillage of bile; using accepted techniques to actively avoid dissecting in the vicinity of the common bile duct; and avoiding cautery in certain situations. Now, I discuss findings more directly related to action, those connecting perceptual expertise to action afforded in a surgical situation.

PERCEPTUAL EXPERTISE AND AFFORDANCE FINDINGS

Quantitative Findings

In looking at the 99 total statements coded with the perceptual expertise variable, a large number of them fall into two categories. First, 32 of them cited perceptual information and accompanying inferences about pathophysiology, or disease progression, in the operative area. For example, "She obviously has stones, you can see them, see the surgeon kicking them with the sucker," and "The artery looks large: it could be due to acute cholecystitis, more inflammation increases blood flow" (see Dominguez, 1997). Second, 31 statements described perceptual information as tied to various actions, goals, or states afforded. It is these affordance statements that are the focus here.

Not all surgeons seem to perceive and understand potential for action in the same manner. One of the most striking aspects of the affordance statements was the high numbers of them made by staff surgeons. In general, staff surgeons made more perceptual expertise statements than residents did: staff averaged 6.5 per transcript, whereas residents averaged 3.5. These numbers were found to be significantly different with a one-way ANOVA, $p < .05$. However, staff made more than three times as many affordance-related statements as residents did (staff made 24; residents, 7). Only 5 of the 10 residents made these statements, as opposed to 9 of 10 staff surgeons. These numbers suggest that understanding and verbalizing affordances inherent in a situation develops with experience and is potentially an essential part of expertise.

Qualitative Findings

What surgeons told us about information they perceived and what it meant to them supports the view that surgeons perceive the potential for action. My colleagues and I categorized the affordance statements according to what action the surgeon indicated was afforded. An example of each of the 11 types of affordance statements seen is given in Table 17.1. For instance, there were seven total statements concerning whether the gallbladder wall would be graspable (see number 4 in Table 17.1 for an example). In three of the statements, surgeons indicate that certain cues (thicker ligaments, how the edema fluid looks, the swollen gallbladder) translate into a higher likelihood of injury (see number 6 in Table 17.1). The surgeons are "preparing their mental workspaces" (Xiao, 1994) to move more carefully to mitigate these risks. This concept ties back to the *prediction* variable statements anticipating difficulty in dissection and identification. Awareness of these constraints is the first step to dealing with them appropriately through cautious control actions.

TABLE 17.1
Sample Statements of Perceptual Expertise

Affordance	*Statement from Surgeons' Transcripts*	*Rating*
1. Dissection	If we see the patient within 2 days, the inflammation usually isn't that bad, after 2 days it's harder to dissect (S)	4.5
2. Exposure	Stiffness of fat, rigidity (pull it away and it's still there) indicates edema, we're concerned about exposure further down (S)	4.5
3. Laparoscopic Procedure	If I can palpate the gallbladder, we're probably not going to be able to take it out laparoscopically, but will try (S)	5.0
4. Grasping	If patient is sick longer than 48 hours, harder to grab GB without it falling apart (S)	4.5
5. Insufflation	Sick old people do better open because CO_2 (insufflation) interferes with respiration, venous return, coronary blood flow (S)	3.0
6. Risk of Injury	Surrounding ligaments much thicker due to edema and inflammation makes it easier to injure duct or artery (R)	4.5
7. Tactile Perception	Operates two-handedly, feel with left hand what the right hand is doing, can tell if pulling too hard (S)	4.5
8. Time (does not afford short procedure)	Thickened omentum and inflammatory exudate cause longer procedure, which is bad for an 80-year-old (R)	4.5
9. Tissue Manipulation	Fat is more stiff, like a stiff rubbery jello you can scrape and pull down, it bleeds/oozes (S)	4.0
10. Visualization	After 48 hours of inflammation the gallbladder gets thick, hard, you can't see planes as much (S)	3.5
11. Visualization and dissection	Wouldn't do this laparoscopically because it's an acute gallbladder, dissection will be difficult, planes won't be easy to find (R)	3.0

Note. Sample statements coded with the Perceptual Expertise variable that reveal affordances, or potential for action, related to cues. (S) indicates a staff surgeon made the statement; (R) indicates a resident. Ratings are an average of the agreement ratings of two independent surgeons on a scale of 1 to 5, 5 indicating *maximum agreement.*

Thus, what surgeons told us about information they perceived and what it meant to them in terms of information, affordances, and control supports the view that surgeons perceive the potential for action. They also develop an understanding of how their own perceptions can be obscured or misled by how tissues respond to disease.

DISCUSSION

The findings of this research are consistent with research from other domains, but add some important contributions. As already noted, the anticipation and prediction behavior surgeons described is similar to the feedforward and preparatory behavior Xiao documented in anesthesiologists. But with these data, it is possible to trace a theoretical line between prediction, affordances, and how the decision to convert is made, in an attempt to integrate these concepts.

First, prediction statements showed how surgeons prepare for risk, anticipating that it will arise soon, and then realize the potential negative outcomes of a risky situation. Making these predictions seems to be a way that surgeons can prepare their minds to encounter problems and decide ahead of time how to deal with them, as well as a way to remind themselves of the risk of negative outcome.

The second major category of predictions, those predicting higher risk of danger to the patient, occurred with greater frequency as the case progressed. For instance, surgeons made statements predicting that the gallbladder would tear open, and predicting postoperative complications (abscesses or infection) stemming from the infected gallbladder. More often, surgeons predicted that there was risk for severing the common bile duct, a major complication.

One prediction statement, made by a resident, included elements of both of the two most common types of predictions I have discussed here: "If you get your landmarks confused you're going to cause the patient harm." The transcript excerpt in Table 17.2 elaborates on this resident's self-acknowledged inability to assess the situation and resultant fear of injury.

Affordance statements are another way to view a prediction. Comprehending an affordance is in essence making a prediction, predicting that a situation will afford doing what you want to do. For instance, the first affordance in Table 17.1 is "If we see the patient within 2 days, the inflammation usually isn't that bad, after 2 days it's harder to dissect." This is an experience-based prediction as to whether the patient will be difficult to dissect based on how many days the patient has had infection symptoms. The second affordance is similarly a prediction as to whether proper exposure will be afforded based on the stiffness of the fat when it is pulled away. Many of the example affordances can be considered in this light and were analyzed under the Prediction variable. These affordance statements are simply surgeons explaining what a piece of information means to them in terms of whether they will be able to take actions that are necessary in order to safely

TABLE 17.2
Two Common Predictions

Speaker	*Question and Response*
Interviewer:	Do you have any concerns at this time?
Surgeon:	Um, my concerns at this time are mainly where are we, ... *I think being in that position you would have a sense that you're lost and you really don't know what's going on* and the only way to overcome that is to open up.
Interviewer:	What errors would an inexperienced surgeon be likely to make in this situation?
Surgeon:	I think progressing on, from what I've seen now, progressing on would probably be the biggest error and any other errors that result would be because of making that first decision. I mean obviously you could progress on and misidentify and ((inaudible)) a structure that you think is the cystic duct and in fact it is the common bile duct or something like that. And *if you get your landmarks confused, you're going to proceed along and it's going to cause the patient a great deal of harm.*

remove this patient's gallbladder. The striking difference in how frequently residents made these statements as compared to staff surgeons indicates that understanding how information relates to potential action may develop with experience and may be a fundamental component of expertise.

The critical assessment surgeons need to make is how all of these prediction statements and affordances combine to feed into an overall comfort level. A surgeon needs to assess continually whether a procedure is either safe enough or too risky and, hence, should be converted to an open procedure. In between these levels of "safe enough" and "too risky" lies an area of risk mitigation, in which surgeons are aware of risks but apply known tools and techniques to be able to continue laparoscopically. Each prediction of risk or danger has its own relative weight; not being able to insufflate the patient is a show-stopper, but lack of proper exposure is a challenge that can be met by trying different techniques. Not surprisingly, predicting difficulty and predicting negative outcome (danger for the patient) are skills that were seen more with the expert group in this study.

Both Xiao (1994) and Amalberti and Deblon (1992) showed how experts used their knowledge to predict how a situation may evolve and to head off dangerous situations using "feedforward control" methods. Similarly, surgeons in this study showed that they also combine knowledge with environmental information to look ahead, to predict difficulty, to assess risk, and to draw conclusions as to what actions will be possible. It may be possible at this time to tie together several domains with similar characteristics into a coherent model of how experts in complex systems exemplify and apply psychological constructs of prediction, affordances, and feedforward control.

CONCLUSIONS

Many of the theories and findings on expertise discussed in the previous section converge on a single concept, that of being able to understand the meaning of patterns of information. For a surgeon, meaning often is synonymous with what course of action would have dangerous consequences and what course would be safe. I have discussed this concept as related to a surgical field of safe travel (Gibson & Crooks, 1938) elsewhere (Dominguez, 1997). To judge what is a safe course, a surgeon learns how visual information and "feel" of the tissue signifies the progression of disease and yields cues for identifying structures. Also important is what this information means in terms of how the tissue can or should be retracted, dissected, or avoided during surgery (affordances). The large staff–resident differences in both prediction/anticipation and discussing potential for action in the research case given indicate (a) the importance of these abilities, (b) the fact they are developed through experience, and (c) the potential for improving performance if these skills could be directly trained. These differences also highlight the importance of analyzing affordances for design. In developing instruments, displays, and other medical devices, designers should expand beyond a user-centered approach to one that accounts for integration across users, the environment, and the work itself (Flach & Dominguez, 1995). An understanding of how surgeons use perceptual information to predict, anticipate, and make decisions about appropriate actions is a starting point to such an analysis.

To put these findings in perspective, examining the prediction and affordance statements from this research is but one lens on understanding the meaning of patterns of information. Many psychological constructs could be (and have been) applied to better understand how to help the medical community design technology, processes, and organizations, which in turn could lead to improvements in patient safety. Applying these constructs across diverse domains such as surgery, anesthesia, and air combat highlights unity in development of expertise in complex environments.

REFERENCES

Amalberti, R., & Deblon, F. (1992). Cognitive modelling of fighter aircraft process control: A step towards an intelligent on-board assistance system. *International Journal of Man-Machine Studies, 36*, 639–671.

Chase, W. G., & Simon, H. A. (1973). Perception in chess. *Cognitive Psychology, 4,* 55–81.

Cuschieri, A. (1995). Whither minimal access surgery: Tribulations and expectations. *The American Journal of Surgery, 169*, 9–19.

de Groot, A. D. (1965). *Thought and choice in chess.* The Hague, The Netherlands: Mouton. (Original work published 1946)

Dominguez, C. O. (1997). *First, do no harm: Expertise and metacognition in laparoscopic surgery.* Unpublished doctoral dissertation, Wright State University. Dayton, OH.

Endsley, M. R. (1995). Toward a theory of situation awareness in dynamic systems. *Human Factors, 37*, 32–64.

Flach, J. M., & Dominguez, C. O. (1995). Use-centered design: Integrating the user, instrument, and goal. *Ergonomics in design.* July, 19–24.

Gibson, J. J. (1986). *The ecological approach to visual perception.* Boston: Houghton-Mifflin. (Original work published 1979)

Gibson, J. J., & Crooks, L. E. (1938). A theoretical field-analysis of automobile driving. *The American Journal of Psychology, 3*, 453–471.

Klein, G. A., Calderwood, R., & MacGregor, D. (1989). Critical decision method for eliciting knowledge. *IEEE Systems, Man, and Cybernetics, 19*, 462–472.

Klein, G. A., & Hoffman, R. R. (1993). Seeing the invisible: Perceptual-cognitive aspects of expertise. In M. Rabinowitz (Ed.), *Cognitive science foundations of instruction* (pp. 203–226). Hillsdale, NJ: Lawrence Erlbaum Associates.

Means, B., Salas, E., Crandall, B., & Jacobs, (1993). Training decision makers for the real world. In G. A. Klein, J. Orasanu, R. Calderwood, & C. Zsambok (Eds.), *Decision making in action: Models and methods* (pp. 306–326). Norwood, NJ: Ablex.

Olsen, S. E., & Rasmussen, J. (1989). The reflective expert and the prenovice: Notes on skill-, rule-, and knowledge-based performance in the setting of instruction and training. In L. Bainbridge & S. A. R. Quintanilla (Eds.), *Developing skills with information technology.* London: Wiley.

Rasmussen, J. (1983). Skills, rules, and knowledge: Signals, signs, and symbols, and other distinctions in human performance models. *IEEE Transactions on Systems, Man, and Cybernetics, SMC-13*, 257–266.

Rasmussen, J., Pejtersen, A. M., & Goodstein, L. P. (1994). *Cognitive systems engineering.* New York: Wiley.

Sanderson, P. M., & Harwood, K. (1988). The skills, rules, and knowledge classification: A discussion of its emergence and nature. In L. P. Goodstein, H. B. Anderson, & S. E. Olsen (Eds.), *Tasks, errors, and mental models.* London: Taylor & Francis.

Xiao, Y. (1994). *Interacting with complex work environments: A field study and planning model.* Unpublished doctoral dissertation, University of Toronto, Toronto.

18

Driving Proficiency: The Development of Decision Skills

Helen Altman Klein
Eric John Vincent
Judith J. Isaacson
Wright State University

Driving is a potentially dangerous activity. Every day, tens of millions of drivers navigate potentially lethal vehicles on highways and neighborhood streets. In 1997, 41,967 people died and more than three million people were injured in crashes (National Highway Traffic Safety Administration, 1998). Nearly seven million motor vehicle crashes were reported to the police and untold more went unreported. We know that better automotive design, for example, crumple zones and features such as air bags, can reduce casualties. Attention has also been devoted to roadway design, signage, surface improvements, and so on. Although improvements in vehicle and roadway design contribute to increasing safety, much of the responsibility for accidents will always remain with the drivers.

Driving is dangerous, but it is also nearly an entitlement for U.S. adults. Although aviation, fire fighting, and surgery have highly selective recruitment and extensive training programs, anyone of a specified age in the United States, depending on the state, can receive a driving learner's permit. Driving is a domain in which skills are learned almost exclusively through experience rather than extensive training. In many states, mandatory instruction time prior to licensing is as little as 8 classroom hours and 6 hours of in-car supervision. Those more than 21 years of age are not required to have any instruction prior to licensing. A permanent license is granted based on a typically perfunctory driving test. In short, while driving can be a dangerous activity, in the United States it has been open to virtually every adult.

The majority of past research on drivers has focused either on the perceptual, motor, and reaction time demands of driving or on risk-taking characteristics. Young adults, on the average, have the best vision (Llaneras, Swezey, Brock, & Rogers, 1993; Shinar & Schieber, 1991), the best motor skills (Laux & Brelsford, 1990), and the fastest reaction times (Salthouse, 1985). If perceptual abilities, physical capacities, and reaction times alone determined performance, we would expect monotonic decreases in driving performance with increases in age. Research also tells us that risk taking or sensation seeking is at its highest among drivers who are 16- and 17-years old and then decreases with age. These trends would suggest that crash rates start very high, and then decrease as drivers reach their late teens. Crashes would then increase slowly from their twenties, as drivers suffer age-related physical decrements. Crash rates do drop from high levels during the teens, but they remain low until the drivers are in their late sixties (NHTSA, 1997). While risk-taking research does predict the poor performance of young drivers, data on perception, motor skills, and reaction times have not explained the good performance of experienced drivers through their sixties.

Beyond the perception, motor skills, and reaction time decrements of aging, and even the risk taking of teenagers and young adults, we must look at the cognitive and decision-making skills of drivers to understand performance. A proficient driver must monitor vehicle functioning, compensate for roadway conditions, and anticipate and react to the irregular behaviors of other drivers. Although much of driving is routine, the drunken motorist, the highway repair zone, or the patch of ice can cause crashes. Regardless of car design, signage, and laws, drivers are sometimes faced with difficult decisions. When driving and your vehicle hits an ice patch, for example, should you hit or pump the brakes? Is a tailgating vehicle making an emergency stop dangerous? Should you steer into the skid to gain control without stopping? Does the berm's width provide a safe escape? Is there an approaching vehicle? These and other questions are common, and drivers sometimes have to decide and respond in milliseconds.

The research reported here was initially designed to capture the cognitive skills of older experienced drivers. We wanted to look at characteristics such as strategies, perceptual cues, tactics, and knowledge of limitations that may keep drivers safe even as their physical abilities decline. We contrasted the skills of long-term drivers with those of newer drivers. The goal was to generate a model of driving that captured cognitive differences. This model would reflect development in a domain that lacks the intense training typical of other potentially dangerous domains such as flying. Such a model could help identify approaches for augmenting performance.

Naturalistic Decision Making (NDM) methodologies have been used to identify the cues, strategies, and other important aspects of successful decision makers. The methods have been used to explore the complex cognitive skills of airline pilots (Waag & Bell, 1997), fire fighters (Klein, Calderwood, & Clinton-Cirocco, 1986), and surgeons (Dominguez, 1997). Like driving, these domains require recognition skills and effective decision-making in complex, dynamic, and time pressured environments. The present research used an NDM approach to explore how people make decisions in driving. We examined the decision-making processes of long-term and newer drivers to attempt to capture how they face the challenges of driving and how they differ from each other.

METHOD

To learn about the nature of driving skill, we interviewed 18 relatively young drivers and 17 long-term drivers. The newer drivers were university students, ages 18 to 23 years, who had driven for one to five years. We selected this age range to ensure that the drivers would have a concept of the driving task, and yet would remember their earlier training experiences. We could expect that by age 18 years they would also be past the peak of risk taking. The long-term drivers were alumni of the same university, whose ages were 64 to 84 years. These drivers had between 46 and 68 years of driving experience. We expected a range of experience and skill in both groups of drivers, and we found it. One of the younger drivers, for example, had over one hundred thousand miles of experience. One of the long-term drivers, a farm woman, had minimal experience. Generally, the newer drivers had considerably less experience than the long-term drivers.

A Cognitive Task Analysis (CTA) interview was used to study the decision making of the drivers. We used an interview guide based on eight identified knowledge areas and appropriate common language probes (see Table 18.1). The guide was based on Knowledge Audits used in earlier CTA research with neonatal nursing (Crandell & Calderwood, 1989) and with housing adaptation in the elderly (Vincent, 1997). The interview focused on the participant's understanding of the driving task and changes in that understanding with driving experience. We also asked about the use of equipment, including mirrors and antilock brake systems (ABS). Newer drivers were asked about training experiences and driver education classes.

In the interviews, we explored a range of hypothetical driving situations. These included typical roadway challenges such as merging, four-way stops, and congested interstate highway travel. In order to present these hypothetical driving situations, a simulation was used. The interviewer marked a large table top with roadway features. Matchbox™ cars and trucks were placed and moved about the drawn roadway to represent the different driving situations. To assess decisions required for merges, for example, a merge situation was presented on a table with roadways, ramps, and lanes marked and cars in place. Participants took the perspective of a driver in one of the cars as they described what they were watching, where threats might come from, and the decisions required. Participants moved the cars as they explained their responses. Variations in

TABLE 18.1
Cognitive Task Analysis Interview Framework

Knowledge Area	*Common Language Probes*
Challenges	Sometimes people find certain driving situations are more challenging than others. In past interviews drivers have mentioned merges, (four-way stops), (congested interstates). Have you ever found these or other areas to be challenging?
Near Misses	Occasionally, during routine driving, we experience near misses, situations where we almost have an accident. Do you have any memorable near misses? (use matchbox cars to simulate).
Using Experience	As we gain more experience as drivers, many of us develop strategies for dealing with driving situations that can sometimes be an effort. Other drivers have told us (unfamiliar locations). Do you have any strategies?
Difficulties	Sometimes driving can be very challenging, requiring a lot of effort. Have you encountered any situations that you find difficult? For example, road signs that are difficult to interpret, (intersections that are challenging).
Future/Hypothetical	How do you expect your driving abilities to change during the next few years? Do you expect them to improve? remain the same, or decline? Why?
Capacities	Capacities change as we gain more experience or our abilities decline. Have you noticed any changes in your capacities? How are the changes affecting driving performance?
Crystal Ball	I'm looking into a crystal ball and I see you (the participant) have had an accident that was your fault. Where do you think this accident would most likely occur and how would it happen?
Rules of Thumb	Many of the drivers we have interviewed have described tricks that they use while driving. Do you have any rules of thumb?

roadway challenges were probed by moving cars and eliciting the participant's analysis. Interviews lasted about $1^1/_2$ hours.

Each interview was audiotaped for transcription, scoring, and analysis. The scoring procedure was developed using an iterative process typical for CTA research (Crandall & Getchell-Reiter, 1993; Dominguez, 1997; Vincent, 1997). Five coders participated at different stages of the process. They included the interviewers and others. Initially, we coded the transcripts for five performance characteristics. These performance characteristics were used while coding for performance levels.

RESULTS

Performance Characteristics

The initial analysis of the interview transcripts provided the driving performance characteristics of cues, tactics, knowledge of limitations, strategies, and changes

Cues.

Cues were what the driver was looking for in other drivers, in the roadway, and in vehicle conditions. Cues provided meaningful information about the driving situation. Newer drivers reported, "I watch for a safe distance in front of me," and "I watch to see when I should shift to the next gear." The long-term drivers were more likely to report complex cues. They, for example, watched for people who were "drunk, or talking on a cell phone." Sometimes these cues reflected situational knowledge. For example, "When I see a big truck riding low, I know that I can't count on it to stop or even decelerate easily." Some reported features of the flow of traffic. "I'm uneasy when all the trucks slow to the speed limit. They must know something I don't know." Each statement acknowledges the cues used by the driver.

Tactical Rules.

Tactical rules were specific IF/THEN statements for dealing with simple road situations and were often linked to cues. *Tactics* are the decision rules for managing immediate challenges encountered on the road. A newer driver reported that during her early months of driving, she would go very slowly when she was in traffic so she would be constantly prepared to stop. Typical examples from long-term drivers were, "When I see a car riding too close to me, I tap my brakes to get them to back off," and "When a semi is near me, I get away from it as quickly as I can." These statements of tactics reflect the drivers' decisions for action in the face of specific cues. They were based on knowledge of rules and social conventions for negotiating the road.

Knowledge of Limitations.

Knowledge of limitations reflected the drivers' understanding of how their own characteristics or the characteristics of their vehicles might challenge their driving safety and effectiveness. Some drivers indicated that fear was a limiting factor for their performance. Newer drivers said, "For my first months of driving, I was afraid of merging," and "The radio was distracting. I just couldn't drive with it on." Knowledge that vision presents a limitation on performance was common for long-term drivers: "I just can't see at night any more." Long-term drivers often understood the role of attention for their performance. We heard, "I know that I get tired in the middle of the afternoon." Drivers also knew the limitations imposed by their car. For example, "My car just doesn't have the power to pass on hills." These acknowledgments reflect information that can be used in driving decisions.

Strategies. Strategies were linked to many statements describing limitations. Strategic rules, in contrast to the simple tactical rules, were complex IF/THEN decision rules for dealing with the perceived long-term limitations. Strategies expressed a rich array of planning changes such as avoidance of night and rush-hour driving. The newer driver who was afraid of merges reported, "I avoided the interstate for my first months of driving because I was afraid of merging." Several long-term drivers talked about stopping more often on long trips and driving for shorter distances each day. Some were using passengers as navigators so their attention would be free to concentrate on the roadway. One long-term driver explained how she took a circuitous route to the mall because left turns across busy highways felt dangerous to her. Another reported, "I double check before I head out of my driveway because I know I'm not as fast as I used to be." These describe decisions for attempting to avoid perceived risks.

Changes Over Time. Changes over time reflected participant awareness as they monitored their performance through years of driving. Participants reported increases or decreases in confidence, patience, risk taking, and comprehension of traffic flow and of other drivers. Newer drivers reported changes, for example: "I'm not nervous about the freeway anymore," "[now] I understand what the other drivers are trying to do. I understand the way the lights and lanes are set up," and "I remember how scary it used to be to go through an underpass near my house. I was never quite sure my car would fit. Now I don't really see how I had that problem." Examples from long-term drivers include: "I'm just not as impatient," and "I don't take reckless risks now, like I used to when I was younger. I'm not sure how I lived through my first years on the road." One driver reported, "I see much more than I used to." Drivers, both newer and long-term, could identify changes over time.

This initial review of interviews suggested that the driver's awareness and decision-making skills changed in an orderly fashion. Consistent with our interest in systematic changes with experience, we then looked for qualitative changes in performance. We noted that the newer drivers offered simpler cues and tactical statements as the basis of their decision making. They saw straightforward characteristics such as red lights and vehicle separations. They applied tactical rules learned as part of their driver training instruction such as "stop at stop signs" and the "2-second rule." Not all tactical rules were learned in driver's education. We heard several times about the "No cop, no stop" rule for stop signs. Some strategic rules were compensatory. "I didn't drive with the radio on." Some rules were linked to the risk taking of teens. One new driver described a merging strategy we labeled "the kindness of strangers." She accelerated down the ramp and closed her eyes as she neared the roadway. She explained that other drivers have always let her in.

More of the long-term drivers reported more knowledge of the road, more complex pattern detection, and more varied tactics and strategies. They were more likely to report attending to subtle characteristics of the driving environment such as acceleration of approaching vehicles, cellular telephone users, and the long stopping distance of a heavily loaded truck. Their tactical decisions were responsive to these cues. Their merging techniques utilized the gaps anticipated based on the flow of traffic, time of day, and road conditions. They gave extra space to potentially dangerous vehicles. These more advanced drivers were attuned to their own limitations and developed strategies for managing these limitations. We heard strategies that compensated for rather dramatic reductions in capacity.

Performance Levels

Consistent performance levels emerged from the initial analysis of cues, tactics, knowledge of limitations, strategies, and changes over time. These levels were declarative knowledge, control, automaticity, broad view, and anticipation. Declarative knowledge was expressed by tactical statements describing either laws of the road, such as "the car on the right has the right of way" or constraints imposed by vehicles or roadways, such as "it takes longer to stop a car on gravel." Although there were some flaws in understanding car and equipment functioning, most participants showed reasonable declarative knowledge for most important knowledge about driving. They knew the meanings of road signs, the recommended vehicle separation—the "2-second rule," the right of way at stop signs, and the like. For this reason, declarative knowledge was not coded.

Guidelines were developed for coding the four higher performance ratings. A four-point scale for each of these levels was developed based on subjects' reported difficulties and range of awareness. See Table 18.2 for these coding criteria. A rating of '1' was given when there was no indication of the subject's having mastered any aspect of that level of performance, whereas a '4' indicated near or complete mastery. Coders were blind to participant demographics. All interview transcripts were coded using this performance scale. This procedure initially resulted in an interrater reliability of $r = .74$. Differences were discussed between coders, consensus reached, and scale descriptions refined. Finally, all interviews were coded independently for performance levels. The final ratings had an interrater reliability of $r = .92$.

Control.

Inexperienced drivers often described working to keep the car between the lines. Newer drivers reported: "I really had to work the wheel to stay in my lane," and "I . . . just make sure I'm by the yellow line to stay away from the curb." Other inexperienced drivers said, "At first, if I took my eyes off the road for even an instant, my car would start to drift," and "I would turn to see what was next to me and I would go off the road." Several mentioned their early difficulty with left hand turns and with curves on narrow roads. Drivers who scored low in control used a limited range of cues and had inflexible decision rules.

TABLE 18.2
Definitions for Qualitative Dimensions of Performance Levels

Control of Car	
1	Can hardly keep car on the road at any speed—frequently off-track, extremely effortful and takes all available attention
2	Control is a problem in many situations of normal driving
3	Control only a problem in one or two situations, e.g., narrow bad roads or drastic curves or significant difficulty with ice but little else
4	Control not a problem (minimal problems in adverse conditions)
Automaticity	
1	No automaticity
2	A few driving situations are automatic, e.g. tracking
3	Most driving situations are automatic, only a couple, e.g., merging, still require careful control
4	Motor control (tracking, parking, etc.) fully automatized, don't have to think about it anymore
Broad View	
1	Tailpipe in front
2	Only one car ahead and behind, some awareness of blind spots
3	A few cars around, blind spots
4	Cars far around, off-road happenings, blind spots of everyone
Anticipation	
1	No anticipation apparent—self-centered, rule-oriented
2	Understands concept of anticipation even if not much in practice. Initial anticipation in a few (low workload) situations, e.g., stop signs, parking lots
3	Anticipates in many (high workload) situations and often acts to avert trouble
4	Always anticipating trouble and working to avoid it

Automaticity. Participants reported needing little attention for the basics of driving such as staying between the lines, shifting, or finding the headlight switch. One experienced driver, reflecting on his early efforts reported, "It took me a long time before I could look away from the road right in front of me. Thinking back, I'm surprised I never hit anything." A newer driver noted, "When I first started driving, I wouldn't have my radio on or talk to passengers. Now I can change the stations." One long-term driver captured automaticity with: "Sometimes I lose track on the way to work and just find myself at the entrance." For drivers who had achieved a level of automaticity, control decisions were made with little effort or attention.

Broad View. Participants attended to vehicles and potential hazards not only in close proximity but also far ahead and far behind. One long-term driver introduced the notion of "managing the road." This meant monitoring vehicles both near and far to better plan for the next actions. Decisions were made using a wide range of cues. In contrast, newer drivers often reported being surprised by vehicles appearing out of their blind spots. A typical remark from a newer driver about the frequency of this phenomenon was: "Once or twice each time I drive . . . they appear out of nowhere." An experienced driver contrasts this by mentioning "someone that hits a bicyclist is someone that wasn't paying attention." Drivers with broad views are rarely surprised. Sometimes, this broad view compensates for limitations: "My right eye never developed vision. Because of that, when I drive, I'm always trying to be aware of where everything is around me. I constantly monitor the rear view mirrors." Drivers at this level have a big picture of their driving. They have a great deal of information to use in their decisions.

Anticipation. Participants showed anticipation as they generated hypotheses about probable future actions of others based on their broad view. An example of anticipation in a newer driver was: "Sometimes you can see cars who don't have their turn signals on but they start drifting over here and I watch for that." One long-term driver reported, "I'm always uneasy when I see a car weaving behind me. I figure they may continue, so I slow down and let them pass me and get out of my way." Another long-term driver said, "When I see orange barrels, I just expect drivers to start cutting in front of me. I'm real careful." The driving decisions of these drivers are based not only on what is actually seen but also on what is projected based on knowledge of the world. Taken together, broad view and anticipation are captured by the notion of managing the road.

The interviews, across levels, did present several interesting anomalies. We observed that environmental attributes could alter performance level. Faced with icy roads for the first time, even skilled drivers reported losing automaticity and having to focus attention on keeping the vehicle on the road. They reported ignoring surrounding traffic that, in better conditions, would have been monitored. We also found that some components of skill spanned different levels of performance. Scanning behaviors, for example, seemed to reflect a broad view or big picture. We found that competent scanning appears first at stop signs where the newer drivers, stopped and freed from the attentional demands of their own vehicle, could observe other vehicles to decide which should go first. Only later did drivers typically attain a broad view for scanning competently at merges.

Group Differences in Performance

Comparisons between the groups for performance levels used the means from the four-point rating scale described earlier. They are presented in Fig. 18.1. Long-term drivers were rated higher than newer drivers on both control (3.75 and 3.15,

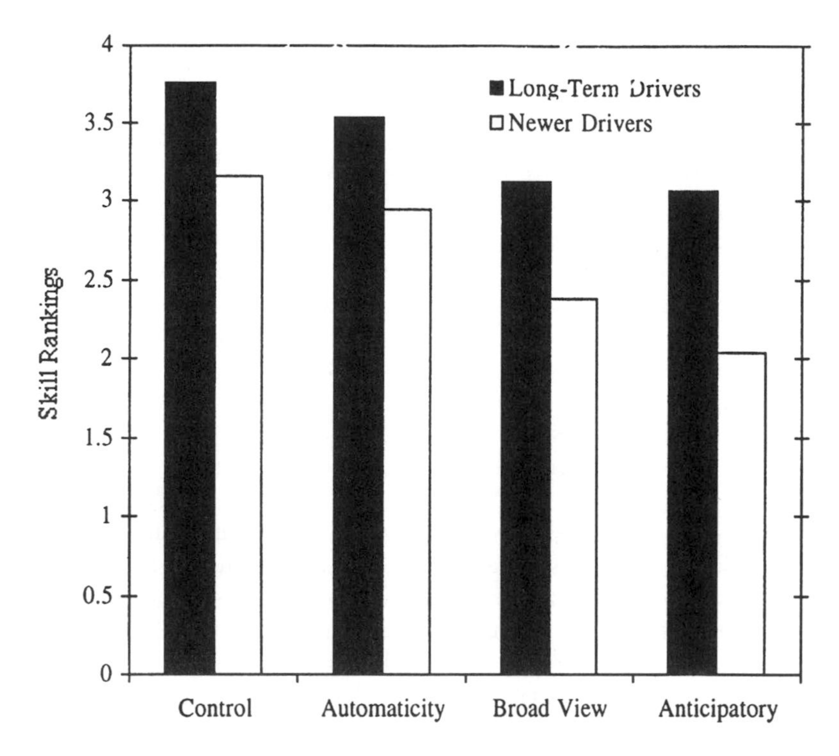

FIG. 18.1. Skill rankings for older and newer drivers.

respectively) and automaticity (3.53 and 2.94, respectively). Although both groups of drivers included a range of skill levels, broad view and anticipation were far more common with long-term drivers than with newer drivers (3.12 and 2.38, respectively, for broad view and 3.06 and 2.03, for anticipation). The lack of broad view in newer drivers was captured by a parent who described his teen as ". . . Not seeing any further than the tailpipe of the car ahead." A frightening number of newer drivers, even after several years of driving, seemed to lack not only anticipation and a broad view but even automaticity in some situations.

Our interviews revealed qualitatively distinct levels of performance, each building on the previous stage. At initial levels of control, drivers were making decisions about how to stop the car, how to maintain lanes, and the like. There were many decisions that required their attention, and they used easy-to-assess cues. Strategic planning was limited to reactive actions. Drivers at automaticity reported that when they decided to turn a corner, no additional decisions were needed. They did not report awareness of the continuous monitoring and adjustment needed by drivers struggling with control. As drivers reached the higher performance levels, they became aware of threats and challenges. They could actively plan for potential troubles. Our interviews were generally consistent with a stage analysis because of the orderly sequence of performance stages. We did not see drivers who had mastered a higher level of performance of broad view, but who had not mastered a lower one of automaticity. Drivers with a lot of experience had achieved automaticity in most areas, but not all experienced drivers had attained the higher levels of driving skill. Our next task was to seek a model that would provide a framework for describing driving and for suggesting ways to augment the decision-making skills of drivers.

DISCUSSION

A Developmental Model of Cognitive Skill Acquisition

Our interviews demonstrated support for distinct levels of driving cognition. Newell's (1990) SOAR and Anderson's (1993) ACT are models of skill acquisition. These models rely on rule-based knowledge, chunking, and practice as the foundation of skill acquisition or expertise. The conversion of declarative knowledge to procedural knowledge and the automatization of those procedures through practice, however, do not seem to account for the qualitative change in decision making we saw in drivers. Dreyfus and Dreyfus (1986; Dreyfus, 1997) presented a model of orderly, directional performance changes with experience. Their model looks at sequential change and qualitatively distinct characteristics of five ascending stages of skill development from the novice to the expert level. Initial performance is characterized as inflexible and rule-governed. Performance then reflects recognition of recurring patterns as experiential knowledge is integrated with rule-based knowledge. Next, performance shows complex planning in the use of information and decreased reliance on rule-governed strategies. The use of accumulated experience to intuitively recognize problems precedes the final level of intuitive decision making and action selection. Although Dreyfus and Dreyfus provide a picture of qualitative differences, their model, as the ones mentioned earlier, does not provide help in understanding why or how these changes occur.

Campbell and Di Bello (1998) suggested that developmental models may be the most appropriate in ill-defined domains in which skills show qualitative rather than simply quantitative changes. In such domains, the processes underlying skill acquisition are more important than descriptions of the end point. We need to know why some skills are mastered before others, and we need to know the conditions under which driver performance degrades. Because drivers show great variability in skill, we need to understand the full range of driving skill levels, not just the highest levels. Developmental models are concerned with the nature and mechanisms of development, not just with descriptions of levels.

Developmental models may allow us to describe levels of decision making and also the processes needed to be a good driver. The similarity of the progression of the decision making of drivers to development in children is striking.

Children and drivers both show qualitative as well as quantitative changes in capacities. In both children and drivers, we see the parallel and sequential increases in competencies with active involvement in the environment. Children, like drivers, show some areas of staggered but parallel development. In the same way that children realize the conservation of number before the conservation of volume, drivers scan at stop signs before they scan at merges. Different degrees or aspects of the same cognitive ability emerge at different times. The type of model used for understanding cognitive development in children may also be useful for describing the contextual and experiential contributions to driving and to the advancement of skill.

Piaget and others who have followed provide a model for understanding development in children (Flavell, Miller, & Miller, 1993; Piaget, 1950, 1970). Changes are developmental in that they are *orderly* and *directional*. At each new level, the competencies of earlier levels provide the basis for qualitatively new approaches to thinking. Developmental changes occur with *interaction* with the world. Research supports the role of interactions in the cognitive development of children. Developmental changes are *adaptive* to the demands of the specific complex domain. Finally, development is marked by *equilibration*—sequences of equilibrium, disequilibrium, and then a higher equilibrium with increases in skill. The first two attributes are components of earlier models of skill acquisition, the second two go beyond.

Orderly and Directional. Within Piaget's framework (Flavell, 1963; Piaget, 1950, 1970), competencies of earlier levels are not lost but rather are integrated into qualitatively new ways of thinking and knowing. The interviews were consistent with this developmental model of orderly and directional changes with experience. Progress is marked by increased integration of motor, sensory, and cognitive functions. We see this increased integration in driving skills when we look at stopping at stop signs. Initially, simply stopping the car is challenging. The new driver knows that the pedal has to be pressed but not how hard. As drivers automatize stopping, they are able to watch the arrival time of other cars at stop signs. Now they can watch the arrival order of other vehicles and know who should go first. In time, they are even able to predict that a certain vehicle is likely to run the stop sign. It is logically necessary to attain the early level before the later one can be mastered.

Early driving was characterized as inflexible and rule-governed. Control for the newer drivers was difficult and required a great deal of attention. Many inexperienced drivers reported struggling to keep their car in the lane. They practiced in empty parking lots before driving on a road with lines and approaching cars. They were unable to fully develop the skills necessary for skilled lane maintenance until earlier skills were mastered. Basic control was followed by increases in automaticity as simple perceptual and motor responses became integrated. Now, the drivers could lift their gaze from the controls. These drivers began to recognize recurring patterns. Eventually, broad view and anticipation allowed drivers to engage in more complex planning. Proficient drivers build on past skills to master more complex maneuvers in demanding contexts.

Interactive.

Consistent with Piaget, we consider a driver's functional understanding of driving as an active adaptation of the driver to the complex roadway environment. A new driver may have years of passive experience in a car and in the mental simulation of driving. Until that knowledge is put into practice, however, the novice cannot make good decisions on the roadway. At the beginning, simple declarative rules are transformed as we adjust to variations encountered in the driving environment. Our early interactions with straight roads, turns, and stop signs develop a new cognitive structure for these challenges. Performance improves as we actively search and interpret the driving environment. Because drivers construct rich mental models, they can search differently from before. Similarly, even a proficient U.S. driver might encounter difficulties when driving in London. Faced with a reversal of familiar traffic patterns, long automatized sequences might be lost. The U.S. driver might be reduced to focusing on rules and control during early interactions with the London roadways.

Adaptive.

Our interactions with the roadway form the basis of adaptation. Piaget's model of development requires an iterative process of adaptation. This adaptation has two aspects: assimilation and accommodation. In *assimilation*, we apply our current knowledge structures to new environmental information. We take our knowledge of hills and use it for decisions about an upcoming hill. This assumed similarity dictates our decision about appropriate motor response. In *accommodation*, we adjust our cognitive structures to the characteristics of the environment. When we are surprised by our speed of ascent up a particularly steep hill, we accommodate our cognitive structure to this new dimension, adding complexity to our understanding of hills.

The broadness of our view at any point in development reflects the adaptation at that point. Drivers who have accommodated to many surprising experiences and have assimilated many experiences using their understanding of the task have broader views. They can "see" elements of the driving situation that are within their adaptive level. In this framework, situational awareness is an assimilation of the driving context. Situational awareness is limited by a driver's adaptation to contextual variables. Adaptation allows action that appears intuitive because it is grounded in experience with related exemplars.

Equilibration.

Within Piaget's theory, equilibration describes the continuous movement between states of equilibrium and disequilibrium. *Equilibrium* is a steady, comfortable state during which we can use already formed mental models to interpret the world. We practice successful patterns and explore situations to which our cognitive structures can be applied. Assimilation is successful because we can interpret the world with existing cognitive structures. Accommodation is less important during equilibrium. We are in equilibrium when we drive the same roads, in the same car, with the same road conditions.

In contrast, *disequilibrium* occurs when interactions in the world present challenges that are not well managed with existing cognitive structures, where our mental model fails. Di Bello (1999) suggested that the discomfort of failure is a powerful pressure for change and development. Failures force us to recognize the limits of our cognitive structures. We must accommodate our cognitive structures to include the new information inherent in the failure or difficulty. With each resolved disequilibrium, we incorporate more of the real world in our cognitive framework, and we can function at a higher level. We experience disequilibrium when we first skid on ice, pull a trailer, or drive in London.

The cycle of equilibration produces an upward spiral in development as we meet each challenge with cognitive reorganization. Progress is discontinuous with periods of equilibrium between periods of disequilibrium. Young drivers who start out in large empty parking lots can practice stops while attending to this task alone. The first venture on the road reveals the limitations of the supposedly successful stopping strategy. The demands of the roadway make it necessary to accommodate the additional information presented by the oncoming traffic and lane markers. This is often a distressful experience, but it is necessary for appropriate adaptation.

Training Driving Skills

This developmental model provides a picture for understanding and training driving skills. New efforts need to incorporate our understanding of cognitive development into training. Our data lend support for extending the training period prior to permanent licensing. Graduated licensing procedures being implemented in many states mandate this extended training period for young drivers (Williams, 1997; NHTSA, 1994). New drivers need driving time to develop control and automaticity. Adaptation requires repeated opportunities to accommodate existing cognitive structures and assimilate familiar exemplars. Driving skills emerge from the complex and extended interaction of the driver with driving situations. Experience alone, however, does not ensure higher levels of driving skill. We must provide training that matches the adaptation level of the driver and creates useful disequilibrium without overwhelming the driver. Our research provides several recommendations, discussed next.

Initial Focus on Declarative Knowledge.

Initial training must begin with declarative knowledge because all drivers must follow the same rules and must master basic knowledge. This is what the classroom portion of driver's education courses cover. Some declarative knowledge is learned even before formal driver's education. The acquisition of declarative knowledge should continue as new technologies are introduced. Although antilock brake systems hold the promise of improved safety, many drivers do not know what system their car has nor do they know how to use their brakes to achieve the safest stop. Efforts to improve skills in long-term drivers should start with an assessment of declarative knowledge.

Foster the Development of Control. Even if novice drivers understand the rules and theory of driving, they must practice extensively to gain control of the vehicle. Practice on a simple driving course or in an empty parking lot allows repeated opportunities for adaptation. The new driver assimilates familiar exemplars and accommodates to unfamiliar ones. Activities such as staying in a single lane or making right turns require extensive practice to achieve control and automaticity. The novice driver's lack of control is easier to understand when the six on-road hours of driver's education are compared to the hundreds of hours devoted to training of pilots and the thousands of hours spent by athletes and musicians. Many drivers, even experienced ones, benefit from practice in trouble spots. A brief session of intentional skids in an empty parking lot can refresh a driver's winter vehicle control skills.

Train Within Limits. We found that the pace of development is inherently limited by the requirement for sequential adaptation. We neither expect 6-year-old soccer players to play like Pélé nor train them as if they could, yet beginning drivers are often instructed well above their current abilities. Although providing information for the next developmental level is important, providing it far ahead of capacity can be useless and even counterproductive. The newer driver lacks the adaptive structure to function in a wide range of situations. If inexperienced drivers need most of their attention just to keep between the lines, then directing them to watch far ahead will be dangerous. Training needs to be appropriate to the level of the student.

Mirror use provides an example of training limits. Some driver training programs recommend scanning rituals for the use of mirrors from the beginning of training. These consist of a scanning pattern of rear view, right side, rear view, and then left side. The pattern is to be repeated every minute. We interviewed drivers who have been taught this approach and reported using it. They were, nevertheless, repeatedly surprised when they were passed on an interstate. They did not understand the field displayed in the mirror nor how to use that information. When mirror use is taught early in training, many students cannot tap the information. If the extensive use of mirrors is taught after basic driving tasks have been automatized, mirrors become an important information source for decision making rather than a momentary distraction.

Build on Past Level. Only when drivers have a feel for how the brake system reacts can they begin to brake appropriately in challenging roadway situations. Only after the acquisition of control can training usefully stress automatization of skills. When automaticity is achieved, the many techniques available to expand view, including mirror use, become helpful to the driver. Each advance in adaptive level should be built on past levels and should provide the base needed for further advancements. If people do not have the necessary adaptive structure, then they cannot accommodate the new challenges. Disequilibrium is useful only if the driver is advanced enough to discover a solution. Di Bello (1999) noted the importance of failures as preparation for abandoning a prevailing mental model and moving to a richer one.

Encourage a Broader View. Automaticity gives the driver the freedom to devote attention to the development of a broad view. Consistent with the developmental analogy, more complex notions of "managing the road" should be introduced when automaticity is mastered. Here, the driver needs encouragement to monitor cars approaching from the rear and to avoid danger. This monitoring also allows the driver to modulate speed based on peripheral cues or traffic pattern changes. As anticipation and broad view becomes possible, "behind the wheel" coaching can focus the driver on these capacities.

A broad view helps older adults who are experiencing capacity changes. They can recognize potential problems and find solutions before their driving becomes hazardous. Courses in defensive driving aim to foster a broader view by way of classroom instruction. Lund and Williams's (1985) review of defensive driving courses found that research did not support the courses' effectiveness in reducing crashes. In contrast to classroom courses, coached driving has the potential for advancing the cognitive level of performance for drivers encountering new limitations. Coached driving allows active involvement with opportunities for adaptation and disequilibrium.

Train the Trainers. Coaching of new drivers usually falls to parents. Parents, however, are often not aware of the implications of the novice driver's limited control capacities. Thus, it is important that coaches understand the capacity limits of the student driver. Coaches also need guidelines for deciding when they can safely move the new driver toward the next level. With the increase in the number of states that are mandating graduated licensing procedures, much more effort should be devoted to educating the parents who will be coaching new drivers.

Our Next Questions

Our next questions address initial skill development, the development of advanced capacities, and the identification of expert driving. Longitudinal studies of the early processes of adaptation and equilibration would show how early driving skills develop. This knowledge of cognition could help enhance driver training. Maintaining driving proficiency depends on developing advanced capacities. New capacities are needed, for example, by the Texan driving on ice, the U.S. driver in London for the first time, and people first using a standard transmission. Drivers developing specific advanced competencies may provide a window into the processes needed for continued skill growth. Finally, we wish to look at highly skilled and expert drivers. Highly skilled drivers from the ranks of professional drivers, such as state patrol officers, truck drivers, and delivery drivers, should help us learn about the development of advanced driving skills.

Driving and the Naturalistic Decision Making Model

The research discussed here demonstrates an application of Naturalistic Decision Making (NDM) methodologies to the complex and dangerous domain of driving. Cognitive task analysis interviews with newer and long-term drivers delineated different patterns of decision making as skill levels progressed. Indeed, novice and proficient drivers seem to be driving in different worlds—the novice must decide how hard to push the brake pedal or who arrived at an intersection first whereas the proficient driver might consider whether the car one fourth mile back is going to continue passing people. This application of NDM methodology has allowed us to describe the performance levels and processes of decision making in driving. It has provided a developmental model with implications for improving the training of drivers. NDM's value has been amply demonstrated in a wide range of specialized domains of expertise. This research shows its promise in a domain in which expertise is rare but in which poor performance can impact any one of us nearly every day.

ACKNOWLEDGMENTS

This research was sponsored in part by a Wright State University Pruet Seed Grant. The authors would like to thank Robert V. Muldoon and Janelle Larson for their valued help.

REFERENCES

Anderson, J. R. (1993). *Rules of the Mind.* Hillsdale, NJ: Lawrence Erlbaum Associates.

Campbell, R. L., & Di Bello, L. (1998) Studying human expertise: Beyond the binary paradigm. *Journal of Experimental & Theoretical Artificial Intelligence, 8*, 277–292.

Crandall, B., & Calderwood, R. (1989). *Clinical assessment skills of experienced neonatal intensive care nurses* (Final report prepared for the National Center for Nursing, NIH under contract No. 1 R43 NR01911 01). Yellow Springs, OH: Klein Associates.

Crandall, B., & Getchell-Reiter, K. (1993). Critical decision method: A technique for eliciting concrete assessment indicators from "intuition" on NICU nurses. *Advances in Nursing Science, 16*, 42–51.

Di Bello, L. (1999). *Iterative methods of technology implementation and performance evaluation.* Paper presentated at the R&T Professional Capability Taskforce Mini-Conference, Washington, DC.

Dominguez, C. O. (1997). *First, do no harm: Expertise and metacognition in laparoscopic surgery.* Unpublished doctoral dissertation, Wright State University, Dayton, OH.

Dreyfus, H. L. (1997). Intuitive, deliberative, and calculative models of expert performance. In C. E. Zsambok & G. Klein (Eds.), *Naturalistic decision making* (pp. 17–28). Mahwah, NJ: Lawrence Erlbaum Associates.

Dreyfus, H. L., & Dreyfus, S. E. (1986). *Mind over machine: The power of human intuition and expertise in the era of the computer.* New York: The Free Press.

Flavell, J. H.(1963). *The developmental psychology of Jean Piaget.* Princeton, NJ: Van Nostrand.

Flavell, J. H., Miller, P. H., & Miller, S.A (1993). *Cognitive psychology* (3rd ed.). Englewood Cliffs, NJ: Prentice-Hall.

Klein, G. A., Calderwood, R., & Clinton-Cirocco, A. (1986). Rapid decision making on the fire ground. *Proceedings of the Human Factors Society 30th Annual Meeting* (pp. 576–580). Dayton, OH: Human Factors Society.

Laux, L. F., & Brelsford, J., Jr. (1990). *Age-related changes in sensory, cognitive, psychomotor and physical functioning and driving performance in drivers aged 40 to 92.* Washington, DC: AAA Foundation for Traffic Safety.

Llaneras, R. E., Swezey, R. W., Brock, J. F., & Rogers, W. C. (1993). Human abilities and age-related changes in driving performance. *Journal of the Washington Academy of Science, 83*, 32–78.

Lund, A. K., & Williams, A. F. (1985). A review of the literature evaluating the defensive driving course. *Accident Analysis and Prevention, 17*, 449–460.

National Highway Traffic Safety Administration (NHTSA). (1994). *Research agenda for an improved novice driver education program.* Washington, DC: U.S. Government Printing Office.

National Highway Traffic Safety Administration (NHTSA). (1997). *Traffic safety facts 1996* (U.S. Department of Transportation Rep. HS 808 543). Washington, DC: U.S. Government Printing Office.

National Highway Traffic Safety Administration (NHTSA). (1998). *Traffic safety facts 1997* (U.S. Department of Transportation Rep. HS 808 806). Washington, DC: U.S. Government Printing Office.

Newell, A. (1990). *Unified theories of cognition.* Cambridge, MA: Harvard University Press.

Patel, V. L., & Groen, G.J. (1991). The general and specific nature of medical expertise: A critical look. In K. A. Ericsson & J. Smith (Eds.), *Toward a general theory of expertise* (pp. 93–125). Cambridge, England: Cambridge University Press.

Piaget. J. (1950). *Introduction to genetic epistemology.* Paris: University Press.

Piaget. J. (1970). Piaget's theory. In P. H. Mussen (Ed.), *Carmichael's manual of child psychology* (Vol.1, pp. 703–732). New York: Wiley.

Salthouse, T. A. (1985). Speed of behavior and its implications for cognition. In J. E. Birren & K.W. Schaie (Eds.), *Handbook of the psychology of aging* (2nd ed., pp. 400–426). New York: Van Nostrand Reinhold.

Shinar, D., & Schieber, F. (1991). Visual requirements for safety and mobility of older drivers. *Human Factors, 33*, 507–519.

Vincent, E. J. (1997). *Naturalistic decision making in older adults: The process of remaining independent.* Unpublished master's thesis, Wright State University, Dayton, OH.

Waag, W. L., & Bell, H. H. (1997). Situation assessment and decision making in skilled fighter pilots. In C. E. Zsambok & G. Klein (Eds.), *Naturalistic decision making* (pp. 247–254). Mahwah, NJ: Lawrence Erlbaum Associates.

Williams, A. F. (1997). *Graduated licensing and other approaches to controlling young drivers.* Arlington, VA: Insurance Institute for Highway Safety.

19

Tool Retention and Fatalities in Wildland Fire Settings: Conceptualizing the Naturalistic

Karl E. Weick
University of Michigan

I want to focus on a stubborn problem in wildland fire fighting. The problem is this. When a wildland fire explodes and threatens to overrun a crew of firefighters, the crew's ability to outrun the fire improves if they drop their packs and tools so they can run faster, cover more ground, and escape to a safety zone. Given this relatively clear means to mitigate the risk of being burned, why is it then that, since 1990, 23 firefighters in four separate incidents,[1] refused to drop their tools when ordered to do so, were overrun by fire, and died with their tools beside them? They died within sight of safety zones that they could have reached had they been lighter and moved faster. For example, at the South Canyon disaster outside Glenwood Springs, Colorado, 14 firefighters were killed on July 6, 1994, when they failed to outrun a fire that exploded through a flammable stand of Gambel Oak just below them. At the time the bodies were being recovered, a site and thermal analysis was written for each person recovered. Part of the analysis for firefighter #10 reads, "was still wearing his back pack. . . . Victim has chain saw handle still in hand with chain saw immediately above right hand. Saw blade is parallel to firefighter #9's left leg." The body of firefighter #10 is approximately 250 feet below the safety of the ridge above, a distance that could have been covered had this person exerted the same amount of energy but dropped his pack and saw 5 minutes earlier.

---

[1]Six died at the Dude fire, 14 at South Canyon, 2 at the California fire, and 1 at the Buchanan fire.

In an effort to accomplish three things, I want to focus on this puzzle of why people keep their tools and lose their lives.

First, I want to prime our thinking about properties of events in the world by discussing a problem that fulfills eight criteria of a naturalistic decision-making setting (Orasanu & Connolly, 1993). The wildland firefighting puzzle is a good means to do that because it involves time pressure, high stakes, experienced decision makers, inadequate information, ill-defined goals, rich context, dynamic conditions, and team coordination (Klein, 1998, p. 4). This problem also fits Zsambok's (1997, p. 4) short definition of Naturalistic Decision Making because it is focused on "the way people use their experience to make decisions in field settings."

Second, I want to prime our conceptualization of naturalistic events by suggesting ways in which the problem of tool retention connects with several issues in NDM theorizing. I conclude this section with the suggestion that NDM theorizing would be strengthened if researchers paid closer attention to improvisation as a process that runs parallel to naturalistic decision making.

Third, I want to add a vivid metaphor to the repertoire of images that capture the nature of crisis decision making. The reluctance to drop tools is not just something that happens in wildland fire fighting. Karl Wallenda, the tightrope walker, lost his life when he fell off the wire and continued to clutch his balance pole all the way down to the ground, rather than drop it and grab the wire on which he had been walking. Fighter pilots whose planes become disabled lose their lives when they hold onto what they call "the cocoon of the cockpit" rather than face the harsh conditions of an ejection. Engineers on the Challenger project failed to drop their launch routines in the face of increasingly severe blowby in O-rings and approved the launch that killed seven astronauts. Naval personnel told to remove their steel-toed shoes before abandoning a sinking ship, refuse to do so and die when they jump off the ship and sink to the bottom or punch holes in life rafts when they board them. Scuba divers who get into trouble in deep water refuse to drop their weight belts to speed their ascent.

TOOL DROPPING IN THE SOUTH CANYON DISASTER

The most recent example of fatalities caused in part by a failure to drop tools when ordered to do so, is the South Canyon disaster. This disaster is representative of the other three wildfire disasters where people died with their tools at hand. It has striking similarities to the Mann Gulch disaster in 1949, made famous in Norman Maclean's (1992) book *Young Men and Fire*, where people also were reluctant to drop their tools.

The South Canyon disaster occurred July 6, 1994, near Glenwood Springs, Colorado, at 4:00 p.m. on a dry, windy, hot afternoon when a mixed crew of smokejumpers and hotshots were constructing a fireline downhill on the east slope of a valley near Storm King Mountain. The fire they were trying to stop circled around them on the south and started up the west slope of the valley. Portions of it spotted across to the east slope underneath them and overran them with flame heights of 150 feet while they were retreating up to the ridge top on the east slope. A group of firefighters already on the ridge top yelled at them to speed up and drop their tools, but they didn't. Why not?

There are several plausible reasons that have been culled from interviews, court depositions, field observation by humans who retraced the route of the victims under similar conditions of heat, weight, exhaustion, and hunger, examination of amateur photographs, analysis of accident investigations, and computer simulations.

Some of the plausible reasons are obvious. First, the exploding fire was so loud that the crew may not have heard the order to drop their tools. Second, they were strung out in single file, and because the fire was behind them and they did not turn around, when they were told to drop their tools, in their eyes there was no obvious reason why they should do so. Third, because the people who told them to drop their tools were not familiar and were from a rival specialty and because the incident command had been flawed all day, the retreating crew had no reason to trust the order even if they heard it.[2] Fourth, because people sometimes survive fires by deploying and crawling inside a fire shelter, and because these shelters are safest when deployed in an area that is cleared of underbrush, they need some of their tools to clear a safe area. Fifth, firefighters were tired, hungry, dehydrated, and had ingested considerable carbon monoxide, all of which made it more difficult for them to think clearly no matter what they heard. And, sixth, firefighters have little knowledge of what their equipment weighs, and even less experience converting a reduction in weight into a gain in speed. Post hoc calculations made by experts suggest that if people at South Canyon had covered 6 to 9 more inches per second when they started to march out, they would have made it to safety. But people in the crew had no way of knowing this. Nor, when events are escalating swiftly, massively, and deafeningly, are people likely to believe that changes this small can make a big enough difference to matter. Thus, there is evidence that the obvious factors of deficiencies in hearing, rationale, trust, control, physical well-being, and calculation may have made people unwilling to drop their tools.

But there are other reasons that are a little less obvious. First, people may take their chances and keep their tools because the alternative of dropping those tools

[2]The Prineville Hotshots did not know Haugh, Erickson, Hipke, Thrash, or Roth, all five of whom were smokejumpers, and all five of whom told the hotshots either to run or deploy shelters. Hotshots are trained to wait for orders from their own leaders. Six of the hotshots who died still had their fire shelters in their cases, presumably because no order had been given to deploy them.

and deploying a fire shelter seems even riskier. Firefighters get almost no practice deploying fire shelters. Furthermore, it is tough to open a shelter while running, in turbulent winds, with gloves on, and while looking for a clear flat area in which to lay down. Fire shelters resist heat radiation and preserve cooler oxygen, but they cannot deflect direct flames. The meaning of shelters in a "can do" culture is occasionally likened to "wearing a life jacket while sailing on the QE II." By this people mean, why should I carry a four-pound shelter all the time for a contingency that probably will never happen. In at least one instance at South Canyon, a fire shelter case was found to contain a rain poncho rather than a shelter, which may have been an intentional or accidental substitution.

A second less obvious reason people may keep their tools is that, to drop one's tools is to admit failure, to keep one's tools is to retain the possibility of success. To keep one's tools is to reaffirm that one is still in it, that the danger will pass, and that everything will work out.

Related is the third possibility that a tool is comforting in the same way a panic button (Glass & Singer, 1972) is comforting. People exposed to stressful conditions appear to cope better if they have control over the situation in the form of a panic button whose activation is able to shut off the stressors. When people are given a panic button, they seldom push it. But its mere presence is enough to calm them down and improve performance. Retention of familiar tools may serve the same function for firefighters.

A fourth possibility is that people may hold onto their tools as a simple result of social dynamics when their crew is lined up single file and marching up a trail. If the first person in a line of people moving up an escape route keeps his or her tools, then the second person in line who sees this may conclude that the first person is not scared. Having concluded that there is no cause for worry or that it would be too embarrassing to go back to camp as the only person without tools, the second person also retains his or her tools and is observed to do so by the third person in line, who similarly infers less danger than may exist. Each person individually may be fearful, but mistakenly concludes that everyone else is calm. Thus, the situation appears to be safe except that no one actually believes that it is. The actions of the last person in line, the one who feels most intensely the heat of the blowup, are observed by no one, which means it is tough to convey the gravity of the situation back up to the front of the line.

Thus, there are the less obvious reasons of fear of unfamiliar shelter technology, reluctance to admit failure in a "can do" culture, eagerness to retain a panic button, and perception that fear is not widespread.

Finally, there are a set of reasons that sound not just less obvious but downright absurd. For example, there is evidence that people at South Canyon did not know how to drop their tools. At South Canyon, Quentin Rhoades, who survived by running perpendicular to the fire, describes running but being slowed because he was trying to find a place to put down his saw so it would not be burned. In his words, "at some point, about 300 yds. up the hill . . . I then realized I still had my saw over my shoulder! I irrationally started looking for a place to put it down where it wouldn't get burned. I found a place I it (sic) didn't, though the others' saws did. I remember thinking I can't believe I'm putting down my saw." Smokejumper Tony Petrelli at South Canyon recounts the following sequencing of events: "Shelton said he was putting his sigg pack down [holds gasoline for chain saw]. This was approximately 150 feet above the lunch spot. It was then I realized I still had my saw. I put my saw down beside the sigg pack. I knew this wasn't the best place to lay the saw, but putting down jacked the pucker factor up one more notch" (U.S. Forest Service, 1994, p. A5-69).

The same pattern occurred at Mann Gulch. In his testimony during the Mann Gulch accident investigation, Walter Rumsey mentioned that even though he was running for his life, he saw that Eldon Diettert was carrying a shovel. Rumsey grabbed it, but then carefully leaned it against a tree. At the Dude fire in Payson, Arizona, the crew chief yelled at his crew to drop their tools. They did so. But, when the crew chief ran out, he stopped and picked up the tools that had been discarded and carried them out, barely escaping the exploding fire. When he got to the escape road and jumped on a moving pickup truck that was trying to outrun the fire, he threw the tools in the pickup bed. One of the shovels was on fire and had to be thrown out to keep from burning firefighters who had just escaped the blowup.

People who have been trained to carry out whatever equipment they carry in to a fire or whatever is dropped to them, people who hear repeatedly how much equipment costs (e.g., a fire shelter costs $23, a parachute costs $600), and people who practice carrying heavier and heavier loads, faster, for longer periods, on steeper slopes, might be at a disadvantage when, without any prior experience of what it feels like or how to do it, they are told to drop their tools and their packs. From what we know about the effects of stress on overlearned behavior (e.g., Weick, 1990), the safest prediction would be that firefighters under pressure would regress to what they know best, which in this case would be keeping their tools. If somehow they could override that tendency, they still might try to protect these expensive tools by putting them down carefully away from potential flames, which still eats up precious time.

It may seem odd to think that people keep their tools because they do not know how to drop them, but it will seem even odder to you when I raise the possibility that the firefighters did not drop their tools because they did not have any. I have assumed all along that tools and people are distinct, separable, dissimilar entities. But fire suppression is an activity done with capabilities, not just bodies. Firefighting tools, such as the Pulaski, are often named after famous firefighters and are designed solely for fire fighting. Their skillful use is the mark of a seasoned firefighter and central to that person's identity. The fusion of tools with identities means that, under conditions of threat, it makes no more sense to drop one's tools than it does to drop one's pride or one's sense of self. Tools and identities form a unity without seams or separable elements.

Listen to Norman Maclean's (1992) reflections on firefighter identity at Mann Gulch:

1. "When a firefighter is told to drop his firefighting tools, he is told to forget he is a firefighter and run for his life" (p. 273).
2. "When fire fighters are told to throw away their tools, they don't know who they are anymore, not even what gender" (p. 226).

What is striking is the apparent fusion of tools with identities. When I first posed the question, why don't firefighters drop their tools, I assumed a separation between firefighters and tools that may not be their circumstance at all. Their circumstance may be one of equipment, projects, and action in context, these being concerns that shift as their needs shift. This is not at all the same as the subject-object world that I impose as a detached observer. I take this to be one of the lessons implied by Heidegger's (1962) discussion of three modes of engagement adopted by people in the world. He distinguishes among *ready-to-hand* engagement where people are absorbed in interrelated projects and deal with equipment in use, not discrete tools; *unready-to-hand* engagement where interruptions of projects show shortcomings of contextualized equipment that can be surmounted; and, detached *present-at-hand* engagement where equipment is decontextualized into discrete entities and viewed in terms of context-free properties, such as weight and mass, that can be measured. The detachment needed to separate people from their tools may require that they shift from ready-to-hand coping to present-at-hand detachment. This switch may be easier for novices than experts, because novices are never very far from a subject-object grasp of new technology and procedures (Dreyfuss, 1997). Thus, those who die with their tools beside them may have more expertise than those who survive. If Heidegger is taken literally, a crew leader who yells "drop your equipment" may gain more compliance than one who yells, "drop your tools." That assumes a crew of experts. In a crew of novices, the opposite prediction would hold. It is in pursuit of possible explanations such as these that I now turn to a sampling of ideas associated with NDM and examine the degree to which they help explain tool retention.

CONCEPTUAL CONNECTIONS OF TOOL DROPPING WITH THEMES IN NDM

The purpose of this section is to sample theoretical issues associated with NDM that seem to have a bearing on the problem of tool dropping. The intent of this sampling is to suggest, in the style of hermeneutic inquiry, that details of tool dropping enrich NDM concepts, and NDM concepts in turn let one see more in the events themselves. Said differently, this section treats ideas associated with NDM as an alternative vocabulary (Rorty, 1989) to interpret the puzzle of tool retention and asks the question, Does this vocabulary help us gain a better understanding of this incident?

Initially, it is important to revisit the earlier assertion that this event exemplifies eight criteria for a naturalistic decision setting. That assertion represents a selective rendering of the event and imposes a figure–ground structure on it that suggests what may be more and less important about the event. The event is *time-pressured* because of the speed with which a fire blowup burns and sweeps over exhausted people running uphill. The event has *high stakes* because, escaped fires not only kill people and destroy property, but also how those fires are fought affects the reputations of crews and their chance to get good assignments in the future. The firefighters in these incidents are *experienced* in the sense that they have been on numerous wildland fires before, but they have had less experience dropping tools, facing a blowup, retreating without clear escape routes and safety zones, or deploying fire shelters. The event is one of *inadequate information* because crews lack an accurate weather briefing, clear instructions from dispatchers as to what backup resources are on the way and when they will arrive, information about the flammability of surrounding vegetation, and knowledge of the attack plan of the incident commander. The event is characterized by *ill-defined goals* because it is not clear whether they are protecting structures, containing the fire, trying to extinguish it directly, preparing for a backfire, or marking time until the fire does what it is going to do anyway. The event has rich *context* although that context is richer at the sharp end of the arrow than at the blunt end, a discrepancy that protects senior management but may preclude learning and change. The event has *dynamic conditions* because winds are freshening, embers are being thrown ahead of the flames into unburned fuel, and radio traffic is increasing while radio discipline is decreasing. Finally, *team coordination* is a growing issue because some firefighters are advancing and others retreating, crews from different specialties and regions are mixed together and unacquainted, and crew leaders are unsure of where people are, what those people understand the situation to be, or how severe the problem is.

If rather than Orasanu & Connolly's (1993) list, we use those crafted by Cannon-Bowers, Salas, and Pruitt (1996) or the adaptation of Cannon-Bowers et al. used by Martin, Flin, and Skriver (1997), the conclusion remains essentially the same. People who die in wildland fires with their tools beside them suffer this fate under conditions associated with other crises in NDM.

The closest direct parallel I have found between the South Canyon setting and issues examined in NDM is a set of problems mentioned by Klein (1997) and by Orasanu, Martin, and Davison (1998; chap. 12, this volume). Klein (1997, p. 19) described a potential family of decisions that involve "abandonment of a routine plan." He includes as examples of such decisions, the decision to abort a takeoff, make an emergency landing, shut down a petrochemical plant, and "the decision to abandon a fire line (because of a wind shift that increases the vulnerability of the wildland firefighters)" (p. 19). What is prescient about this fire line speculation is that "freak winds" were contributing factors to all four incidents where tools were found alongside bodies. The hesitation to abandon the plan of downhill line construction at South Canyon while the fire advanced, possibly encouraged by de minimus misjudgments (Klein, 1998, pp. 66, 69–70) of cues that the fire was growing more intense, closed off the option of a safe retreat with tools and made it almost impossible to run to safety even without tools. Notice that there is close resemblance between the scenario just described and the phenomenon of "flying behind the plane" (Klein, 1998, p. 155) in which there is bad situation awareness and people are unable to anticipate.

Orasanu et al. (1998) find that the most common kind of decision error in aviation accidents occurs when the crew continues with the original plan of action in the face of cues that suggest changing the course of action (p. 5). Generalized to fire fighting, whether the plan underway is an orderly retreat or digging a longer fireline before the winds hit, its continuation means that people either fail to do something that should have been done (e.g., re-position lookouts to get a clearer vantage point to watch fire spread) or they take actions out of the ordinary that enable them to continue with the plan (e.g., keep digging line downhill in the belief/hope that aircraft would soon drop water and cool the area ahead of them that was beginning to smoke). These plan-continuation errors appear to be caused both by bounded rationality (e.g., firefighters lacked information and were also "rusty" because they had not been on a big fire in some time) and by error-inducing contexts. *Error-inducing contexts* occur when people are under the influence of organizational and social pressures that shape their interpretation of ambiguous cues in high risk situations in ways that lead them to underestimate risk, overestimate competence, and misestimate the consequences of planned actions (Orasanu et al., 1998, pp. 6–9). Details about firefighters at South Canyon already presented suggest that their situation fits the profile of an error-inducing context. For example, firefighters memorize a list of 10 fire orders and 18 watch out situations that are signs of pending danger, but they often violate these rules without negative outcomes. Sometimes, these rules need to be violated in order to suppress fires quickly. Several of these rules were being violated at South Canyon (Weick, 1995b), a situation that is not unlike Orasanu et al.'s error-inducing contexts where an experienced pilot lands yet again in Alaska with "bad" conditions only to crash when he or she continues with an unstable approach and this time, loses the gamble.

If firefighters are caught in a potential plan-continuation error, then the leader has to intervene to change the project convincingly. If firefighters are unable to see beyond their project of fire suppression, then the leader has to create a defining moment and take countermeasures against the group's own fixations. The leader has to stop the suppression project cold, confirm that they face an exploding fire, and reset the project clearly and firmly as a race. If the project of a race replaces the project of suppression, then speed and lightness and rapid movement toward a safe zone become the new relevancies. Anything that interferes with winning the race now becomes visible as an obstacle and is discarded. Tragedy may occur when the moment is perceived as less than defining, and there is a holdover from the prior project of suppression that creates ambivalent mixed action.

Although much about the tool retention problem fits a NDM framework, there are portions of it that resemble classical rational decision making and the search among options. For example, the question of which escape route to take (up the side of the mountain or back up the fireline that has already been cut) is similar to the question of which airport a pilot should divert to when there are weather problems at the destination (Klein, 1998, p. 93). Furthermore, analytical decision methods may be necessary when it is difficult for people to build up expertise (Klein, 1998, p. 282). That becomes relevant in the case of wildland firefighters because they are often seasonal employees who do not fight fire in winter. Furthermore, major fires are relatively rare, and crew leadership may change from fire to fire (e.g., smokejumpers treat the first person out the door of the aircraft as the Jumper-In-Charge, as they did at South Canyon, and this may lead to a less expert firefighter being in charge of those who are more expert).

IMPROVISATION AS AN INFRASTRUCTURE FOR NATURALISTIC ACTION

Given the several assists from the vocabulary of NDM just reviewed, I want to suggest that there is a common theme and a common form found in many of the episodes of recognitional, intuitive, ready-to-hand decision making and problem solving that are in the NDM data base. This common theme and form centers on the process of improvisation. People who discuss NDM are not strangers to the notion of improvisation. Klein (1998), for example, discussed improvisation as crucial in the recovery of Apollo 13 (p. 137), as a capability of experts expressed in their use of counterfactuals (pp. 149, 154), as something that cannot be faked by pretenders to expertise (p. 156), as something visible in blitz moves in chess (p. 163), and as more or less likely to be effective depending on the ways in which commanders express their intent (pp. 222–225). Klein's concern that overly detailed statements of intent shut down improvisation is the same concern I have when firefighters are burdened down with lengthy checklists whose value seems to be largely that it allows administrators to point to some neglected rule in a disaster and to shift blame from the blunt end of the arrow back toward the sharper end.

I want to suggest that improvisation is not just an activity that attaches itself every now and then to episodes of NDM. Rather, I want to argue that it is a recurrent form of action in everyday life, that it is a candidate for the "glue" that ties together the various properties of NDM, and that, at least in the case of fire fatalities involving tools, it suggests an approach to deal with the problem.

Improvisation can be defined as "reworking precomposed material and designs in relation to unanticipated ideas conceived, shaped, and transformed under the special conditions of performance, thereby adding unique features to every creation" (Berliner, 1994, p. 241). Improvisation involves the flexible treatment of preplanned material, but it is not about "making something out of nothing." Instead, it is about making something out of previous experience, practice, and knowledge during those moments when people surface and test intuitive understandings while their ongoing action can still make a difference (Schon, 1987, pp. 26–27). Jazz musician Stan Getz described the preplanning of improvisation using the metaphor of language: Jazz is "like a language. You learn the alphabet, which are the scales. You learn sentences, which are the chords. And then you talk extemporaneously with the horn" (Maggin, 1996, p. 21).

Gilbert Ryle (1979) argued that virtually all behavior has an ad hoc adroitness akin to improvisation because it mixes together a partly fresh contingency with general lessons previously learned. It is this depiction that best illustrates the possibility that improvisation is part of the infrastructure of NDM. "[T]o be thinking what he is here and now up against, he must both be trying to adjust himself to just this present once-only situation *and* in doing this to be applying lessons already learned. There must be in his response a union of some Ad Hockery with some know-how. If he is not at once *improvising* and improvising *warily*, he is not engaging his somewhat trained wits in a partly fresh situation. It is the pitting of an acquired competence or skill against unprogrammed opportunity, obstacle or hazard" (1979, p. 129, italics in original).

In the context of conceptualizing NDM, what seems important about mating NDM with improvisation is that improvisation implies grounding in experience, wary use of recognition, reliance on lessons learned, focus on situation awareness ("thinking what he is here and now up against"), the centrality of intuition ("union of some Ad Hockery with some know-how") and embellishment of modest structures. When people improvise, they act in order to think, which imparts a flavor of retrospective sensemaking (Weick, 1995a) to their activity. Ted Gioia (1988), writing about jazz improvisation, put it this way: unlike an architect who works from plans and looks ahead, a jazz musician cannot "look ahead at what he is going to play, but he can look behind at what he has just played; thus each new musical phrase can be shaped with relation to what has gone before. He creates his form retrospectively" (p. 61). The person who builds form retrospectively builds something that is recognizable from whatever is at hand, contributes to an emerging structure being built by the group with which he or she is working, and creates possibilities that can be taken up by other members. Gioia's description suggests that in improvisation, as in ready-to-hand engagement, intention is loosely coupled to execution and more tightly coupled to creation and interpretation.

Improvisation is grounded in experience, something that people often miss when they describe it as making something out of nothing. "This simplistic understanding of improvisation belies the discipline and experience on which improvisers depend, and it obscures the actual practices and processes that engage them. Improvisation depends, in fact, on thinkers having absorbed a broad base of musical knowledge, including myriad conventions that contribute to formulating ideas logically, cogently, and expressively. It is not surprising, therefore, that improvisers use metaphors of language in discussing their art form. The same complex mix of elements and processes coexists for improvisers as for skilled language practitioners" (Berliner, 1994, p. 492). A person who improvises successfully is more accurately described as a highly disciplined "practicer" (Berliner, 1994, p. 494) than as a practitioner.

Finally, it is important to note that improvisation does not materialize out of thin air. Instead, it materializes around a simple melody, formula, or theme that provides the pretext for real-time composing. Some of that composing is built from precomposed phrases that become meaningful retrospectively as embellishments of that melody. And some comes from elaboration of the embellishments themselves. The use of precomposed fragments is an example of Ryle's "wary improvisation" anchored in past experience. The elaboration of those emerging embellishments is an example of Ryle's opportunistic improvisation in which one's wits engage a fresh situation. Thus, improvisation is guided activity whose guidance comes from minimal ongoing structures and patterns discovered retrospectively.

Let's look quickly at an example of how improvisation works. Isenberg (1985, pp. 178–179) argued that, on battlefields, commanders often "fight empirically" in order to discover what kind of enemy they are up against. "Tactical maneuvers will be undertaken with the primary purpose of learning more about the enemy's position, weaponry, and strength, as well as one's own strength, mobility, and understanding of the battlefield situation. . . . Sometimes the officer will need to implement his or her solution with little or no problem definition and problem solving. Only after taking action and seeing the results will the officer be able to better define the problem that he or she may have already solved!" (pp. 178–179). Commanders essentially hold a diagnosis lightly and tie their understanding to activity. This is akin to a simple melody that is embellished until a more appropriate melody emerges from the embellishments. A hunch held lightly is a direction to be followed, not a decision to be defended. It is easier to change directions than to reverse decisions, simply because less is at stake.

What is striking about military improvisation is a subtle difference between sense making and decision making that is implicit in the phrase, "hunches held lightly." This difference comes out clearly in a discussion I had with Paul Gleason, reputed to be one of the five best wildland firefighters in the world. Gleason said that when fighting fires, he prefers to view his leadership efforts as sense making rather than decision making. In his words, "If I make a decision it is a possession, I take pride in it, I tend to defend it and not listen to those who question it. If I make sense, then this is more dynamic and I listen and I can change it. A decision is something you polish. Sensemaking is a direction for the next period" [Personal Communication, June 13, 1995].

When Gleason perceives himself as making a decision, he reports that he postpones action so he can get the decision "right" and that, after he makes the decision, he finds himself defending it rather than revising it to suit changing circumstances. Both polishing and defending eat up valuable time and encourage blind spots. If, instead, Gleason perceives himself as making sense of an unfolding fire, then he gives his crew a direction for some indefinite period, a direction which by definition is dynamic, open to revision at any time, self-correcting, responsive, and with more of its rationale being transparent.

With these ideas as background, let's return to the problem of how to deal with firefighters who die with their tools in hand. If improvisation is a meaningful analogue of their world, then it ought to give some guidance about how to make that world safer.[3] Think again about what it means to be a firefighter. Firefighting subcultures often enforce norms that encourage people to be macho, tough, proud. Firefighters like to think of themselves as "can do" people who get the job done regardless of the obstacles. In their words, "you're not professional if you run from something."[4] Elite hotshot firefighters, for example, describe themselves as "the go-to person in overtime" who sinks the winning basket when the going gets toughest. What tends to get missed in all this "can do" rhetoric is the infrastructure that makes it possible. To get the job done, people have to be able to improvise using whatever resources they have at hand. Thus, the people best able to sustain a can do identity, may be those who are best able to "make do."

Those who have most success at making do tend to have both deep experience and good "melodies" to embellish. In the case of firefighters, a possible melody available for improvisation in the face of danger is a flexible set of four guidelines for escape that read like this:

1. Build a backfire if you have time;
2. Get to the top of the ridge where the fuel is thinner, where there are stretches of rock and shale, and where winds usually fluctuate;
3. Turn into the fire and try to work through it by piecing together burned-out stretches; and

[3]To examine lapses in fire safety using the perspective of improvisation suggests that firefighters are not practicing their equivalent of scales, are not enlarging their repertoire of knowledge and skill, have poor melodies to embellish, do not know how to combine skills and experience with melodies, and they are punished for improvisation and rewarded for compliance. For example, an increasing number of wildland firefighters come from urban rather than rural backgrounds. This means that they have less experience being woodsmen, have less familiarity with tools and with burning stubble off fields and with how winds swirl in the wild, know less about how to survive and to pace oneself in the wilderness or how to move on uneven ground, are less able to infer directions and time of day, and are less informed about how heat interacts with terrain and foliage. This knowledge for firefighters is the equivalent of knowledge about scales and chords for jazz musicians. It is the material that is available for recombination. If it is unavailable, then it needs to be made available through survival training, a rural–urban buddy system, longer training before fire season, or training away from a firebase locale.

[4]Quotation from firefighter interview at National Convention of Hot Shots.

4. Do not allow the fire to pick the spot where it hits you because it will hit you where it is burning fiercest and fastest. (Maclean, 1992, p. 100)

That list, which has considerable elasticity to accommodate local fire conditions, was used by firefighters until the mid-fifties[5] when it was replaced by lists that were longer, more detailed, more constrained, and that were harder to adapt to conditions that fell outside their specifications (e.g., fires at the urban-wildland interface). Those lists now exceed 48 items (Weick, 1995b, p. 65), and contain injunctions such as (a) fight fire aggressively but provide for safety first, (b) ensure instructions are given and understood, and (c) retain control at all times. I would argue that the longer lists, written to mitigate the risk of fire fatalities by *anticipating* potential risks, may actually have increased risk because they removed both a guide that could be embellished and tacit approval for improvisation as an accepted way to deal with unexpected events. Long checklists shut down improvisation. That shut down of improvisation is often just what is desired in hierarchical, rank-conscious, seniority-driven, deference-based, organizations that pay more attention to anticipation of events than to resilience and that view the top echelon as the source of assertions rather than questions.

If firefighters were trained[6] to become more skilled at improvising, then they might see the danger in an escalating fire sooner and disengage or reposition themselves or change their suppression tactics more quickly. The reason their situation awareness might improve is that when people increase their capability to improvise, they should increase the size of their response repertoire. The repertoire should get bigger because greater skill at improvisation makes it easier for people to recombine old skills and knowledge in new ways to deal with unexpected

[5]It appears likely that this list saved Wagner Dodge's life as is suggested by the following excerpt from the transcript of the original board of review conducted September 26–28, 1949, after the Mann Gulch fire of August 5, 1949. The relevant exchange is as follows:

MAYS: I would like to ask if dodge had ever done that before—lighted an escape fire?

DODGE: No, I never had to use an escape fire before. I have been run off of big fires.

MAYS: It was just your knowledge of fire behavior that told you that was a good thing to do?

DODGE: Yes. The type of burning on the advancing main fire necessitated going back through 250–300 feet of practically solid flame to get to the burned area.

MAYS: Had you ever been instructed in setting an escape fire?

DODGE: Not that I know of. It just seemed the logical thing to do. I had been instructed if possible to get into a burned area. (U. S. Forest Service, 1949, p. 123, Mann Gulch transcript).

[6]An example of the "basics" in improvisation training would be detailed exposure to the fire triangle, the fact that to produce the flame and combustion of a fire, you need three things: heat + oxygen + fuel. Escape from injury occurs when one of these is absent. Thus, when Wagner Dodge lit his escape fire at Mann Gulch, he removed fuel and lowered the heat in the area that was burned away relative to the heat generated by the blowup, with the result that when he laid down in the ashes of the escape, the blowup passed above him rather than through him.

events. It is the potential richer recombination that enlarges the repertoire. After firefighters have a larger repertoire of things they can do, then they should notice more details because now they are in a better position to do something about whatever they notice. In Westrum's (1988) words, "a system's willingness to become aware of problems is associated with its ability to act on them." An increase in a generalized capability to act on problems should allow people to see more problems.

Firefighters who see an escalating fire in more detail should develop a more differentiated view of what to do, ranging from dropping all tools, or keeping only a fire shelter and a water bottle, or dropping packs but keeping pulaskis, or keeping everything and facing into the fire, or keeping a radio to call in a retardant-dropping aircraft, or starting a back fire, or starting a full fledged race[7] or even accepting that one's time has come. If firefighters refine a "can do" identity to include an enhanced capability to "make do," this should enlarge what they see when they act holistically in a ready-to-hand mode of engagement. If endangered people see more, then there is a better chance that they will enact a safe option.[8]

CONCLUSION

Let me conclude by returning to my original three goals.

First, I wanted to prime our thinking about the properties of naturalistic events by describing a concrete problem in wildland firefighting that fit both Klein's list of boundary conditions for NDM and Zsambok's short definition of NDM. The explanation and possible solution that were reviewed tended to become both conceptually less plausible and pragmatically more plausible as they moved from a detached present-to-hand perspective toward an engaged ready-at-hand perspective.

Second, I wanted to prime researchers' conceptualization of naturalistic events by discussing a sample of ideas from NDM that seemed to provide insight into the wildland puzzle. The outcome of this sampling was an interpretation of the wildland fire setting as an error-inducing context that had special impact on plan-

[7]Maclean (1992, p. 272) noted that "there are no classes in how to run from a fire as fast as possible." There are also no disengagements from a fire that get labeled as "a race." If firefighters who were operating in a ready-to-hand mode understood that they were in a race and that the goal was to win the race, then it would be obvious that they have to be as light as possible in order to run as fast as possible. Dropping one's tools in order to win would now be natural and fitting. Recall Mosier's (1997, p. 322) observation, "once the situation is understood, the appropriate course of action is obvious."

[8]Examples of novel but safe options in the face of blowups include Wagner Dodge's invention of an escape fire at Mann Gulch and the smokejumpers' choice of an escape route that ran upward from the launch spot at South Canyon rather than along the path of the fireline that was being constructed.

continuation decisions. Decision errors of commission and omission were seen to postpone abandonment of the fire suppression plan and to compromise the commitment to a new plan of disengagement. This pattern of results was then reconceptualized as a possible deficiency in a more basic capability for improvisation. It was suggested that the structure of naturalistic decision making resembles the structure of improvisation. To improve one's skills at improvisation is to improve one's situation awareness, wary recognition, and response repertoire. These improvements enable people to see more and see it more quickly, to intuit a stronger case for discontinuing an inappropriate plan, and to improvise a better one on the spot.

And third, my goal was to suggest a vivid metaphor that captures the nature of crisis decision making. To discard one's comforting tools while fleeing a wall of fire is a defining moment. My colleague Lance Sandelands (personal communication, May 9, 1998, paraphrased) has posed the issue this way: "The exquisite puzzle of crisis decision making seems to be just how to reconcile a limited person with an overwhelming circumstance, particularly when the circumstance further limits the range and flexibility of that person's thought. It seems there are two possible solutions. One is to try to anticipate those limits by overlearning thoughtless actions that might save the person. Another is to fight those limits by consoling the person with a simple but open framework that enables that person to improvise and meet the circumstance intelligently. Probably you'd want to do both somehow. But when the crisis hits, one solution occurs at the expense of the other and I'd hate to be wrong."

Firefighters hate to be wrong.

And so do researchers. We researchers need to be mindful of which tools slow our progress and need to be dropped so that we become faster, lighter, more agile analysts. We need not fear that if we drop our favorite analytical tools we are necessarily left naked and empty-handed because we still have our intuitions, feelings, stories, experience, ability to listen, presence, shared humanity, being in the moment, introspection, capability for fascination, awe, vocabulary, and empathy to trigger lines of questioning whose answers are available for reflection. To face people without our tools is not always a bad thing, nor is it debilitating to thought, although our identity as social scientists may momentarily take a hit. If it does, we will probably survive.

ACKNOWLEDGMENTS

I am grateful to Kathleen Sutcliffe, Gary Klein, Lance Sandelands, Karen Weick, Ted Putnam, Raanan Lipshitz, Stephen Small, Catherine Augustine, and Eli Berniker for their help with various drafts of this argument.

REFERENCES

Berliner, P. F. (1994). *Thinking in jazz: The infinite art of improvisation.* Chicago: University of Chicago Press.

Cannon-Bowers, J., Salas, E., & Pruitt, J. (1996). Establishing the boundaries of a paradigm for decision-making research. *Human Factors, 38*, 193–205.

Dreyfuss, H. L. (1997). Intuitive, deliberative, and calculative models of expert performance. In C. E. Zsambok & G. Klein (Eds.), *Naturalistic decision making* (pp. 17–28). Mahwah, NJ: Lawrence Erlbaum Associates.

Gioia, T. (1988). *The imperfect art.* New York: Oxford University Press.

Glass, D. C., & Singer, J. E. (1972). *Urban stress: Experiments on noise and social stressors.* New York: Academic Press.

Heidegger, M. (1962). *Being and time.* New York: Harper & Row.

Isenberg, D. J. (1985). Some hows and whats of managerial thinking: Implications for future army leaders. In J. G. Hunt & J. D. Blair (Eds.), *Leadership on the future battlefield* (pp. 168–181). Washington, DC: Pergamon.

Klein, G. (1997). The current status of the naturalistic decision making framework. In R. Flin, E. Salas, M. Strub, & L. Martin (Eds.), *Decision making under stress* (pp. 11–28). Brookfield, VT: Ashgate.

Klein, G. (1998). *Sources of power.* Cambridge, MA: MIT Press.

Maggin, D. C. (1996). *Stan Getz: A life in jazz.* New York: Morrow.

Maclean, N. (1992). *Young men and fire.* Chicago: University of Chicago Press.

Martin, L., Flin, R., & Skriver, J. (1997). Emergency decision making—A wider decision framework? In R. Flin, E. Salas, M. Strub, & L. Martin (Eds.), *Decision making under stress* (pp. 280–290). Brookfield, VT: Ashgate.

Mosier, K. L. (1997). Myths of expert decision making and automated decision aids. In C. Zsambok & G. Klein (Eds.), *Naturalistic decision making* (pp. 319–330). Mahwah, NJ: Lawrence Erlbaum Associates.

Orasanu, J., & Connolly, T. (1993). The reinvention of decision making. In G. Klein, J. Orasanu, R. Calderwood, & C. E. Zsambok (Eds.), *Decision making in action* (pp. 3–20). Norwood, NJ: Ablex.

Orasanu, J., Martin, L., & Davison, J. (1998, May). *Errors in aviation decision making: Bad decisions or bad luck?* Paper presented at Fourth Conference on Naturalistic Decision Making, Warrenton, VA.

Rorty, R. 1989. *Contingency, irony, and solidarity.* New York: Cambridge University Press.

Ryle, G. (1979). Improvisation. In G. Ryle, *On thinking* (pp. 121–130). London: Blackwell.

Schon, D. A. (1987). *Educating the reflective practitioner.* San Francisco: Jossey-Bass.

U. S. Forest Service. (1949). Mann Gulch Transcript. Washington, DC: Author.

U. S. Forest Service. (1994). Report of the South Canyon Fire Accident Investigation Team. August 17, 1994. Washington, DC: Author.

Weick, K. E. (1990). The vulnerable system: An analysis of the Tenerife air disaster. *Journal of Management, 16*, 571–593.

Weick, K. E. (1995a). *Sensemaking in organizations.* Thousand Oaks, CA: Sage.

Weick, K. E. (1995b). South Canyon revisited: Lessons from high reliability organizations. *Wildfire, 4*(4): 54–64.

Westrum, R. (1988). *Organizational and inter-organizational thought.* Paper presented at the World Bank Conference on Safety Control and Risk Management, Washington, DC.

Zsambok, C. E. (1997). Naturalistic decision making: Where are we now? In C. E. Zsambok and G. Klein (Eds.), *Naturalistic decision making* (pp. 3–16). Mahwah, NJ: Lawrence Erlbaum Associates.

20

Puzzle-Seeking and Model-Building on the Fire Ground: A Discussion of Karl Weick's Keynote Address

Raanan Lipshitz
University of Haifa

When Gary Klein invited me to discuss Karl Weick's presentation, I felt at the same time honored and apprehensive. After all, Weick's Social Psychology of Organizing (Weick, 1979) is one of the most thoroughly read and re-read books in my library. However "Dooty is dooty," as Long John Silver used to say cheerfully, and because opportunities of this kind do not come my way too often, I graciously agreed.

Weick's presentation is, as usual, evocative and thought provoking. For one thing, it is constructed like polyphonic music: Different themes progress simultaneously in unexpected directions, often ending in suggestive implications rather than explicit assertions. I cannot hope to do justice to this rich tapestry in the time allotted to me. Hence, I just touch on some of the "rough spots in the mindset of naturalistic decision making" that Weick highlights. These I grouped under three headings: "What Is NDM?" "What Areas Should Be Studied in NDM?" and "How to Study NDM?"

WHAT IS NDM?

In a draft of the presentation, Weick suggested that "to be serious about NDM we need to understand, as fully as possible, what it means to live forward in the face of delayed understanding, and what it means to experience the world as ready-to-hand." To be absolutely frank, I find Heidegger's terminology more than a bit opaque. To the best of my understanding, Weick suggests that NDM should focus on nondeliberate automatic action, that is, using Rasmussen's (1983) terminology, skill-based behavior. This, is a radical suggestion, considering that at present NDM focuses on rule-based and knowledge-based behavior. I agree that many real-world decisions are made partly or wholly out of awareness. I also agree that there is much to gain from studying the schemas and assumptions that drive this behavior and from alerting decision makers when taken-for-granted assumptions become dysfunctional: All decision makers encounter instances in which they better drop tools that generally serve them well and which are dear to them. Nevertheless, I disagree that NDM should change its focus. Dewey (1933) suggested that real-world inquiry (i.e., deliberate problem solving and decision making) begins with the encounter of the type of problematic, difficult, or important situations that Orasanu and Connolly (1993) identified as the arena of NDM. Taking Dewey and Orasanu and Connolly seriously, thus, clearly entails that NDM should focus on unready-to-hand mode of engagement.

WHAT AREAS SHOULD BE STUDIED IN NDM?

I take Weick's list of hypotheses regarding the South Canyon disaster to suggest, above all, that thorough understanding of NDM requires one to study numerous factors. From a descriptive point of view, these factors include: physical setting (e.g., what can or cannot be seen or heard); social setting and dynamics (e.g., socialization and trust in one's coworkers); the limited utility of general rules in low probability situations (e.g., hang on to your tools); bounded rationality (e.g., the inability to calculate correct rates of advance and the use of hard-to-refute heuristics such as "small changes cannot make big enough differences in massive events"); and last, but not least, the nonrational aspects of decision making (e.g., the difficulties posed by admitting defeat and working against deeply ingrained values and images of ideal self). From a prescriptive point of view these include training in improvisation. I presume that all of us agree with this list. Moreover, we can probably point out that NDM made some interesting, however modest, contributions in some of these areas. For example, a contribution that came to my mind was Marvin Cohen's Recognition/Metacognition model (Cohen, Freeman, & Wolf, 1996), which captures the need—and experts' ability—to depart from simple pattern-matching rules. To wit, observe the similarity between Schon's (1983) characterization of improvisation as "making something out of previous experience, practice, and knowledge during those moments when people surface and test intuitive understandings while their ongoing action can still make a difference"(p. 23) and Cohen's (1989) definition of assumption-based-reasoning as "filling gaps in firm knowledge by making assumptions that (1) go beyond (while being constrained by) what is more firmly known, and (2) are subject to retraction when and if they conflict with new evidence or with lines of reasoning supported by other assumptions" (1989, p. 260). I suspect that some of the items in Weick's list are neglected because we do not quite know how to study them, or, at any rate, do not know how to do it in ways that are acceptable to journal editors, funding agencies, or prospective clients. In the final analysis, it is safer and easier to study decision makers' cognitive processes than their equally important emotions and values. This brings me to the third question raised by Weick's presentation—How to Study NDM?

HOW TO STUDY NDM?

A cursory glance suffices to show a strong affinity between Weick's work and NDM: Sense making and situation awareness are virtually synonymous. Nevertheless, it is safe to assume that no journal referee will mistake a manuscript submitted by Weick for an NDM piece of work. Weick and NDM have distinctly different styles of approaching the same basic question: How do people operating in their natural habitats commit themselves to a certain course of action? Because Weick's analysis of the South Canyon case study concerns firefighters, it offers a unique opportunity to compare his approach with that of Gary Klein, a central and representative figure in NDM.

Table 20.1 presents some salient differences between Weick's case study and Klein's research on the decision making of firefighting unit commanders (Klein, 1989). Although the table pertains to two particular studies, I think it presents a fairly representative sample of the distinctions between Weick's work and NDM in general. Going over the table, we note first that Weick chooses to analyze a single case (constructed from numerous sources), while Klein chooses to analyze multiple cases (each coming from a single source). Both choose to study extreme situations that challenge decision makers' competencies. Klein, however, finds situations in which decision makers surmount the challenge to be informative. Weick finds situations in which competencies fail to be informative. Both Weick and Klein have cognitive orientations, but, as the next two pairs of items show, they have very different conceptions of what cognitive processes drive action and how they do it. Klein concentrates on what decision makers see and retains the traditional separation between cognition and action. Weick focuses on what decision makers make of whatever they see and advocates a radical notion that cognition and action are not distinct entities. Decision makers do not think and act but think by taking action, which they do more or less thinkingly (Weick, 1983). Having dealt with the data that Weick and Klein collect and the processes that they study, we can move down the table to look at their analytical methods. Klein looks for stable patterns, which he summarizes in the form of a flowchart model. Weick looks for puzzles to which he offers several alternative interpretations.

TABLE 20.1
Styles of Studying Decision Making In Action

Weick	*Klein*
Single case	Multiple cases
Demonstration of break down in competency	Demonstration of effective competency
Cognition intertwined with action	Cognition drives action
Meaning making	Pattern matching
Search for puzzles	Search for stable patterns
Divergence seeking	Convergence seeking
Argue via salient demonstration	Argue via replication
Evocative	Definitive
Ideal/target: The reflective practitioner	Ideal/target: The domain expert

Klein, thus, tries to converge on a single explanation (or model) that can account for several cases and argues for the significance of his study and the plausibility of his findings by pointing to the stability of his model across several representative (albeit challenging) situations of practice. Weick follows a different strategy. First, he looks for several plausible explanations to account for a single, extremely atypical case without making a particular effort to establish one of them as the most plausible. In fact, instead of asking which explanation is most plausible in a given set of explanations, he adds yet another explanation by reflecting on the assumptions underlying the original set (in this case, for example, that firefighters and tools are separate identities). Finally, Weick relies on a metaphor (as distinct from data) to argue for the general significance of his analysis. That is why I think that Klein strives to be definitive by converging on a single explanation and Weick strives to be evocative by looking for diverging, albeit plausible, explanations. In a way, the two styles embody alternative images of the ideal decision maker under uncertainty. In Weick's case, he is a reflective practitioner (Schon, 1983) who copes with equivocality in overdetermined situations by representing them, consistent with Ashby's requisite variety, with complex mental model. In Klein's case, he is a domain expert who copes with uncertainty in noisy situations by detecting patterns of critical signals hidden in them.

Having concentrated on the differences between Weick's and Klein's styles of doing research, let me turn to a predicament that they share: Both need to achieve rigor without the paraphernalia of laboratory experimentation.

Naturalistic decision making research relies primarily on observation. According to Weick (1968, p. 358), "the term 'observational methods' is often used to refer to hypothesis-free inquiry, looking at events in natural surroundings, nonintervention by the researcher, unselective recording, and avoidance of manipulation in the independent variable." These features are as different from the conditions required for satisfying laboratory-type rigor as possible. To understand how Klein and Weick solve the problem of rigor requires, therefore, a revision in conceptions of rigor that are suitable for (and informed by) laboratory experimentation. To perform this task, I apply to their methodologies three guidelines for achieving rigor that I draw from two example of evolutionary studies (as evolution achieved spectacular success by means of observational methods). The first example, presented by de Geus, comes from the work of Allan Wilson:

> In the late nineteenth century, milkmen left open bottles of milk outside people's doors. A rich cream would rise to the tops of the bottles. Two garden birds common in Great Britain, titmice and red robins, began to eat the cream. In the 1930s, after the birds had been enjoying the cream for about 50 years, the British put aluminum seals on the milk bottles. What happened? By the early 1950s, the entire estimated population of titmice in Great Britain, from Scotland to Land's End, had learned to pierce the seals. The robins never acquired that skill.
>
> Why did the titmice gain the advantage in the inter-species competition? . . . [according to Wilson, two conditions are necessary for] learning to take place in a population, numerous mobile individuals, some of them innovative, and a social system for propagating innovation. The red robins lacked such a social system. Of course . . . they can communicate. But they are fundamentally territorial birds. Four or five robins live in any garden, and each has its own small territory. There's a lot of communication among them, but what they usually have to say to one another is, Get out. Titmice . . . live in pairs in May and June. By the end of June and July, you see the titmice in flocks of 8, 10, and 12. They fly from garden to garden, and they play and feed. Birds that flock learn faster. (de Geus, 1997, pp. 56–57)

The second example is taken from Maynard Smith:

> Flatfish, for example, can change both their color and pattern in a few minutes; such changes are produced by the movement of minute granules of pigment within the cells of the epidermis. When the pigment is concentrated at the center of the cell the animal looks pale, and when it is spread throughout the cell the animal looks dark. Still more rapid changes in color are possible to squids and to octopuses; pigment is contained in small bags which, being elastic, take up spherical form, in which condition they cover only a small part of the surface, leaving the animal pale in color. However, each bag can be flattened into the shape of a disc by a series of radially arranged muscles, each supplied by a nerve fiber. When flattened the bags of pigment cover a large part of the surface, so that the animal appears dark by color. (Maynard Smith, 1993, p. 31)

I regard the credibility of Wilson's and Maynard Smith's explanation to rival that of most empirical studies published by *Psychological Review* and *The Journal of Applied Psychology*. Both Wilson and Maynard Smith rely on Mills' (1936) principle of concomitant variation to explain one phenomenon (adaptive learning of titmice and change of color of flatfish and squids) by another (social structure and spread of pigments, respectively). Using Toulmin's (1958) framework, which includes a *Claim*, *Grounds* (i.e., the data supporting the claim), and a *Warrant* (underlying assumptions that legitimize bringing the grounds as support for the claim), the structure of Wilson's and Maynard Smith's arguments is presented in Fig. 20.1.

The power of the first argument lies in the reliability of the two phenomena that constitute its ground and claim components—it is fair to assume that any observer of titmice and robin behavior would have made the same reports—and on the plausibility of the assumptions underlying the causal link relating them. The second argument, which basically has the same features, rests, in addition, on a solid background theory that establishes pigments as determinants of organisms' color. Rigorous observation, thus, implies three principles: Study phenomena, obtain reliable data, and base inference on transparent, plausible assumptions.

Study Phenomena

Essentially, this principle suggests that one frames research questions and collects and analyzes data in terms of phenomena as opposed to abstract variables detached from the phenomena that they presumably capture or explain. Thus, Mintzberg (1979, pp. 585–586) suggested that "Probably the greatest impediment to theory building in the study of organizations has been research that violates the organization, that forces it into abstract categories that have nothing to do with how it functions . . . [instead of] measuring things that really happen in organizations."

Obtain Reliable Data

According to Gellert (1955, p. 194, quoted in Weick, 1985), "the fewer the categories, the more precise their definition, and the less inference required in making classifications, the greater will be the reliability of the data."

Base Inference on Transparent, Plausible Assumptions

The first two principles are compatible inasmuch as both advise researchers to "stay low in the ladder of inference" (Argyris, 1982), that is, as close as possible to observable data. Because research is a form of argumentation, researchers

Wilson

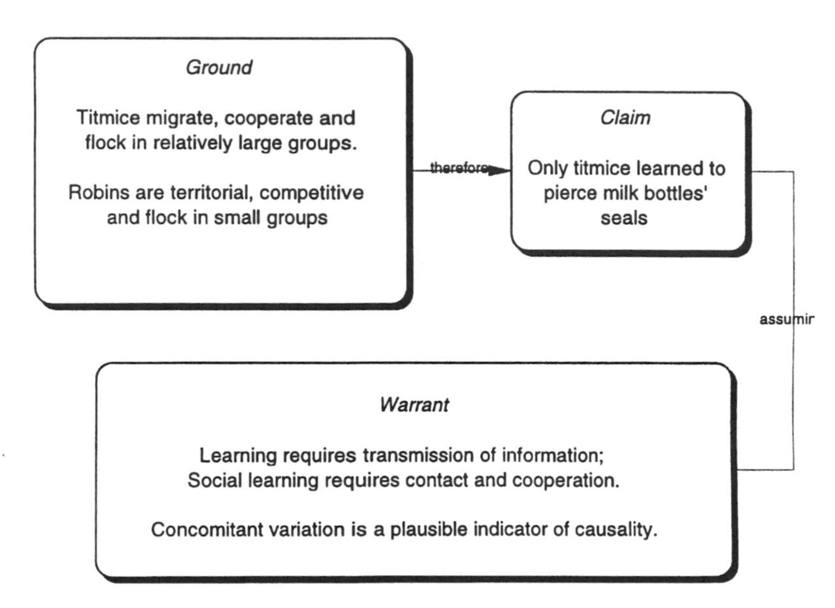

Maynard Smith

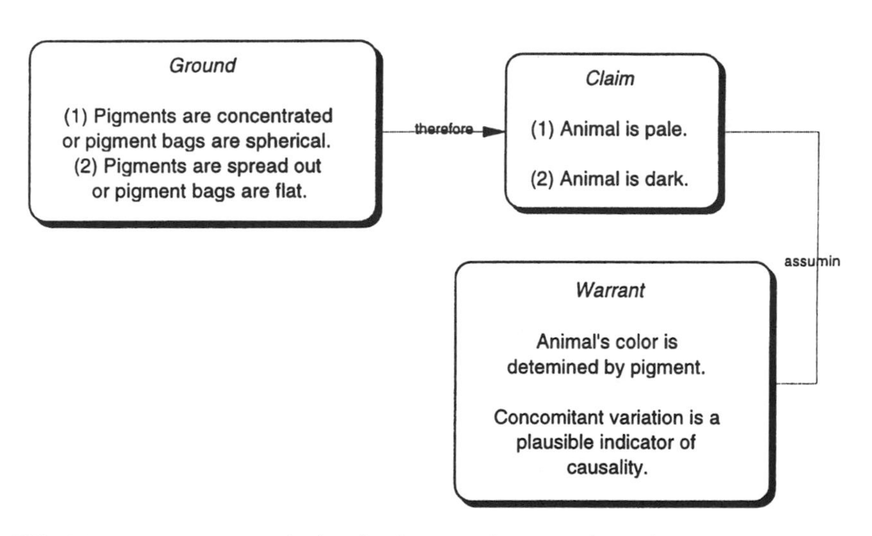

FIG. 20.1. Argument analysis of Wilson and Maynard Smith.

must, sooner or later, make inferences and rely on assumptions. The third principle is based on Weick's suggestion that "To deal methodically with validity is to make explicit one's status on those dimensions that are central to the assessment of credibility as defined by different audiences" (Weick, 1985, p. 604).

Klein and Weick apply a different mixture of the three guidelines that I proposed for rigorous naturalistic research. Both apply the first guideline, studying phenomena (e.g., how firefighters make decisions on the fire ground) and assiduously avoid the language of variables. (Klein also avoids as much as possible hypothetical constructs such as "schemata" where others use them, because they are not directly observable; Lipshitz & Ben Shaul, 1997). Klein, who is a model builder, relies in addition mostly on the second guideline: The Critical Decision Making method (Crandall & Getchell-Reiter, 1993) is a careful interviewing technique that is designed to elicit tacit expert knowledge. Being a puzzle seeker, Weick in addition relies mostly on the third guideline, presenting his basic assumptions as follows:

> Inquiry in general is based on the assumption that the paths to understanding in nature may be infinite and characterized by unique problems, but that all of these paths lead to *a* goal *an* understanding of *one* nature . . . Problem-solving seems to be favored by opposing sets of propositions, *both* of which are correct on some occasions . . . The investigator who retains opposed conceptual orientations will be open to comprehending a larger portion of the referent event. (Weick, 1979, pp. 29–30, italics in original).

I did not present the two styles of studying naturalistic decision making in order to recommend one of them. Both are legitimate, and each, I suspect, is useful under different circumstances and for different purposes. For example, Klein's methodology lends itself to designing domain-specific decision skills training. Weick's methodology lends itself to designing more general reflection skills training. Inasmuch as most of us rarely use Weick's approach, his work as much as his words points to a "rough spot" in the "NDM mindset."

REFERENCES

Argyris, C. (1982). *Reasoning, learning and action.* San Francisco: Jossey-Bass.

Cohen, M. S. (1989). A database tool to support probabilistic assumption-based reasoning in intelligence analysis. *Proceedings of the Joint Director of C-2 Symposium* pp. 260–264, Ft McNair, VA.

Cohen, M. S., Freeman, J. T., & Wolf, S. (1996). Meta-recognition in time stressed decision making: Recognizing, critiquing and correcting. *Human Factors, 38*, 206–219.

Crandall, B., & Getchell-Reiter, K. (1993). Critical decision method: A technique for eliciting concrete assessment indicators from the intuition of NICU nurses. *Advances in Nursing Science, 16*(1), 42–51.

De Geus, A. (1997). The living company. *Harvard Business Review, 75*(2), 51–59.

Dewey, J. (1933). *How we think*. Boston: D.C. Heath.

Gellert, E. (1955). Systematic observation: A method in child study. *Harvard Education Review, 25*, 179–195.

Klein, G. A. (1989). Recognition primed decisions. In W. R . Rouse (Ed.), *Advances in Man-Machine Systems Research, 5*, (pp. 47–92). Greenwich, CT: JAI.

Lipshitz, R., & Ben Shaul, O. (1997). Schemata and mental models in recognition-primed decision making. In C. Zsambok & G. A. Klein (Eds.), *Naturalistic decision making* (293–304). Mahwah, NJ: Lawrence Erlbaum Associates.

Maynard Smith, J. (1993). *The theory of evolution*. Cambridge, England: Cambridge University Press.

Mills, J. S. (1936). *A system of logic*. London: Longman.

Mintzberg, H. (1979). An emerging strategy of "direct" research. *Administrative Science Quarterly, 24*, 582–589.

Orasau, J., & Connolly, T. (1993). The reinvention of decision making. In G. A. Klein, J. Orasanu, R. Calderwood, & C. Zsambok (Eds.), *Decision making in action: Models and methods* (pp. 3–20). Norwood, NJ: Ablex.

Rasmussen, J. (1983). Skills, rules and knowledge: Signals, signs, and symbols, and other distinctions in human performance models. *IEEE Transactions on Systems, Man and Cybernetics, 13*, 257–266.

Schon, D. A. (1983). *The reflective practitioner*. New York: Basic Books.

Toulmin, S. (1958). *The uses of argument*. Cambridge, England: Cambridge University Press.

Weick, K. E. (1968). Systematic observational methods. In G. Lindzey & E. Aronson (Eds.), *The handbook of social psychology*, (2nd ed.) (Vol. 2, pp. 357–451). Reading, MA: Addison-Wesley.

Weick, K. E. (1979). *The social psychology of organizing*. Reading, MA: Addison-Wesley.

Weick, K. E. (1983). Managerial thought in the context of action. In S. Srivastva (Ed.), *The executive mind* (pp. 221–242). San Francisco: Jossey-Bass.

Weick, K. E. (1985). Systematic observational methods. In G. Lindzey & E. Aronson (Eds.), *The handbook of social psychology*, (3rd ed.) (Vol. 2, pp. 567–644). Reading, MA: Addison-Wesley.

V

Teams

21

Learning in the Context of Incident Investigation: Team Diagnoses and Organizational Decisions At Four Nuclear Power Plants

John S. Carroll
MIT Sloan School

Jenny W. Rudolph
Boston College Carroll School of Management

Sachi Hatakenaka
MIT Sloan School

Theodore L. Wiederhold
MIT

Marcello Boldrini
MIT Sloan School

Decision making is a construction one places on behaviors and events. There is a dominant ideology of choice in the Western world that reveres the expression of individual preferences and perceived control over outcomes (Sethi & Lepper, 1999). This ideology biases managers in the direction of interventions that affect local, short-term, and observable outcomes, rather than distant, long-term and harder to detect results (Carroll, 1998; Repenning & Sterman, 2000). Residents of the United States expect people to have preferences, to find choice desirable and motivating, and to be skilled at making choices (e.g., Langer, 1975; Lepper, Greene, & Nisbett, 1973). The reality is more complex. Simon (1955) pointed out that people avoid choices by taking the first acceptable option, and subsequent theorists have focused research attention on predecisional processes such as framing (Tversky & Kahneman, 1981), sensemaking (Weick, 1995), and recognition via pattern matching (Klein, 1998). Most recently, attention has focused on postdecision processes such as learning (e.g., Argyris & Schon, 1996; Levitt & March, 1988) and failures to consider feedback (Sterman, 1989). In short, decision making can be placed into a larger systemic view as one among many psychological, social, organizational, and cultural processes that shape action.

In this chapter, we present a research project that examines organizational decisions about improvement efforts in the nuclear power industry from the viewpoint of an organizational learning cycle. Decisions are made and constrained at numerous points in this cycle, and we approach the behaviors and events from a multilevel perspective (cf. Rousseau, 1985) on individual, group, and organizational processes.

ORGANIZATIONAL LEARNING IN NUCLEAR POWER PLANTS

High-hazard or high-reliability production systems (LaPorte & Consolini, 1991; Perrow, 1984; Roberts, 1990), such as nuclear power plants, face unusual demands for error-free operation, due to the potentially catastrophic outcomes associated with mistakes and the intense attention directed to their performance by regulators and publics. High-hazard organizations, therefore, cannot afford to learn solely through reactive, trial-and-error processes (Weick, 1987). Substitutes for trial-and-error learning include imagination, vicarious experiences, stories, simulations, analytical models, other symbolic representations, and the pooling of interpretations across observers (March, Spoull, & Tamuz, 1991; Rasmussen, 1990; Weick, 1987). Organizations can learn from success alone, but this may lead to premature closure on a suboptimal set of procedures and to complacency (Herriott, Levinthal, & March, 1985; Sitkin, 1992). Small failures are less risky and less threatening, yet often contain the kernels of larger failures and may provide the most desirable mode of learning (Sitkin, 1992). For example, information needed to prevent the Three Mile Island event was available from similar prior incidents at other plants, recurrent problems with the same equipment at the same plant, and engineers' critiques of operator training; yet this information was not incorporated into operating practices (Marcus, Bromiley, & Nichols, 1989).

Archetypal learning activities enact a generic feedback cycle (e.g., Argyris & Schon, 1996; Daft & Weick, 1984; Kim, 1993; Kolb, 1984; Schein, 1987), often represented by four ordered processes: (a) observing—noticing, attending, heeding, tracking; (b) reflecting—analyzing, interpreting, diagnosing; (c) creating—imagining, designing, planning, deciding; and (d) acting—implementing, doing, testing. Ideally, this full learning cycle takes place at individual, group, organizational, and institutional levels as various kinds of work activities are carried out, such as self-checking and quality control, daily planning meetings, incident investigations, postjob critiques, exchanges of good practices, and so forth. These activities draw upon various resources such as time, money, knowledge, authority, credibility, and procedures. In our research, we have focused on incident investigations as a particularly interesting learning activity shaped by conscious and unconscious decision heuristics.

LEARNING AND DECIDING DURING INCIDENT INVESTIGATIONS

Every nuclear power plant has a regulatory responsibility to report serious events that involve plant shut down, failure of safety equipment, injury, or release of radioactive or toxic materials, and so forth. However, plants are also expected to carry out internal analyses of both serious and less serious incidents in order to address current problems and avoid future events. Consider, for example, the following incident ("Fall From Roof") investigated by one of the teams in the research we describe shortly:

> An electrical maintenance worker climbed onto the roof of a shed inside the hot machine shop, an area used to decontaminate equipment with radiological residue. His goal was to replace burned-out fluorescent lights. As he crawled along the 1.5 inch steel frame of the roof (the roof is constructed of a steel frame and thin panels set in the frame), he slipped off the frame and fell through the roof to the floor 10 feet below. His injuries included 5 fractures and severe lacerations.

Characteristic of many naturalistic decisions, incident investigations involve multiple and sometimes conflicting goals, ill-defined structure, multiple parties with distinct areas of expertise, time constraints, and consequences that play out over time (Orasanu & Connolly, 1993). How much effort should be put into this incident investigation? Is it more important to fix this problem right now or to connect this incident to other past and potential incidents? Should one focus attention on the worker, the worker's training or supervision, the work culture, the design of the shed, incentives, management expectations, or organizational structure? Is it important that this worker had never done this job before, but was with an experienced coworker who warned the worker that the roof panels would not support a person's weight? Is it important that this job was considered routine and was routinely performed in violation of the accident prevention procedure that requires wearing a protective harness when working aloft? How much weight should be given to what the maintenance manager thinks, what the union thinks, or what the regulator thinks?

Incident investigations are part of a deliberate feedback system that includes: (a) categorization of incidents and evaluation of severity, (b) methods for gathering information from personnel and instruments, (c) aggregation of incidents by type of problem and type of cause for tracking over time, (d) in-depth analyses of important incidents to determine "root causes" and generic lessons, (e) recommendations for corrective actions, and (f) tracking of corrective action implementation and resultant improvements. Less serious incidents result in simple fixes, but also in tracking and trending of the incident categories. More serious incidents also are analyzed for apparent causes of the problem, and these are also tracked and trended. Even more serious incidents or undesirable trends are subject to more formal and thorough analysis to seek the underlying or "root" causes and ways to address them. Finally, the most important, complex, or puzzling incidents and trends are analyzed for root causes by a multidiscipline team.

Despite regulatory requirements, dissemination of best practices, and other mechanisms that serve to establish and improve incident review programs, each plant seems to carry out incident investigations in its own way. For example, there are regulatory requirements for reporting more serious incidents, but the threshold for reporting, documenting, and analyzing less serious incidents and whether this is carried out by special staff, regular line employees, individuals, or groups, is left up to each plant. Thus, each incident investigation program has its own theories of organization and management embedded in its structure and procedures, definitions, and examples used for illustration.

In previous research, Carroll (1995, 1998) showed that typical incident review programs are based on narrow frameworks that focus on particular kinds of causes and corrective actions. For example, at one nuclear power plant, administrative documents describe the incident review program as a search for "root cause": "the primary or direct cause(s) that, if corrected, will prevent recurrence of performance problems, undesirable trends, or specific incident(s)." Although this seems sensible, on close examination the very concept of "root cause" focuses attention on a single cause rather than on an exploration of multiple causes or chains of events. This has been called "root cause seduction" (Carroll, 1995) because the idea of a singular cause is so satisfying to one's desire for certainty and control.

It is interesting that at this plant there was a difference of opinion regarding the proper number of causes to include in a root cause report. The majority believed that there should be only two or three root causes, at most, in order to avoid "diluting" the impact of the report. The minority believed that a more extensive list of causes would be helpful for learning and improvement. This issue was expressed even in the computer system that recorded root cause reports for tracking and analysis, which limited data entry to a single root cause. Decisions about what counts as a root cause may limit learning about interactions among multiple root causes. For example, whereas the designated root cause in the Fall From Roof case was "worker's tunnel vision" (narrow focus on the task), our research team thought the plant could have considered the role of architectural design (lights in a hard-to-reach place), supervisory performance (supervisors routinely skip prejob briefs and work site visits), and learning orientation (this situation was known but not reported).

Incident investigations give some kinds of causes more attention than others, based on how well they are understood. Decisions about what to include in reports are also constrained by what investigators' cognitive frames tend to include and exclude: "believing is seeing" (Weick, 1995). For example, at the same plant, the incident review program documents had extensive detail regarding equipment and human error issues, but very little on organizational, programmatic, and cultural issues. In incident investigation reports, the causes found are those "familiar to the analysts. . . . There is a tendency to see what you expect to find; during one period, technical faults were in focus as causes of accidents, then human errors predominated while in the future focus will probably move upstream to designers and managers" (Rasmussen & Batstone, 1991, p. 61).

Incident review systems in nuclear power plants tend to focus on causes that are proximal to the problem (at the "sharp end" of systems, Reason, 1990), typically involve human agents, and have available solutions that can be implemented predictably (Carroll, 1995). This sharp-end focus provides a powerful decision heuristic for what causes to pursue and what corrective actions to suggest. Sharp-end focus also tends to constrain learning from events by focusing attention on "single loop" learning strategies (Argyris & Schon, 1996) that involve compliance with rather than questioning of existing routines and heuristics. Incident investigators are inclined to overlook or ignore the organizational values and the biases created by professional training that may shape individual action (Perin, 1995), the informal work culture (Rudolph, 1997), incentive systems, and management's role in setting expectations and allocating resources.

Additionally, the management review and approval process forces the incident investigation team to negotiate with line management over the content of the report. The virtue of this requirement is that, because line managers are supposed to take responsibility for implementing change, they should therefore have opportunities to provide input and commit to the new actions. Everyone recognizes that it is easier to produce reports and analyses than it is to create effective change (Langley, 1995). However, the danger of this approach is that needed change may get buried in politics: the power resides in line managers, who may fail to acknowledge issues that reflect badly on them, diminish their status or power, or have solutions that are risky to their own agendas. The anticipation of resistance from line managers can lead to sanitizing the report, which is a kind of "acceptability heuristic" (Carroll, 1995; Tetlock, 1985). Consider that, in some states, the public utility commission does not permit plants to recover costs that arise from management mistakes. Intended to be fair to ratepayers and deter further mistakes, the actual result is that problems are rarely if ever blamed directly on management.

This narrow focus is consistent with U.S. managers' desires for certainty and action, as well as the fundamental attribution error (Nisbett & Ross, 1980) that biases managers toward individual characteristics rather than systemic relationships. People are not encouraged to look at the linkages among problems and issues, because it creates a feeling of overwhelming complexity and hopelessness in the face of pressure to do something quickly. For example, one engineering executive at a U.S. nuclear power plant commented that, "it is against the culture to talk about problems unless you have a solution." The question is whether this approach works successfully with complex, ambiguous issues that lack ready answers, with diffuse organizational and cultural processes that are poorly understood, or whether a different approach is needed.

In any industry, well-intentioned, commonplace solutions can fail to help, have unintended side effects, or even exacerbate problems. When complex, interdependent problems are understood in linear cause–effect terms that result in a search for "fixes," it is common to find a "fixes that fail" scenario (Senge, 1990). Consider that a plant has an increased number of equipment breakdowns and that these are attributed to poor quality maintenance. It is typical to "fix" this problem by writing more detailed procedures and monitoring compliance more closely in order to ensure the quality of work. The more detailed procedures usually result in fewer errors on that particular job; however, the increased burden of procedures and supervision can be perceived by maintenance employees as mistrust and regimentation. This may result in loss of motivation, blind compliance to procedures that may still be incomplete, malicious compliance when workers know the right thing to do but also know that only rote compliance is safe from disciplinary action, the departure of skilled workers who find more interesting work elsewhere, and, ultimately, more problems.

In the "Fall From Roof" case, if the reaction had been only to discipline the worker, the workforce might have become angry with management and less likely to cooperate in other ways that could have produced more problems. Although the corrective actions included counseling the people involved, they also included appointing a full-time safety person, having managers communicate and reinforce expectations on industrial safety, and providing more detailed job guidance on working aloft. But the report did not address management behavior in any detail, the role of training, the relative priority of safety and production, or attitudes toward compliance with rules.

TWO STUDIES OF INCIDENT INVESTIGATION TEAMS

We studied incident investigation teams whose members were taken off their regular assignments in order to gather information, analyze causes, and make recommendations regarding the most serious incidents and undesirable trends. The resulting reports and recommendations were then handed to responsible managers for corrective action. Plants commission a small number of such teams each year to investigate problems that appear to extend beyond the domain of any one functional specialty or organizational unit. These teams are an opportunity to examine cross-functional cooperative team decision making and learning that is intended to become organizational learning and change.

For clarity, but not to be taken too rigidly in its details, we offer the following framework for understanding the relationships among individual, team, and organizational characteristics and activities (see Fig. 21.1). Drawing on models of team performance (e.g., Ancona & Caldwell, 1992; Cannon-Bowers & Salas, 1998; Hackman, 1987), we view team learning as building a shared mental model of the problem and its impact on the plant (expressed in written and oral reports), as well as a shared mental model of the team itself and how to work together. Team learning emerges from team composition or the separate viewpoints and expertise brought by individual members, the team process that partitions and coordinates the work, the task difficulty or new knowledge required, and the resources from managers and others needed to support the team. Team learning then affects broader organizational learning through formal and nonformal processes such as meetings, reports, and conversations. Team learning also influences individual learning as the team members develop their personal knowledge, skills, and styles, which then may diffuse to other organization members and activities. Organizational learning affects individual learning, and individual learning has impact on organizational learning as well (Kim, 1993). This is a dynamic process that unfolds over time within the team's activities and across future teams as organization members reconfigure and carry new organizational knowledge and routines with them.

This framework suggested interrelated research questions at the group, individual, and organizational levels. The group level question is "What is the relationship of incident investigation team composition, in terms of expertise, hierarchical level, and cognitive style and complexity, on the quality and depth of the review and the resulting team learning, considering team process and task complexity?" A substantial contribution of the incident review process is that people from diverse functional groups, hierarchical levels, and cognitive styles get to talk, share viewpoints, establish connections, and so forth. The direct benefit of this diversity on the review is presumably to build a better team mental model and produce a more comprehensive and creative report. Given a process

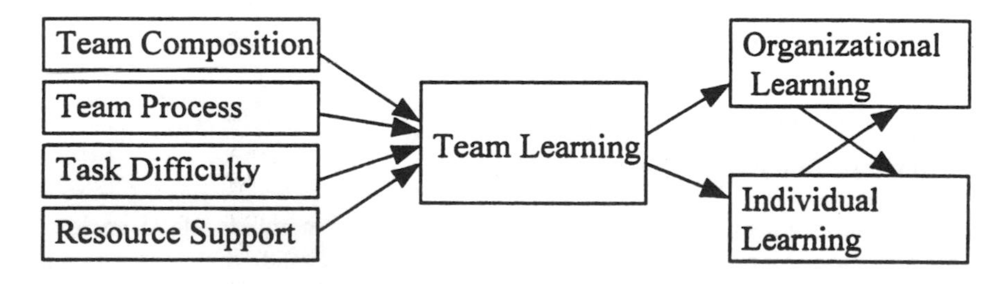

FIG. 21.1. Team, individual, and organizational learning.

that achieves openness and inquiry, the diverse team will produce a richer report. However, if the process devolves into conflict or exercise of authority, the team will be no better, and possibly worse, than a less diverse team. We would also expect that team members with more categories and dimensions and more complex mental models would be able to consider more causes and combinations, side effects and delayed effects, as well as systemic and dynamic causal patterns.

At the individual level, the research question is "In what ways does participation on an incident investigation team change mental models and cognitive style?" Serving on an incident review team should provide an important source of individual learning: participation is sometimes viewed by organizations as a form of training. Ex-team members should have an increased sensitivity to cross-functional problems and systemic issues. Their own mental models of performance and organization should be enriched and represented by more cognitive complexity (cf. Tetlock, 1983; Torbert, 1994) and changes in cognitive style (e.g., Myers & Myers, 1980). They should also have more respect for others' viewpoints, less self-serving attributional bias, and a greater ability to work in a diverse team.

The organizational question is "How does plant improvement or organizational learning depend upon team and individual learning?" At the organizational level, knowledge from the incident review process is supposed to influence the plant. This presumably takes place when participants in incident reviews share their new knowledge and viewpoints with others at the plant (and outside the plant). We suspect that more people hear about incidents and analyses through informal discussions and storytelling (Boje, 1991; Weick, 1995) than from reports and formal training. Involvement of plant personnel in root cause analysis training acts to improve plant performance apart from any "answers" to "problems" that the reviews are ostensibly about (Carroll, 1998). Success also consists of improving sensitivity to problems in general, encouraging more inquiry, challenging assumptions and making the undiscussible more discussible (cf. Argyris & Schon, 1996), mobilizing support for self-analysis and change, improving communication throughout the organization, preventing a wide range of future problems, and finding ways to improve.

In this chapter, we report results from two studies: (a) a retrospective questionnaire and interview study of 25 incident investigation teams at two nuclear power plants, and (b) an observational study of incident investigation teams at two other nuclear power plants.

Study 1: Team Learning and the Handoff to Implementation

We collected questionnaire and interview data from members of 25 incident investigation teams at two nuclear power plants and the management sponsors/customers of these reports and also obtained the reports for content coding. These were the most serious incidents and most troubling trends during the previous 2 years. Teams were defined as having at least three members who participated broadly in team activities and shared a perception of themselves as team members (there are other people who participated in a more limited way and were not counted as team members).

The team member questionnaires asked both closed- and open-ended questions about the incident or team task, team composition, available resources, team process, the resulting report, management reactions, organizational changes, and personal learning (see Table 21.1 for examples of composite scales). There were also measures of cognitive complexity and style as well as demographic information. Most team members took approximately 1 hour to complete the questionnaire, although some who were members of multiple teams took considerably longer. The response rate was 83% for team members (85% for Plant A, 81% for Plant B). The management questionnaire asked a relevant subset of questions about the report and resulting changes. Managers self-identified as being sponsors or customers, so there is no way to compute a response rate.

We also collected the team reports and developed coding categories to characterize the reports and to compare them against the highest safety and learning standards in the industry. We held several day-long meetings with Dr. William Corcoran, an experienced industry consultant and specialist in incident investigation programs, to

TABLE 21.1
Study 1 Selected Measures

	Individual Level Measures	*Team Level Measures*	*Organizational Level Measures*
Team Member Questionnaire	Self-report of: • Cognitive complexity • Cognitive style • Age, education, experience, etc. • Personal learning from being on the team	Aggregated self-reports on: • Task complexity • Team composition • Team dynamics • Communications • Management support • Report quality	Aggregated self-reports on: • Implementation of plant changes • Assessment of outcomes
Management Questionnaire	• Personal learning from reading the report	• Management support • Trust in team • Report quality	• Implementation of plant changes • Assessment of outcomes
Incident Reports		Coding based on: • Causal analysis • Corrective actions • Learning focus • Narrative style	

review our coding protocol, code pilot incident reviews, and revise our coding criteria. Thirty codes were generated, grouped into four dimensions (see Incident Reports-Team Level Measures in Table 21.1): clarity and depth of causal analysis (including approaches to prevention and defense), adequacy of corrective actions, learning focus, and narrative style. In three rounds of pilot coding, inter-rater agreement on the 30 codes improved from 60% to 88%. All reports were coded by two of the authors, one a nuclear engineer, the other an organizational theorist; differences of opinion were resolved by discussion.

We focus our primary analyses on the three measures of team learning or report quality deriving from the team members' assessment of the "excellence" of their report, managers' corresponding assessment, and our own coding of the reports on the above four dimensions. On average, team members and managers moderately agreed that the teams had produced excellent reports (means of 1.88 and 2.20 on a 6-point scale, with 2 = *moderately agree*), however, these judgments correlated weakly with each other ($r = .08$, $n = 21$, n.s.). Consistent with the apparent differences in their reactions to the reports, the pattern of relationships with our coding of the reports differed between team ratings and manager ratings. Team ratings of reports were significantly predicted by our own coding of Cause, and the composites of questionnaire items from the teams' coding that we labeled Internal and External Communication, Team Dynamics, and Management Support. In short, teams seemed to believe they had produced a better report when they experienced better team process, open communication within the team, better support from outside the team, and a deeper causal analysis. In contrast, managers' ratings were significantly predicted by our coding of Corrective Action, a composite from the management questionnaire that we labeled Trust in the team, and team members with more prior experience in incident investigations and Myers-Briggs cognitive style scores (Myers & Myers, 1980) that were more Sensing and more Thinking. In short, managers were more impressed with reports from a team they could trust and whose members thought concretely and analytically and identified corrective actions that were logical and capable of addressing deeper problems. There was no evidence that variation in team diversity, whether age, cognitive style, or department, had an impact on team process or report quality.

Our own coding of the reports found a disappointing level of depth and completeness, insight and clarity. Causal analyses rarely went very deep, averaging 2.29 in our coding system (where 2 is *less than adequate* and 3 is *adequate*). For example, the root cause of the "Fall From Roof" incident was described as "tunnel vision," or the worker's narrow focus on the task. When we asked one plant general manager about these results he said, "Our culture focuses on the individual, the sharp end. We don't know how to focus on the engineer or vendor. . . . Systems thinking is foreign to our culture, our organizational chart, and American culture. Our view of our own success is 'how successful is my department?' Silo thinking." Shallow analyses also reinforce the sharp-end heuristic by embedding these cause labels in training materials, trend reports, and supposedly objective data bases.

Our analysis of the reports also found that corrective actions were sometimes misaligned with the supposed causes, either leaving root causes without corrective actions, or introducing corrective actions with little prior foundation (average score of 2.16). Few reports were well-written—they were sometimes confusing, often redundant, usually in passive voice, and almost always avoided acknowledging ambiguity (average Story = 2.21 and Learning = 2.04). We speculate that it is hard for plant personnel to learn from a written document that is difficult to comprehend and remember. We also found no overall positive relationship among qualities of the report: of the six pair-wise correlations among Story, Cause, Corrective Action, and Learning, four are negative, suggesting that it is unusual to present a complex and deep analysis and a good story at the same time.

We note from the open-ended comments that intentional and unintentional messages from management appear to affect the process, content, and outcomes of the incident reviews. Many comments focused on two themes: having enough time to devote to the investigation, that is, being relieved of competing responsibilities by one's manager, and manager receptiveness and nondefensiveness. Team members reported feeling more comfortable psychologically, operationally, and politically when addressing technical or procedural issues. For example, one member of the "Fall From Roof" team wrote, "When it was becoming apparent what the real problem was, I think the group became (temporarily) unsure where to go—what to do—it looked like a big step." However, team members also raised questions about the value of this "shallower" approach: "[A weakness of the report is] blaming the worker, who made the decision to perform the work, instead of 'why did management allow the design problem to exist for so long?'"

The comparison of team members' and managers' views of incident investigations presents an apparent paradox. On the one hand, team members need good contacts with managers in order to have the resources to do a good report, and managers react more positively to reports from people they trust. On the other hand, managers want team members to present reports with clear guidance for correcting problems. Thus, the deeper the report goes into uncharted territory and fundamental issues of human systems and organizational analysis, the more difficult it is to create conditions for learning about these deeper issues. More diverse teams with more unfamiliar and complex stories to tell may find themselves in a difficult spot. It would seem that skills for managing diverse mental models are needed both within teams and between teams and managers.

The open-ended questionnaire comments and other interviews conducted at the plants emphasize the difficulty of the handoff between the incident investigation teams and the managers who will commission various change efforts. Some team members reported not knowing what had happened after the report was given to management. Several suggested that management had been defensive about initial drafts and, therefore, reports had pulled their punches. One team member from Fall From Roof wrote, "If top level managers aren't willing to listen to the people doing the work, and respond to their findings, it all becomes a waste!" If an acceptability heuristic sanitizes reports, then learning and improvement may be undermined. Yet, there may be benefits to tailoring reports to what managers will accept to get commitment of resources. In interviews, managers complained about long lists of causes and corrective actions that undermined the impact of the report and seemed to yield little value for the investment. Teams that understood management concerns appeared to have made their reports more actionable. The more important issue for learning is whether these decisions are being made covertly or whether conversations allow some examination of and learning about conflicting goals. In general, the results reinforce the importance of the boundary spanning function of teams (Ancona & Caldwell, 1992)—their ability to negotiate resources and goals with management and to establish strong relationships with other groups in order to get information and later to sell the report.

Study 2: Observations of Incident Investigation Teams

Study 2 uses direct observation of incident investigation teams as they organize, carry out their work, and deliver their product to management. This serves to complement the retrospective information from Study 1. Although the logistics of such research is difficult, we have so far observed six teams at two plants. For each team, one researcher attended team meetings, talked with team members individually, observed team members' interviews with plant personnel when appropriate, and attended the management review of the report if possible.

Our results tend to reinforce and deepen what we learned from the questionnaire study. Specifically, teams do not always have the resources they need to make good decisions that have impact on the plant. Our discussion focuses on time, management attention, investigative techniques, analytical tools, and leadership skills.

Most teams we observed had difficulty breaking people away from other tasks and/or faced heavy pressure from management to meet a deadline. For investigations on the critical path for plant availability, there were few problems getting time from team members, but there was pressure to finish quickly. For investigations not on the critical path, there was less time pressure (for example, deadlines were extended twice for one of the teams), but team members' managers did not relieve them fully from their other tasks. When plant availability is no longer at issue, team members tend to disappear and it can be difficult to pursue the deeper and longer term aspects of the investigation.

Management seemed to have considerable ambivalence about root cause investigations. On the one hand, those that resulted in bringing equipment back online were highly valued. In addition, there was an expectation that investigations would go deep enough to speak to management's own role in the underlying problems.

However, management could be critical during review of the draft report, very careful about the wording of conclusions, and allow rather vague root causes such as "lack of accountability" and "management expectations need to be reinforced." At one plant where an investigation involved interviews with senior management, team members were very anxious and reluctant to go beyond prepared questions. At both plants, root cause teams were seen by managers and team members as a drain on resources and a "pain" in terms of the level of detail and validation required.

Another indication of the overall lack of resources for investigation was the lack of training, experience, and skills on the part of team members and team leaders. Both plants have a process for qualifying team leaders and a desire to have a wide range of employees receive training and experience in incident investigation. However, the level of skills on the teams we observed was relatively low. They appeared to be improvising and still learning how to assemble and validate facts, to differentiate "failure modes" from "root causes," and to use tools such as causal factor charting. One team seemed to have an initial hypothesis that they did not challenge during the investigation. Nor did the team leaders have much training or experience in leading investigation teams, which made it difficult to coordinate and focus the team effort. One plant typically provided an experienced "mentor" for the teams, but in one case the mentor was reassigned to an urgent investigation, leaving the team to manage on its own. Given these limitations, it is not surprising that team members and managers have such ambivalence and unwillingness to invest even more in the process.

DISCUSSION AND RESEARCH DIRECTIONS

Our work on root cause analysis and incident investigations has suggested that certain kinds of causes are favored, such as those at the "sharp end," those with readily imaginable fixes, and those with political support. This has an impact on how nuclear power plants move through the learning cycle of observing, reflecting, creating, and acting. For example, managers readily observe, analyze, and interpret problems that relate to equipment and procedures, create solutions to address them, and often act as if these are the main problems worth addressing. The plant thus "learns how to learn" to fix these concrete, bounded problems. In contrast, we note the difficulties expressed by managers at dealing with the "human system": they seem almost paralyzed at how to take action even after a "human system" cause has been identified. They express the problem as an inability to know how to decide what to do. They know how to decide whether to maintain or replace a defective pump; it is possible to lay out the decision problem in an orderly manner with known consequences and risks that can be assessed by system engineers or others with experience. However, they are stymied about how to decide whether to "increase accountability" or how to "empower" people. It is unclear, they report, how much you are going to have to spend, during what periods of time, with what results and what risks, and how you would know that you have had any useful impact.

Based on this contrast, we speculate there is a gap in how we characterize causes in our current theories about causal analysis in incident investigation. On the one hand, we have focused or bounded causes that have a physical existence in a place and time. Sharp-end causes such as human error, equipment failure, and procedure inadequacy have this nature. However, some non-sharp-end causes have the same tangible features, such as many involved in "failure mechanisms," incentive systems, and reporting relationships in the organizational chart. These kinds of causes tend to be fixable because they are discrete and component-like, with few residual effects (i.e., after changing the procedure, the old procedure no longer exists, supposedly). On the other hand, we have causes that are diffuse in the sense that they are "hard to put your arms around." These are things like safety culture, accountability, leadership, standards and values, and even "programs." Reengineering and business process mapping are popular in part because they take things that are understood as diffuse and turn them into bounded objects. Diffuse causes tend to be broadly connected to many other things, to have a lot of inertia, to be hard to measure, and to have many side-effects and delayed effects when changes are attempted (Repenning & Sterman, 2000). From the viewpoint of managers, there are no decision rules to apply to tackling such messy or "wicked" problems. Therefore, they often ignore them and focus on short-term visible root causes and corrective actions.

Viewing human and physical processes as separate may make solutions to both less robust, and learning more difficult. It is hard to make decisions about human processes like empowerment if they are not tied to a measurable process—they appear diffuse. Likewise, it is easy to assume, based on professional training, management practice, and common sense, that decisions about physical processes involve no human systems and, therefore, will be easy to tackle. When interventions produce paradoxical or disappointing results (for complex problems, things usually get worse before they get better, Repenning & Sterman, 2000), the outcomes are attributed to human failings (fundamental attribution error, Nisbett & Ross, 1980). The net result is that intendedly rational decisions may subvert the goals of the larger system. If the human problems can be tightly linked to the tangible aspects of production, then the human problems assume boundaries and the consequences of intervention become imaginable.

The difficulty of predicting report characteristics in terms of individual and team characteristics suggests that the theories that describe individual learning may not replicate across levels of analysis. Issues such as breadth versus depth of experience, diversity and learning from conflicting viewpoints, and supportive external relationships, may operate differently for individuals, teams, and organizations. Instead, we may need a cross-level theory that builds on phenomena of different qualities at different levels of analysis. We believe that incident investigations provide an important context for developing and testing concepts of organizational learning at multiple levels of analysis.

We hope that this chapter helps to stimulate connections between decision-making researchers and those concerned with other psychological and organizational processes. A broader context for decision making will help us to reframe our own work in ways that may lead to fresh insights and new kinds of data from which to develop theoretical and practical implications for organizational learning. In the meantime, we are looking more closely at the team process, the interactions of team composition, team process, task complexity, and the handoff from the teams to the implementation of corrective actions and the assessment of effectiveness. We also have started work with a plant that is considered among the best in the industry and plan to compare their reports and teams with those already studied.

ACKNOWLEDGMENTS

Based on a paper presented at the Fourth Conference on Naturalistic Decision Making, Airlie, Virginia, May 29–31, 1998. This research was supported by National Science Foundation Grant # SBR96-1779 and a Marvin Bower Fellowship at the Harvard Business School.

REFERENCES

Ancona, D. G., & Caldwell, D. F. (1992). Bridging the boundary: External activity and performance in organizational teams. *Administrative Science Quarterly, 37,* 634–55.

Argyris, C., & Schon, D. (1996). *Organizational learning II: Theory, method, and practice.* Reading, MA: Addison-Wesley.

Boje, D. M. (1991). The storytelling organization: A study of story performance in an office-supply firm. *Administrative Science Quarterly, 36,* 106–26.

Cannon-Bowers, J. A., & Salas, E. (Eds.). (1998). *Making decisions under stress: Implications for individual and team training.* Washington, DC: American Psychological Association.

Carroll, J. S. (1995). Incident reviews in high-hazard industries: Sensemaking and learning under ambiguity and accountability. *Industrial and Environmental Crisis Quarterly, 9,* 175–197.

Carroll, J. S. (1998). Organizational learning activities in high-hazard industries: The logics underlying self-analysis. *Journal of Management Studies, 35,* 699–717.

Daft, R. L., & Weick, K. E. (1984). Toward a model of organizations as interpretation systems. *Academy of Management Review, 9,* 284–295.

Hackman, J. R. (1987). The design of work teams. In J. W. Lorsch (Ed.), *Handbook of organizational behavior* (pp. 315–342). Englewood Cliffs, NJ: Prentice-Hall.

Herriott, S. R., Levinthal, D. A., & March, J. G. (1985). Learning from experience in organizations. *American Economic Review, 75,* 298–302.

Kim, D. H. (1993). The link between individual and organizational learning. *Sloan Management Review, 35,* 37–50.

Klein, G. (1998). *Sources of power: How people make decisions*. Cambridge, MA: MIT Press.

Kolb, D. A. (1984). *Experiential learning as the source of learning and development*. Englewood Cliffs, NJ: Prentice-Hall.

Langer, E. (1975). The illusion of control. *Journal of Personality and Social Psychology, 32*, 311–328.

Langley, A. (1995). Between "paralysis by analysis" and "extinction by instinct." *Sloan Management Review, 36*, 63–76.

LaPorte, T. R., & Consolini, P. (1991). Working in practice but not in theory: Theoretical challenges of high reliability organizations. *Journal of Public Administration Research and Theory, 1*, 19–47.

Lepper, M. R., Greene, D., & Nisbett, R. E. (1973). Undermining children's intrinsic interest with extrinsic rewards: A test of the overjustification hypothesis. *Journal of Personality and Social Psychology, 28*, 129–137.

Levitt, B., & March, J. G. (1988). Organizational learning. *Annual Review of Sociology, 14*, 319–340.

March, J. G., Sproull, L. S., & Tamuz, M. (1991). Learning from samples of one or fewer. *Organization Science, 2*, 1–13.

Marcus, A. A., Bromiley, P., & Nichols, M. (1989). *Organizational learning in high risk technologies: Evidence from the nuclear power industry*. Minneapolis: University of Minnesota Strategic Management Research Center. (Discussion Paper #138).

Myers, I. B., & Myers, P. B. (1980). *Gifts differing*. Palo Alto, CA: Consulting Psychologists.

Nisbett, R. E., & Ross, L. (1980). *Human inference: Strategies and shortcomings of social judgment*. Englewood Cliffs, NJ: Prentice-Hall.

Orasanu, J., & Connolly, T. (1993). The reinvention of explanation-based decision making. In G. Klein, J. Orasanu, R. Calderwood, & C. E. Zsambok (Eds.), *Decision making in action: Models and methods* (pp. 3–20). Norwood, NJ: Ablex.

Perin, C. (1995). Organizations as contexts: Implications for safety science and practice. *Industrial and Environmental Crisis Quarterly, 9*, 152–74.

Perrow, C. (1984). *Normal accidents: Living with high risk systems*. New York: Basic Books.

Rasmussen, J. (1990). The role of error in organizing behavior. *Ergonomics, 33*, 1185–1190.

Rasmussen, J., & Batstone, R. (1991). *Toward improved safety control and risk management*. Washington, DC: World Bank.

Reason, J. (1990). *Human error*. New York: Cambridge University Press.

Repenning, N. P., & Sterman, J. D. (2000). Getting quality the old-fashioned way: Self-confirming attributions in the dynamics of process improvement. In R. Scott & R. Cole (Eds.), *The quality movement and organization theory* (pp. 201–235). Newbury Park, CA: Sage.

Roberts, K. H. (1990). Some characteristics of one type of high reliability organization. *Organization Science, 1*, 160–176.

Rousseau, D. (1985). Issues of level in organizational research: Multi-level and cross-level perspectives. In L. L. Cummings & B. Staw (Eds.), *Research in organizational behavior, Vol. 7*, pp. 1–37. Greenwich, CT: JAI.

Rudolph, J. W. (1997, August). *Enhancing safety practice in high-hazard organizations: An introduction to single, double, and triple loop safety work*. Paper presented at the Academy of Management Annual Meeting, Boston, MA.

Schein, E. H. (1987). *Process consultation: Vol. 2: Lessons for managers and consultants*. Reading, MA: Addison-Wesley.

Senge, P. (1990). *The fifth discipline: The art and practice of the learning organization*. New York: Doubleday.

Sethi, S., & Lepper, M. R. (1999). Rethinking the value of choice: A cultural perspective on intrinsic motivation. *Journal of Personality and Social Psychology, 76*, 349–366.

Simon, H. A. (1955). A behavioral model of rational choice. *Quarterly Journal of Economics, 69*, 99–118.

Sitkin, S. (1992). Learning through failure: The strategy of small losses. In B. Staw & L. L. Cummings (Eds.), *Research in organizational behavior, Vol. 14*, pp. 231–266. Greenwich, CT: JAI.

Sterman, J. D. (1989). Misperceptions of feedback in dynamic decision making. *Organizational Behavior and Human Decision Processes, 43*(3), 301–335.

Tetlock, P. E. (1983). Cognitive style and political ideology. *Journal of Personality and Social Psychology, 45*, 118–126.

Tetlock, P. (1985). Accountability: The neglected social context of judgment and choice. In B. Staw & L. Cummings (Eds.), *Research in organizational behavior, Vol. 1.* (pp. 297–332). Greenwich, CT: JAI.

Torbert, W. (1994). Cultivating post-formal adult development: Higher stages and contrasting interventions. In M. Miller & S. Cook-Greuter (Eds.), *Transcendence and mature thought in adulthood: The further reaches of adult development* (pp. 181–204). Lanham, MD: Rowman & Littlefield.

Tversky, A., & Kahneman, D. (1981). The framing of decisions and the psychology of choice. *Science, 211*, 453–458.

Weick, K. E. (1987). Organizational culture as a source of high reliability. *California Management Review*, Winter, 112–127.

Weick, K. E. (1995). *Sensemaking in organizations.* Thousand Oaks, CA: Sage.

22

Distributed Cooperative Problem Solving in the Air Traffic Management System

Philip J. Smith
Institute for Ergonomics, The Ohio State University
C. Elaine McCoy
Department of Aviation, Ohio University Airport
Judith Orasanu
NASA Ames Research Center

In the design of complex systems, there is often a tension between a desire to achieve solutions that are globally optimal and the limitations imposed by the cognitive complexity of determining such optimal solutions (Billings, 1997). These limitations can arise for one of two reasons:

1. The individual operators who are trying to make the system function could not deal with the cognitive complexity of the task if they were to fully consider all of the relevant goals and data in arriving at an optimal solution.
2. The designers of technological tools, who intended to make it possible to overcome the cognitive limitations of human operators, are themselves unable to fully model the true complexity of the system.

Because of these two types of limitations, most real system designs rely on simplifications that allow the system to perform well, without trying to achieve optimal solutions. One common approach is to decompose the task of managing the overall system into subtasks, and then assign these subtasks to separate individuals. The hope is that there is sufficient independence among these subtasks, so that when each subtask alone is performed well, the combined effects will produce acceptable (rather than optimal) levels of performance for the system as a whole. Furthermore, because few systems are actually decomposable into fully independent subtasks, it is also hoped that these individuals will interact with one another as needed when the solutions to their various subtasks in fact interact in significant ways.

These considerations about the design of complex systems are discussed in the context of the Air Traffic Management (ATM) system (Hopkin, 1995; Wickens et al., 1997). Four points are highlighted in these considerations:

1. The traditional design of the ATM system has been highly distributed. Tasks are assigned to different individuals in order to limit the amount of information and knowledge that each individual is expected to access and process directly, thus limiting the cognitive complexity of the task for that individual.
2. This traditional decomposition generally produces acceptable performance in terms of safety and efficiency. However, it occasionally results in safety hazards because the decision maker does not have direct access to all of the important data and knowledge and fails to interact with the person who has access to the data or knowledge. It also routinely results in less than optimal performance in terms of efficiency within the ATM system.
3. Efforts to improve efficiency in order to achieve closer to optimal system performance have focused on changing either the frequency and nature of the interactions among different people, or on changing the locus of control. For this latter solution to be effective, however, such changes in the locus of control (which result in significant changes in the overall task decomposition) need to be accompanied by appropriate changes in direct access to the relevant data and knowledge, or by new patterns of interaction between the decision maker and the people who have the relevant data and knowledge.
4. The ATM system has changed in significant ways over the past few years. By studying this evolution, we gain insights into how different control paradigms influence individual and group performance, especially with regard to the degree of interaction among different individuals when the distribution of knowledge and data does not match the distribution of control. In addition, we begin to identify some of the more detailed design features that influence decision making performance.

AN EXAMPLE OF TASK DECOMPOSITION IN AIR TRAFFIC MANAGEMENT

As a specific example of task decomposition, consider the following setting within the air traffic system. In order to reduce cognitive complexity, the overall task of selecting safe routes of flight and of operating these flights is currently decomposed such that each of the participants (pilots, controllers, dispatchers, and traffic managers) has only partial information. In particular, within the current air traffic management system, tactical decisions are made by flight crews and controllers without always having the information necessary to develop the same big picture about weather system developments available to dispatchers and traffic managers. As an example, in cases involving significant reroutes, the flight crew needs to bring the dispatcher back into the loop to ensure that the big picture has been adequately considered. Although this distribution of information and responsibilities generally affords an efficient operation, it is susceptible to occasional errors due to false assumptions about "what the other guy has already considered" or due to incorrect assessments of whether a particular change in route is "significant."

Details of the Scenario

As an illustration of the impact of this task decomposition, consider an actual incident involving a Boeing 727–200 flying from Dallas/Ft. Worth to Miami. As part of his job, the dispatcher responsible for this aircraft was required to provide the pilot in command with information regarding any hazardous enroute weather. In this case, the dispatcher noted a line of thunderstorms that he felt potentially jeopardized the safety of the flight and issued a reroute to the aircraft, with the captain's concurrence. During this process, the captain was briefed on the situation. That reroute was coordinated with Air Traffic Control (ATC) and approved, but as the flight progressed along its refiled route of flight, the receiving center rejected the reroute and put the airplane back on its originally filed route of flight. The sector controller in that center rejected it because that new route was already congested due to flights from Europe and the East Coast, and because he had no access to data that would have informed him about the weather problems in southern Florida. As a result, the aircraft became trapped south of the line of weather.

More specifically, as the aircraft was going across the Florida panhandle, there was a line of thunderstorms from the Tampa Bay area southeastward down to the Miami/Ft. Lauderdale area. At that point, the dispatcher contacted the captain, briefed him on the enroute weather conditions, and recommended a reroute taking the aircraft direct to Ormond Beach and then down the east coast of Florida into the Miami airport from the northeast, ahead of the weather. The captain concurred with the reroute and contacted the appropriate Jacksonville Center frequency to coordinate the reroute. The reroute was approved. The aircraft made a turn to the east and was proceeding directly to Ormond Beach on the Florida east coast. At the point where there was a hand-off made from one controlling center sector to the next center, the receiving center sector advised the captain that, due to traffic along the east coast of Florida, they would not be able to accommodate the reroute and that the aircraft would have to return to the originally filed route of flight. The aircraft made a fairly abrupt turn back to the southwest, got offshore along the west coast of Florida and proceeded down toward the Ft. Myers area. Furthermore, the aircraft was slowed to 180 knots due to traffic, increasing fuel burn.

At that time, the line of thunderstorms was sinking to the southeast, moving down toward Miami/Ft. Lauderdale/Sarasota/Ft. Myers. As the aircraft arrived in that vicinity and was preparing to turn to the east for the final to Miami to land to the east, the weather came across the airport and shut down the operation. As a result, the aircraft entered airborne holding and was given "expect further times from ATC" that continued into the future. Thus, the crew was faced with an indefinite situation as to when they would be released to proceed into Miami.

It was not until this point that the captain contacted the aircraft's dispatcher and advised the dispatcher that the reroute the captain and the dispatcher had agreed upon had been refused by an ATC sector, that the aircraft had ended up back on its original filed route of flight, and that they had encountered airborne holding. The dispatcher's attention had been diverted to another situation and he had not noted the ATC-initiated reroute. Thus, at that point the aircraft was holding with thunderstorms between its position and the intended destination.

What complicated this scenario was that Sarasota, Ft. Myers, Ft. Lauderdale, and West Palm Beach, which were all of the other usable alternate airports for this aircraft, were either unusable due to thunderstorms or were now north of the weather as well. The aircraft was basically trapped south of its intended destination and south of its usable alternates. (This aircraft was not authorized to use the Key West airport.) Consequently, the crew was faced with a situation of being very low on fuel with limited options in terms of available diversion airports. The aircraft finally broke through the line of thunderstorms as the weather passed south of Miami and was able to land at Miami. However, they picked up significant turbulence going through the line of weather, producing a very uncomfortable ride for the crew and the passengers as the aircraft passed through severe turbulence.

It is also important to understand that the dispatcher working this particular flight on this day had about 30 other flights that he was responsible for at that time and felt as though this situation had been resolved and had turned his attention to other situations that required his attention.

Important Features Illustrated by the Scenario

This scenario provides an example of one of the ways in which the air traffic management system has been decomposed into subtasks to reduce the cognitive complexity for individuals. This particular scenario also illustrates one potential weakness of such a decomposition: the reliance on individuals to decide appropriately when there is a need for interaction (i.e., when the decomposition is inadequate).

One response to such an incident would be to attempt to improve judgments about when to interact by improved training or more clearly defined procedures. A second would be to maintain the existing task decomposition, but to give everyone better access to critical elements of the bigger picture (such as weather), so that they could better judge when there is a need to interact with the other system operators. Another would be to develop technological support tools such as an "intelligent" alerting system that would inform the dispatcher when a flight has begun to deviate "significantly" from its original route. A fourth would be to try to integrate decision making, abandoning or partially abandoning the task decomposition strategy. All of these approaches have strengths and weaknesses and merit serious consideration for this specific scenario. The remainder of this chapter, however, focuses on the fourth approach and does so in another ATM context concerned with preflight planning and traffic flow management.

DESIGN OF THE AIR TRAFFIC MANAGEMENT SYSTEM

Historically, traffic flow management (TFM has primarily been a function under the control of the Federal Aviation Agency (FAA), with traffic managers at various facilities making decisions about what routes could be flown by the flights scheduled by the airlines (Odoni, 1987). In recent years, however, there has been an emphasis on giving the airlines greater flexibility, based on the assumption that the airlines have better information about the costs of alternative flight plans and should, therefore, be in a position to make better decisions about the economics of alternative flight plans. In essence, this shift changes the task decomposition as, under such changes, airline dispatchers must consider a much larger set of factors if they are to in fact improve performance. Issues surrounding such a shift are discussed in terms of alternative system architectures for accomplishing it.

Alternative System Architectures

Alternative architectures for the ATM system that change the decomposition of tasks for flight planning can be grouped into three categories (Smith et al., 1997):

1. Management by directive, in which FAA traffic managers simply inform an airline regarding the route that can be flown by a particular flight.
2. Management by permission, in which there is a default flight plan assigned by the FAA, which can be revised if the airline operations center requests an alternative and receives permission from FAA traffic management staff.
3. Management by exception, in which the airline operations center can simply file the flight plan that it desires for a given flight (Sheridan, 1987, 1992). This flight plan is automatically approved, and the route of flight is changed only if a problem is detected while it is enroute.

Over the past several years, the ATM system has been evolving from a system in which management by directive was the predominant form of interaction to a hybrid system including examples of all three forms of interactions.

Control by Permission

The first major change arose in 1992, with a shift in management by directive to management by permission. Specifically, FAA Advisory Circular 90-91 (FAA, 1992) established a formal procedure allowing the airlines to request nonpreferred routes (routes for flights that differed from the FAA assigned preferred routes). Under this procedure, an airline could send a message via teletype to the FAA's Air Traffic Control Systems Command Center (ATCSCC) requesting an alternative route for a particular flight. A specialist at ATCSCC would then evaluate this request, checking with traffic managers at the involved enroute regional air traffic centers and, based on their input, would approve or disapprove the request.

This shift to management by permission gave the airlines a means for improving efficiency, because they had better information for determining the most economical flight plans for their aircraft. It still left the locus of control with the FAA traffic managers, however, as they had to individually approve all requested alternative routes. These approvals were made based on considerations of safety and overall efficiency in traffic flows. Thus, this shift left the basic task decomposition the same, but provided a procedure for increasing the frequency of interactions among traffic managers and dispatchers.

This new paradigm was viewed very positively by both the airlines and the FAA. One airline, for example, reported that in one year, it submitted 15,279 requests for nonpreferred routes and that 75% of these requests were approved. These approvals resulted in an estimated savings of 13,396,510 pounds of fuel. Studies by Smith, et al. (1997) identified a number of factors that appeared to contribute to this success.

Factors Contributing to Success. The first factor concerned matching the locus of control with access to relevant information. The criticism of prior procedures was that, under the management by directive paradigm, FAA traffic managers were making decisions that did not take into consideration the airlines' business concerns. Thus, the claim was that, for any given flight, there could be a number of equally acceptable flight plans from the perspective of safety and overall system efficiency and in such cases the FAA was making the choice without the benefit of any input from the airline about its economic considerations. Under this new paradigm based on management by permission, the ultimate decision was still left up to FAA traffic managers, who had information and experience regarding potential traffic bottlenecks, but it allowed the airlines to indicate their preferences based on economic concerns; safety was ensured while economics were improved.

As with any system architecture, however, supporting arguments based on high-level considerations do not, by themselves, ensure that the architecture will be successful. The details of its implementation are equally important. Three major factors appeared to contribute to the success of this program:

1. Implementation of communication channels that led to the development of a shared understanding of goals, problems, constraints and solutions.
2. The form of the distribution of responsibilities to a number of different individuals.
3. Incorporation of feedback and process control loops.

Regarding the relevance of these hypotheses to naturalistic decision making, the point is twofold. First, an architecture involving control by permission has the potential to ensure that certain important interactions occur without requiring changes in the basic task decomposition or the locus of data and knowledge. Thus, the patterns of behavior during decision making are significantly influenced by a relatively straightforward architectural change. Second, however, is that several other factors need to be considered to make the resulting interactions as efficient and effective as possible.

Thus, in terms of the earlier discussion regarding task decomposition, this procedure maintained the basic decomposition that had previously been used, in the sense that both the FAA traffic managers and airline dispatchers still had to analyze alternative routes from their own perspectives. The routine interactions, however, gave both groups a broader understanding of the factors considered by the other group, resulting in more effective and efficient interactions when they were likely to be productive (i.e., when the task decomposition was inadequate, and there was a need for interactions between both groups in order to determine the best solution).

Limitations of Control by Permission. The primary weakness of this paradigm was that it was manpower intensive (requiring extra staffing to support the additional interactions) and was thought by the airlines to, at times, be excessively conservative in terms of the approval of requests for alternative routes. As a result, the system evolved further in 1995 to give the airlines additional flexibility using a different "architecture."

Control by Exception

Although the use of the "control by permission" architecture was viewed as a significant improvement, its perceived limitations were sufficient to result in a followup program based on "control by exception." This new program, known as the expanded National Route Program (FAA, 1995), allowed the airlines, subject to certain constraints, to simply file the routes that they preferred for particular flights. FAA traffic managers would then monitor conditions, watching for situations (such as severe weather) where the program had to be canceled temporarily for particular portions of the country. Tactical changes by FAA air traffic controllers (as well as by airline pilots and dispatchers with the concurrence of the responsible air traffic controllers) could also be initiated after the flight was enroute. Unlike the earlier shift to control by permission, this architectural change significantly altered the historical task decomposition, requiring airline dispatchers, if they wanted to be fully effective, to now consider factors (such as the prediction of air traffic bottlenecks) that in the past had been handled largely by FAA traffic managers.

To evaluate the impact of this architectural change, two studies were conducted dealing with the impact of the expanded National Route Program (NRP) on fuel consumption. The motivation for this study came from two sources. First, dispatchers at a number of airlines as well as traffic managers at enroute air traffic control centers provided numerous examples of how flights filed under the NRP were sometimes given significant amendments and suggested that some of these changes occurred on a regular basis. In at least some cases, the changes were clearly initiated by the ATC system to deal with traffic congestion. Along these lines, dispatchers made comments such as:

> Under the expanded NRP, it's like shooting ducks in the dark.

> The problem with the expanded NRP is that there's no feedback. Nobody's getting smarter. Someone has to be responsible for identifying and communicating constraints and bottlenecks.

> It used to be the weather that was the biggest source of uncertainty. Now it's the air traffic system.

In short, the dispatchers appeared to be indicating that the shift in their tasks gave them more flexibility but did not give them the information and tools necessary to integrate considerations of air traffic (one of the major factors that used to be handled primarily by the FAA traffic managers) into their decision making.

As a specific example, one dispatcher indicated that NRP flights from Washington National to Cincinnati frequently have a problem because of the strategy used by ATC to deal with crossing traffic:

> It happens to us all the time. We file the flights at altitudes of 35,000 or 39,000 feet and they're held at 23,000, 25,000 and 27,000 feet. They don't tell us ahead of time that it's going to happen.

A second example of how traffic bottlenecks can affect NRP flights was provided by a traffic manager:

> Quite often ... 8-10 extra aircraft are on this northern route to DFW [from Southern California to Dallas flying north of White Sands into the northwest cornerpost at the Dallas-Fort Worth airport] during the noon arrival rush [noon local time]. This causes a sector saturation problem in ZFW Sectors 93 and 47 [two Dallas-Fort Worth (ZFW) air traffic control sectors]. To relieve this volume problem, the ZFW TMU [Traffic Management Unit] moves 5 aircraft back to the south route [south of White Sands] via CME.TQA.AQN.DFW [a sequence of navigational fixes into the southwest cornerpost of the Dallas-Fort Worth airport]. This longer route of flight, plus the fact that DFW is in a south flow (meaning these flights will spend more time flying below 10,000 feet), will reduce fuel savings or negate them altogether for this bank of flights.

Thus, anecdotal evidence suggested that traffic bottlenecks were arising that influenced the efficiency of NRP flights and raised questions about the effectiveness of this new decomposition of tasks. To gain further insights into this concern, two followup studies were conducted. These are described next.

Study 1: Analysis of Predicted Versus Actual Fuel Consumption

To look for evidence of such inefficiencies, we collected data from a major airline on all of their flights filed over a 5-month period. These data were used to compare predicted fuel consumption on NRP routes with both predicted fuel consumption on FAA preferred routes and with actual fuel consumption. In the following discussion, a *flight* is defined to be a particular combination of an origin, destination, Ptime (scheduled departure time), and equipment type. Thus, a given flight could have a new instance filed each day. Predicted and actual fuel consumptions were from wheels-up to wheels-down.

Predicted fuel consumptions were first analyzed, comparing performances on FAA preferred routes with the filed NRP routes. This airline filed 21,334 flight instances under the NRP during this time period. The average predicted fuel savings per day during this time period ranged from 2.3% to 6.0%. The total predicted savings was 17,723,329 pounds of fuel.

Comparison of Predicted Versus Actual Fuel Consumption. Given the anecdotal evidence outlined earlier, however, it seems possible that these predictions overestimate actual fuel savings for some flights, because the computer's predictions do not take into account the new reroutings that might occur as a result of filing an NRP route and then encountering a traffic bottleneck while enroute. Consequently, we also compared predicted with actual fuel consumption for these flights.

To ensure adequate statistical power, only flights with at least 20 instances were considered. There were 267 such flights. A statistical analysis indicated that 94, or 35%, of these 267 flights routinely burned more fuel than predicted ($p < .05$). Of these 94 flights, 21% routinely burned more extra fuel than was supposed to be saved by flying the NRP route instead of the FAA preferred route. As an example, the flight from Dallas-Fort Worth to San Diego scheduled to depart at 1645 Universal Coordinated Time on average burned 1,013 pounds of fuel more than predicted for the FAA preferred route. That flight, which was supposed to save 759 pounds of fuel compared to the FAA preferred route (a predicted 4% savings), actually burned 254 pounds more than the prediction for the FAA preferred route (a 1.3% loss).

Thus, these data indicate that there was some sort of a problem associated with 35% of the flights filed by this airline under the NRP during this time period. One possibility would be an underlying inaccuracy in the prediction model for one or more of these flights, over and above any new problems introduced by use of the expanded NRP. If, however, we assume that the prediction model provides unbiased estimates for the FAA preferred route (and assume no new inefficiencies are introduced for the FAA preferred routes by the use of the expanded NRP), then these data indicate that the actual benefits in terms of fuel consumption from the use of the NRP are less than predicted.

Study 2: A Detailed Observational Study of Los Angeles–Dallas-Forth Worth Flights

These data also indicated that the city pair that most often had flights with regular problems was Los Angeles to Dallas-Forth Worth (LAX-DFW). Seventeen of those flights routinely burned more fuel that predicted. Therefore, we decided to study this route in detail in order to collect more detailed data on the nature of the problems with NRP flights for this city pair and to better quantify the influence of these problems.

Methods. Four students from the Aviation Department at Ohio University collected data from June 22, 1996 to August 23, 1996 on the performances of flights from LAX-DFW. Flights with five different scheduled departure times (Ptimes) were studied (1400, 1415, 1445, 1515, and 1810 Universal Coordinated Time). The students collected data on predicted and actual fuel consumptions and observed each flight instance on an aircraft situation display that showed filed and actual routes in order to record any flight amendments.

Results. The resulting observations quickly made it clear that the underlying problem was the rerouting described earlier. Very briefly, what happens is this:

1. A flight instance is filed under the expanded NRP along a route north of White Sands (special use airspace) to the northwest cornerpost at DFW.
2. While that flight is enroute, the ATM system decides that there is likely to be a sector saturation problem in the Turkey or Falls high sectors when the flight reaches that point as it approaches the northwest cornerpost into DFW.
3. To deal with that problem, the flight or flights with the most southerly routes that are flying to the northwest cornerpost are rerouted south of White Sands to the FAA preferred route so that they will approach DFW via the southwest cornerpost.

Table 22.1 indicates the frequency with which the cornerpost swap occurred for the different flights that we observed. (Keep in mind that this swap usually occurs before White Sands, not as the flights are approaching the airport.) The results indicate that the flights that arrive at DFW for the noon rush (flights that are arriving into DFW around noon local time and that have scheduled departure times or Ptimes of 1400 and 1415 Universal Coordinated Time) are particularly affected. During that time period, 33% to 39% of the flights fell into that category and were rerouted south of White Sands to the FAA preferred route.

Table 22.2 indicates the impact of this rerouting on overall savings for the NRP flights filed at particular Ptimes for those instances where an NRP flight was actually rerouted south of White Sands. All of these flights, on average, burned more fuel than was predicted if they had been filed on the FAA preferred route. On average, for example, it cost an additional 1,502 pounds of fuel each time the flight at 1400 Universal Coordinated Time was rerouted to the southwest cornerpost. A statistical test comparing actual with predicted fuels consumptions for these flights was significant ($p < .05$) for the Ptimes of 1400, 1445, and 1810.

TABLE 22.1
Percentage of Flights Flying the FAA Preferred Route (Pref Route) and NRP Routes with or without Cornerpost Swaps

	Equipment	*Number*	*Route Flown*		
Ptime	*Type*	*Observed*	*Pref Route*	*NRP-No Swap*	*NRP-Swap*
1400	DC10	41	44%	17%	39%
1415	B767	42	48%	19%	33%
1445	MD80	36	50%	44%	6%
1515	MD80	41	51%	39%	10%
1810	DC10	29	38%	52%	10%

Note. Ptime is Universal Coordinated Time.

TABLE 22.2
Expected versus Actual Fuel Savings for Flights that were Rerouted from the Northwest Cornerpost to the Southwest Cornerpost

	Equipment	*Number*		
Ptime	*Type*	*Observed*	*Expected Change*	*Actual Change*
1400	DC10	16	-3.5%	+0.4%
1415	B767	14	-4.5%	+0.3%
1445	MD80	2	-3.4%	+1.9%
1515	MD80	4	-2.3%	+0.1%
1810	DC10	3	-3.0%	+2.7%

Note. Only those flights that were rerouted in this manner are included in this table. Ptime is Universal Coordinated Time. Savings are the percentage reduction or increase relative to the predicted fuel consumption for the FAA preferred route that day.

OVERALL CONCLUSIONS

As described earlier, the ATM system has gone through two major evolutions in terms of the paradigm for controlling preflight planning. Initially, the shift was to a control by permission paradigm, where FAA traffic managers maintained control of the actual decision to approve requests from the airlines to deviate from the FAA preferred route. This procedure was then further modified to a control by exception paradigm, in which the airlines were allowed to file their desired flight plans without permission from FAA traffic managers but these plans were then altered tactically by air traffic controllers (as well as pilots and dispatchers) only as needed while the flight was enroute. As documented in this study, both of these alternatives had a sizable influence on the patterns of decision making and performance.

The control by permission paradigm is particularly interesting because of two factors:

1. It left the basic task decomposition the same, except for providing an impetus to increase interactions between airline dispatchers and FAA traffic managers in order to consider airline requests for individual flights to fly something other than the default (FAA preferred) routes. One implication of this is that there was no need to change the distribution of data and knowledge, because the party with control (the FAA traffic manager) already had the data and knowledge necessary to decide whether a given request for a different route was acceptable from an air traffic perspective, whereas the party requesting permission (the airline dispatcher) already had the data and knowledge to identify route changes that were preferable from an airline business perspective. Thus, without having to shift the locus of data and knowledge, significant changes in decision making were induced through the introduction of a new pattern of interactions.
2. A caution regarding this paradigm, however, was that even though the shift to control by permission was viewed as a significant improvement, it was ultimately replaced with another paradigm. The motivation for this appeared to be a belief that there remained a sort of anchoring or inertial effect when the locus of control was left the same (with the FAA traffic managers). Essentially, this belief was that, for one of several possible reasons (habit, workload, comfort, level of understanding, etc.), traffic managers tended to be more conservative than necessary and that, because they had the final say, airline requested routes were sometimes denied unnecessarily.

In contrast, implementation of the control by exception paradigm shifted the locus of control from traffic managers to airline dispatchers and clearly served to overcome some of this anchoring or bias toward the traditional FAA preferred routes that appeared to continue under the control by permission paradigm. However, use of the control by exception paradigm lead to additional considerations:

1. Even though the dispatchers now had more control to determine the routes to be filed, they were not provided with direct access to the data and knowledge necessary to evaluate alternative routes in terms of the effect of potential air traffic bottlenecks.
2. Because there was no longer a mechanism requiring routine interactions with traffic managers to identify and deal with such problems, the patterns of communication originally induced by the control by permission paradigm were now greatly reduced.

Thus, as a result of these two factors, dispatchers frequently were filing routes that did not achieve the desired improvement in efficiency.

Overall Implications

The case studies reviewed in this chapter serve to illustrate several points:

1. One classic strategy for reducing cognitive complexity in the ATM system has been to decompose the system into subtasks and to assign these tasks to different individuals. Then, in those circumstances where the assumption of independence among these subtasks is inadequate, it is necessary for the responsible individuals to interact with each other.
2. A drawback of such a decomposition strategy is that the responsible individuals may not recognize the need for such interaction. This can result in problems from either a safety or efficiency standpoint, as illustrated by the example of the flight from Dallas to Miami.
3. Another drawback is that, because the assumption of independence made during the task decomposition is at best an approximation and because the "as required" interactions of individuals to deal with inadequacies in this task decomposition typically only partially compensate for these approximations, although overall system performance may be good, it is not likely to achieve its theoretical optimum.
4. Because of these two drawbacks, a variety of alternative architectures or task decompositions are now being explored within the ATM system. Two such architectures, control by permission and control by exception, are illustrated in the context of preflight planning. The first architecture, control by permission, attempts to improve performance by maintaining the traditional task decomposition, but by improving the interactions between traffic managers and dispatchers to cope with the limitations of the decomposition. The second architecture, control by exception, represents a major change in task decompositions.
5. Studies of the use of the control by exception architecture provide cautions about the need to consider fully the impact of alternative task decompositions on information requirements and on the cognitive complexity of the newly defined tasks. These studies caution that, without such considerations, the expected move toward a more optimum level of performance may in fact not be fully achieved.

In short, these studies on the evolution of the ATM system help demonstrate how relatively high-level decisions regarding task decomposition and the locus of control can impact patterns of interaction and decision making in very profound ways.

ACKNOWLEDGMENTS

This work was supported by the FAA Office of the Chief Scientist and Technical Advisor for Human Factors (AAR-100) and NASA Ames Research Center. We would like to express special appreciation to Larry Cole, Eleana Edens, Tom McCloy, Mark Hoffman, Roger Beatty, Joe Bertapelle, Rob Blume, Scott Ridge, Moira Hoban Edwards, and John Tittle.

REFERENCES

Billings, C. (1997). *Aviation Automation: The Search for a Human-Centered Approach.* Mahwah, NJ: Lawrence Erlbaum Associates.

Federal Aviation Administration. (1992). *National Route Program* (Advisory Circular 90-91, ATM 100, April 24, 1992). Washington, DC: Author.

Federal Aviation Administration. (1995). *National Route Program (NRP)* (FAA Order 7110.128, Free Flight, ATM 100, effective Jan. 9, 1995). Washington, DC: Author.

Hopkin, V.D. (1995). *Human factors in air traffic control.* New York: Taylor-Francis.

Odoni, A. R. (1987). The flow management problem in air traffic control. In A. R. Odoni, L. Bianco, & G. Szego (Eds.), *Flow control of congested networks* (pp. 64–79). Berlin, Germany: Springer-Verlag.

Sheridan, T. (1987). Supervisory control. In G. Salvendy (Ed.), *Handbook of human factors* (pp. 146–173). New York: Wiley.

Sheridan, T. (1992). *Telerobotics, automation and human supervisory control.* Cambridge, MA: MIT Press.

Smith, P. J., McCoy, C. E., Orasanu, J., Billings, C., Denning, R., Rodvold, M., Gee, T., & Van Horn, A. (1997). Control by permission: A case study of cooperative problem-solving in the interactions of airline dispatchers with ATCSCC. *Air Traffic Control Quarterly, Vol. 6*, 229–247.

Wickens, C., Mavor, A., & McGee, J. (1997). *Flight to the future: Human factors in air traffic control.* Washington, DC: National Academy Press.

23

The Nature of Constraints on Collaborative Decision Making in Health Care Settings

Vimla L. Patel
José F. Arocha
McGill University

Two basic approaches have governed decision-making research to date: the study of individual decision-making processes in controlled environments and the study of individual and team decisions in naturalistic settings. Decision-making research in naturalistic settings (Klein, Calderwood, & McGregor, 1989) differs markedly from typical decision-making research, which most often focuses on a single decision among a fixed set of alternatives in a controlled environment. In naturalistic settings, decisions are embedded in a broader situational and cultural context and are part of a dynamic decision process. Therefore, decisions are affected by the dynamics of the situation rather than by a single judgment isolated from contextual constraints (Orasanu & Connoly, 1993). These two approaches complement one another—the first by providing insight into the making of decisions after a rational and reflective processes of reasoning, and the second by highlighting the importance of contextual constraints on individual or team decisions. In this chapter, we present two studies on decision-making expertise. These focus on the constraints that complex real-world settings impose on individual and group reasoning and decision making. Two naturalistic settings are investigated: two intensive care units in a hospital and an emergency telephone triage service responsible for 911 emergency calls.

The chapter is organized as follows. First, we present the theoretical ideas that have guided our research in expertise and medical cognition (Patel, Arocha, & Kaufman, 1999). The research has dealt with the nature of problem solving and decision making by experts and less-than-expert persons in a variety of situations in both controlled and naturalistic settings. After describing our theoretical framework, we present a study of collaborative decision making in the Intensive Care Unit (ICU), where we look at the contextual constraints on decision-making strategies employed by ICU team members. In the following section, we examine the decision-making process by nurses as they interact with patients during 911 emergency calls. We conclude with some comments on how contextual factors affect team decision making and how tacit knowledge may shape the type of learning that occurs in naturalistic settings with differing characteristics.

INDIVIDUAL DECISION MAKING

Most medical cognition research has been informed by a problem-solving perspective, which emphasizes the nature and development of expert performance. A consistent finding has been that experts use a form of forward-directed reasoning (from data to hypothesis) to solve routine problems within their domain of expertise (Patel & Groen, 1986; 1991). In the medical domain, forward reasoning is characterized by a chain of inferences from data (e.g., patient's signs and symptoms), leading to an incremental refinement of hypotheses, which finally results in a diagnostic solution. Presented with a routine clinical case in their domain of expertise, expert physicians generate the correct diagnosis based on the recognition of the case findings, typically within the first few minutes of the interview. For instance, an expert physician observing a patient who complains of chest pain and who leans forward to relieve the pain, may immediately generate the diagnosis of pericarditis, without having to evaluate all the findings in the case or considering alternative diagnoses. Research into other domains, such as physics (e.g., Chi, Feltovich, & Glaser, 1981; Larkin, McDermott, Simon, & Simon, 1980) and mathematics (e.g., Hinsley, Hayes, & Simon, 1977), has also revealed forward-directed reasoning to be a hallmark of expert performance. In all variety of experts, forward reasoning is strongly correlated with accuracy.

By contrast, novices (e.g., medical students) and intermediates (e.g., medical residents) typically tend to employ a form of backward reasoning, in which one or more hypotheses are conceived and then tested against the available data (Patel & Groen, 1991) by deriving consequences for each of the hypotheses (as in the hypothetico-deductive method). Observing the same patient complaining of chest pain, a less experience physician may generate a list of possible diagnoses, such as heart attack and pulmonary embolism, and ask questions that allow him or her to decide among these alternatives. This is a less efficient problem-solving strategy, necessitating the running of mental simulations and making heavy demands on working memory. Research (Arocha & Patel, 1995) has shown that second- and third-year medical students generate numerous hypotheses, but fail to evaluate these in a systematic fashion. As their education progresses, medical students learn to test each generated hypothesis more methodically (Arocha, Patel, & Patel, 1993). After graduation, experience in hospital settings allow physicians to make effective use of forward-driven strategies in routine cases, whereas they exploit their knowledge of pathophysiology in difficult cases in a backward-driven manner. How systematically backward or mixed reasoning strategies are employed to make decisions is partly a function of training and experience (Arocha, Patel, & Patel, 1993).

Reasoning strategies are tightly linked to the nature of the domain. In medicine, reasoning strategies operate on a hierarchical and dynamic knowledge base. This is structured into several levels of organization (Evans & Gadd, 1989; Patel, Evans, & Kaufman, 1989; Patel, Arocha, & Kaufman, 1994). An epistemological framework has been used to describe the levels at which clinical knowledge (in the form of a disease schema) is typically organized in medical problem-solving contexts. This is presented in Table 23.1.

The first level consists of *observations*, or units of information that in and of themselves may not be clinically useful, but when combined with other observations, can be recognized as potentially relevant to the problem. The second level consists of *findings,* or observations that have potential clinical significance. Findings are established on the basis of a decision; that is, that a set of observations is significant and must be clinically accounted for. The next level, *facets,*

TABLE 23.1.
Knowledge Levels for the Analysis of Medical Problem Solving

Level	*Example*
Observations: Basic perceptual categories	We observe a person with his hands on his chest and leaning forward as he walks slowly. This can be interpreted as someone looking for something on the floor, by a lay person, as back pain, by a medical novice, or as pericarditis, by an expert cardiologist.
Findings: Clinical significance of clusters of observations	In the case above, the person's hand on his chest are interpreted as "chest pain" and learning forward as "pain relieved by movement."
Facets: Intermediate constructs used in medical reasoning (e.g., in running a mental model underlying a disorder).	The findings above are interpreted or examined in terms of their medical significance. For instance, "chest pain" and "pain relieved by movement" are interpreted as suggesting that the patient has some unspecified heart problem, or a pulmonary disorder.
Diagnosis: A medical category listed under as a distinct disorder, with known signs and symptoms.	"Viral pericarditis," "juvenile diabetes," "myocardial infarction," etc.

involves clusters of findings that suggest prediagnostic interpretations. These serve to divide the clinical problem into manageable subproblems, including general pathological descriptions (e.g., aortic insufficiency) or categories of diseases (e.g., endocrine problem). Finally, at the "highest" level in the epistemological structure, a *diagnosis* is made, which subsumes and explains all levels beneath it. The model is hierarchical—with facets and diagnoses establishing the context in which observations and findings are interpreted and also providing a basis for anticipating and looking for confirming or discriminating findings.

Because effective forward-driven reasoning in medicine depends on the physician possessing an elaborate knowledge hierarchy, it is particularly prone to error in the absence of knowledge. The reason is that forward reasoning is an inductive strategy (i.e., from data to hypothesis), and there are no procedures for checking the validity of the "inferences" made. Disease schemata, slowly constructed on the basis of experience in a given domain, guide physicians to key aspects of a problem and serve to filter out irrelevant information. Thus, when diagnosing clinical cases within their specialty, expert physicians can distinguish significant findings from irrelevant observations, rapidly access appropriate schemata, and delineate a structured problem space mainly consisting of facets and diagnosis.

In contrast, when faced with less familiar problems, experts typically employ a mixed reasoning strategy—forward-oriented reasoning to account for aspects of the problem that can be easily solved and backward reasoning to tie up "loose ends" and anomalies (Patel, Groen, & Arocha, 1990). Research has shown that resolving anomalies is essential for the development of globally coherent explanations, and conducive to learning (Dunbar, 1995; Patel, Kaufman, & Arocha, 2000). When experts are presented with unfamiliar or difficult cases, their reasoning consists of a two-phase process: First, they generate the main diagnosis, based on a set of the case findings, and then they evaluate the hypotheses against the other findings (Joseph & Patel, 1990; Patel, Arocha, & Kaufman, 1994).

In complex real-world settings, such as intensive care units, physicians are often confronted with problems (e.g., multisystem problems) that are at the outer edge of their area of expertise. They need to rely on other people's domain knowledge to treat or manage the patients appropriately. In these circumstances, no single physician possesses a complete schema for the disorder, requiring team collaboration (Patel, Kaufman, & Magder, 1996). Thus, a model of medical reasoning based on individual expertise may not apply in rare cases or in multiply constrained, complex, and dynamic settings, where health professionals need to act in collaboration with one another. The following section summarizes results from two studies of team problem solving and decision making in the complex, uncertain environment of a hospital intensive care unit (Leccisi & Patel, 1997; Patel, Kaufman, & Magder, 1996).

COLLABORATIVE DECISION MAKING IN AN INTENSIVE CARE UNIT

Although the investigation of individual expertise is important for understanding the nature and limits of expert performance, in naturalistic environments experts often rely on other people to accomplish their tasks. We hypothesize that this reliance changes the nature of expert problem solving and decision making in two ways: First, by constraining the expert actions in particular, situation-specific, ways. Second, by allowing the expert to use the environment (other people's expertise) in ways that extends his or her range of action (e.g., managing a patient outside their area of expertise). For instance, the expert in an intensive care unit (ICU) relies on the nurses to collect and interpret basic patient data and on the residents to make day-to-day decisions. This allows the expert to focus solely on the critical decisions. The ICU functions more efficiently this way.

In short, the ICU functions as a team, where each of its members (e.g., nurses, residents, physicians, nutritionists, pharmacists) possess specialized, although somewhat overlapping, knowledge. The team works together toward a common goal in a coordinated way and relies on more than one source of information (Orasanu & Salas, 1993). The following section illustrates some of these issues.

The Nature of the Environment in Intensive Care Units

Hospital ICUs are designed to care for seriously ill patients or high-risk patients who require rigorous monitoring. Among these patients, many suffer from multiple problems and are administered medications, which can produce severe side effects, which necessitates further observation and action. As soon as some improvement is noted, patients are typically transferred to other wards, which offer less intensive and less costly medical care.

We studied two ICU environments—a medical ICU (MICU) and a surgical ICU (SICU). In both environments, the core team consisted of residents, nursing staff, consulting physicians, and an attending physician (MICU) or a chief surgeon (SICU). The attending physician headed the MICU, and the chief surgeon headed the SICU. Other team members included a staff pharmacist to look for adverse drug interactions and to serve as a consultant to the attending physician or the chief surgeon, nutritionists, and laboratory technicians. The team was complemented by residents (ages 26–34 years), who ranged from first year (i.e., just completed their medical training) to fourth year (i.e., about to become board-certified specialists). Aside from their medical degree, the residents possessed an undergraduate science degree (e.g., physiology).

Each member of the team contributed specific kinds of expertise (e.g., in nursing, pharmacology, or nutrition). The attending physician and the chief surgeon had considerable expertise in the ICU environment (more than 10 years of experience) and were frequently required to manage patients with disorders that fell outside their area of expertise (cardiology for MICU and surgery for SICU). This required relying on their generic knowledge of medicine and on the help of specialists in other medical areas, such as neurologists or neurosurgeons.

One key characteristic of ICU settings is that team members rotate frequently. For example, there is a day shift and a night shift for nurses and residents, and typically four or five attending physicians who rotate on a weekly basis. A report is kept of such rotations to minimize information loss and ensure continuity of care. This frequent personnel rotation often complicates the smooth coordination of decisions (Patel, Kaufman, & Magder, 1996). For instance, a physician may receive information about an event from a resident or a nurse who was not present when the event occurred.

In interviews with the attending physicians and the chief surgeons, three general goals common to both MICU and SICU were identified: (a) to stabilize the patient, (b) to identify and treat the underlying problem, and (c) to plan a longer term course of action for the patient. The achievement of these goals required that tasks be effectively coordinated and distributed among the various players on the team. Different data sets are gathered by different team members, and although team leaders are ultimately responsible for making most major decisions, responsibility is allocated to various team members to maximize efficiency and deliver care in the most effective possible manner. Each individual team member carries out task-specific situation assessments, attends to immediate problems, and coordinates information with the person at the next level of the hierarchy.

Observational data were collected at the medical intensive care units of the Royal Victoria Hospital, a McGill University teaching hospital. The researchers, including graduate students and post-doctoral fellows, observed the ongoing activities over a period of time to familiarize themselves with the patterns of behavior and patient care in this work setting. This phase involved observation and note taking of informal conversations with the physicians and residents in charge. Subsequently, two experimenters spent a week in each of the ICU settings collecting data, as follows.

The investigators followed two patients (one in MICU and another one in SICU) from the time they entered the ICU to the time they were transferred to the general medical ward. The MICU patient was an elderly man who had been brought to emergency with Sudden Death Syndrome (SDS) and was given emergency cardiopulmonary resuscitation. The patient received treatment for three days after being transferred from the MICU to another hospital ward. In the SICU, the patient followed was an elderly woman who was admitted to the SICU after having had a liver transplant. The patient had previously suffered from Hepatitis C cirrhosis. She remained in the SICU for a period of 4 days.

The main source of data was the audiotaped recordings of morning rounds, which were complemented by patient charts, recordings of morning lectures, and semistructured interviews with the major participants. Morning rounds are important because the teams visit and evaluate each patient in detail. During these rounds, members of the team provided patient reports, which were discussed to evaluate each patients' status, evaluate the appropriateness of each of the decisions made and of the actions taken, and plan future courses of action. As is customary, the morning rounds were also used as an instructional forum for resident trainees.

The audiotapes were transcribed verbatim. The transcript of each daily round was divided into "episodes," based on the topic of the discussion, where each change in topic signaled a different episode. Each episode (which focused on a particular aspect of patient care and management) was divided into segments. These segments typically referred to a team discussion of one or more issues relevant to the patient state, laboratory data, or various medical measurements. The segments were further divided into propositions (i.e., idea units). Furthermore, the transcripts were coded for the type of inference generated; namely, forward inferences (when data was used to generate a hypothesis) or backward inferences (when hypothesis was used to evaluate data).

To investigate how high-level decisions were made, transcripts from the case discussions between residents and physician or surgeon were analyzed. Decisions made by these team members were categorized in terms of the level of the decisions. Three levels were coded *findings*, *actions*, or *assessments*. *Findings* are defined as statements regarding decisions on some patient-specific information. *Actions* refer to statements regarding decisions about a procedure to be performed, Finally, *assessment* refers to evaluation or deliberations about the benefits or drawbacks of treatment. The two researchers analyzed the transcriptions independently and discussed their differences until an agreement was reached. An expert physician was consulted when the discussion involved unresolved medical matters.

In the next section, we present some results from our study, which focus on how the tasks required by the two clinical settings affected the communication patterns among team members, the reasoning patterns, and the levels of decision made.

Tasks and Environmental Constraints on Decision Making

There were some general differences in the medical and the surgical units that stemmed from the different goals in each of the environments and that affected the way in which decisions were made. Table 23.2 presents a list of the major differences that we identified in conversations with both MICU and SICU members and from observations (note taking, audio recordings of meetings) of day-to-day activities by the team members.

TABLE 23.2
Differences in the Constraints on Decision Making
Between the Medical and Surgical Intensive Care Units

Type of Constraint	*MICU*	*SICU*
Main Task	Problem Solving / Problem Management	Problem Management
Flow of information	More Hierarchical. From nurse to resident to attending physician	Less Hierarchical. All team members interact
Gathering of information	Nurses & Residents	Mostly Nurses
Evaluation of Patient Condition	Attending Physician	Team
Domain Knowledge	Focus on Findings	Focus on Procedures and Facets

The first major difference involved the nature of the tasks the two teams have to carry out. In the MICU, the main task was to identify the patient problem (what the diagnosis is) and to make decisions for implementing the most effective treatment and management plan. The MICU, therefore, had two main goals: to diagnose the patient problem and to manage the patient so that he or she becomes stable (out of a life-threatening situation). Before the patient could be managed, the concern was to determine why the patient had suffered from SDS and to identify the underlying pathology and, therefore, the diagnosis. This type of task is common in the MICU, where patients with unknown disorders are often brought. The task of identifying the problem is exacerbated by the fact that patients have experienced a very serious, life-threatening event, such as heart attack, heart failure, or a serious infectious disease, such as tuberculosis or AIDS, typically in combination with another ailment.

In contrast, in the SICU the nature of the problem was known because it was the result of the surgical operation. The task was to monitor the postoperative problem by treating and managing the patient's condition. As part of problem management, one of the main concerns with the SICU patient was the possibility of infection that often results after a surgery. Dealing with infection requires administrating antibiotics, which involves very aggressive or invasive procedures, such as the use of intravenous central catheters. In this case, the task of the SICU staff is to focus on stabilizing the patient condition by making sure that the patient's airways remain open and that blood circulation is not impaired, while monitoring the patient postoperative evolution and subsequent treatment.

A second major difference between the two was in the pattern of communication among the team members. Figures 23.1 and 23.2 illustrate the flow of information in both the MICU and the SICU. The figures present the pattern of

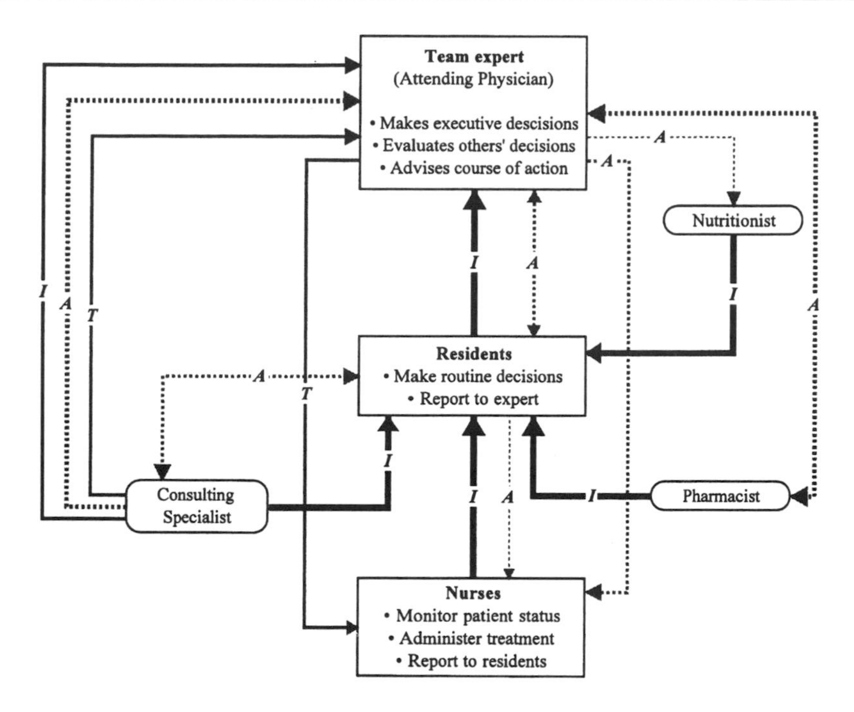

FIG. 23.1. The process of information gathering and decision making during communication among MICU team members, who communicate individually through the day (I), during morning rounds (T), or in advising exchanges (A). The thickness of lines indicates amount of communication, with thicker lines representing more communication exchanges.

communication. The thickness of the lines indicates different amounts of interaction (thicker lines representing more interactions). The SICU team members tended to have a more "democratic," less hierarchical communication, where each member interacted with other members (e.g., nurse-chief surgeon) more often and in various roles (e.g., advising, in unspecific daily interactions, in round discussions). The MICU the flow of information was found to be more linearly organized, where the pattern of interaction was most often between attending physician and resident and between the latter and the nurses, while less communication was observed between nurses and attending physician. In both settings, the consultants (i.e., medical specialists, pharmacists, and nutritionists) only communicated with the physician or the surgeon and the residents, but not with the nurses.

A third difference was in terms of roles that team members played in the decision-making process. In the MICU, the attending physician made all the major decisions directly relevant to the evaluation of the patient condition, although the nurses gathered most of the information (e.g., patient's signs and symptoms, results of tests performed, general characterization of the patient's state). In the SICU, the decisions were often made in discussions with the team,

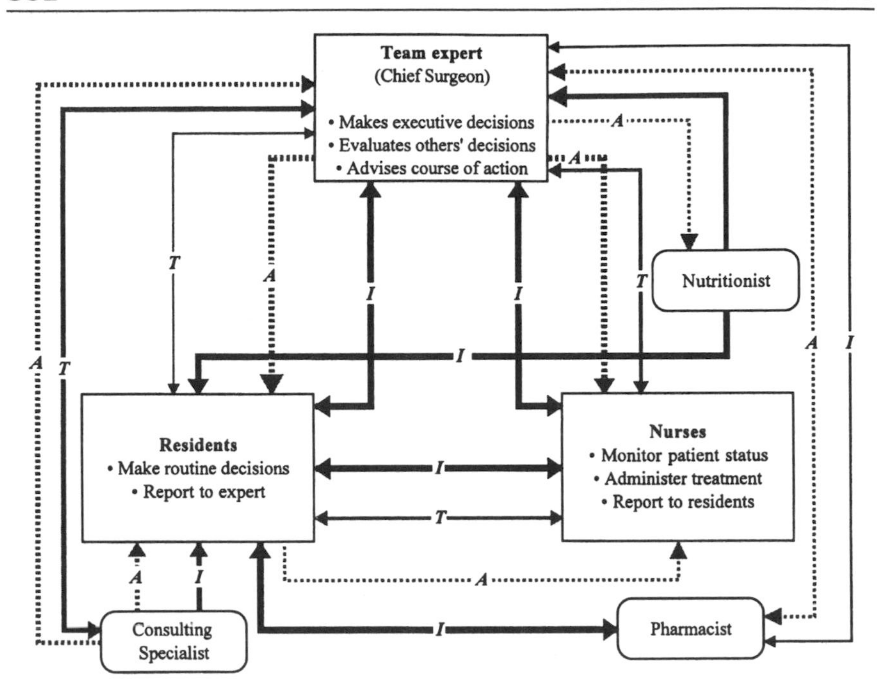

FIG. 23.2. The process of information gathering and communication during decision making among SICU team members, who communicate individually through the day (I), during morning rounds (T), or in advising exchanges (A). The thickness of lines indicates amount of communication, with thicker lines representing more communication exchanges.

involving the intervention of nurses, residents, and chief surgeon. The data that serve as a basis for making decisions were the responsibility of both residents and nurses.

In summary, the goals of the tasks performed in each of the ICUs were different. That is, identifying the patient problem and patient stabilization, in MICU; and monitoring the patient for complications, in SICU. These goals may account for the different patterns of communication among the team members. The clinical case in the SICU was more routine, the goal being that of monitoring the patient for complications. In addition to monitoring the patient's condition, the MICU patient required identification of the problem (i.e., Why did this patient suffer a sudden death syndrome?). The latter is a more complex task, as it requires diagnostic reasoning in a more difficult case (recall that in the MICU most patients present multiple underlying problems).

We hypothesize that in the MICU the pattern of communication required more knowledge of the underlying pathophysiological process, which may explain why there was more communication between attending physician and resident and less communication between attending physician and nurses. The SICU patient required only the implementation of a routine management and treatment plan. Because the nurses play a very important role in this task, there is a great deal of information that they need to convey directly to the chief surgeons. These characteristics of both settings may have resulted in differences in team members' reasoning and decision-making processes, as we see later. The next section deals with how the ICU setting affects the reasoning patterns and the types of decisions made by residents and experts.

Directionality of Reasoning and Levels of Decision Making in ICU Settings

As stated earlier, research has shown that experts tend to use mainly forward-driven reasoning in routine problems and a mixture of forward- and backward-driven reasoning in more complex problems (Patel & Groen, 1986; Patel, Groen, & Arocha, 1990). Table 23.3 shows the percentage of inferences generated by team leaders (i.e., attending physicians and chief surgeons) and residents that were coded as either forward-driven or backward-driven to reflect their directionality (Leccisi & Patel, 1997). Whereas in the MICU there was a higher percentage of backward-driven inferences (59%) than forward-driven inferences (37%), the opposite was found in the SICU (43% and 58%, respectively). This was found to be the case for both experts and residents.

Given that the case in the SICU was a more routine problem than the case in the MICU, it is not surprising that a higher percentage of forward-driven inferences was observed. Similarly, a lower percentage of forward-driven inferences and a higher percentage of backward-driven inferences is consistent with the uncertain nature of the patient problem in the MICU.

The differential difficulty of the cases may have also affected the decisions the team members made. To examine their decisions we categorize them into levels (what these decisions were about). Table 23.4 presents the percentage of decisions that were made at different levels in each ICU setting. As described before, data from the case discussions with team leaders (e.g., attending physicians or chief surgeons) and residents were used to evaluate the frequency and the nature of decisions. The decisions examined were concerned with *findings* (e.g., decisions

TABLE 23.3
Percentage of Forward-Driven and Backward-Driven Inferences Generated by Team Leaders and Residents in Medical and Surgical Intensive Care Units

	Forward-driven inferences		*Backward-driven inferences*	
Team Members	*MICU*	*SICU*	*MICU*	*SICU*
Attending physicians	26	42	37	36
Residents	11	16	22	7

TABLE 23.4
Percentage of Types of Decisions Made by
Attending Physicians and Residents in MICU and SICU Settings

	MICU		*SICU*	
	Attending	*Resident*	*Attending*	*Resident*
Finding	3	28	23	53
Action	39	57	23	32
Assessment	50	12	38	10

regarding the patient's state, signs, and symptoms), *actions* (e.g., management procedures to be followed) and *assessments* (i.e., decisions regarding overall state of the patient).

In the MICU environment, the most common deliberations and decisions were related to securing appropriate course of action for treating and managing the patient. In contrast, the most commonly addressed issues and decisions made in the SICU were related to the nature of the findings (e.g., discussions of the state of the patient, vital signs). In the SICU, patients' signs and symptoms are individually treated, which means focusing on problems at the level of findings, which is in keeping with the goal of stabilizing the patient (e.g., making sure that the patient's vital signs are normal or that the patient is well enough to be transferred to the general ward). This is not completely true for the MICU, although in certain patient conditions that are life threatening, findings do need to be individually attended to.

The results presented here suggest that the two settings impose different constraints on reasoning. The MICU demands a more deliberative form of decision making that ensures adequate action, whereas the SICU requires more procedures that aim at stabilizing the patient after complications arising from surgery. Constraints that influence the decision-making processes are situational—the task that the team has to carry out affects the nature of the strategies used.

As an illustration of these types of decisions, we present an excerpt from a dialogue between the chief surgeon (CS) in the SICU and a resident (R). This excerpt is representative of the type of communication that existed between the chief surgeon and the resident in that it illustrates the procedural nature of much of SICU interaction, where most of the communication focuses on what to do:

R: The [patient's] PO2 dropped to 75 ... 74, then I have to put it back to 45 ...

CS: Yeah, there was no blood gas for some time and she was on bicarb and all kinds of stuff, trying to fix this and that, and I sort of felt uncomfortable, so I checked that and [resident's name] thought in the middle of the night that she wasn't oxygenating better, so he increased the pressure support to get it oxygenating better, which is not the right thing to do, at least not with that kind of data, so the bicarb was stopped yesterday

In contrast, consider the following excerpt illustrating an exchange between the attending physician (AP) and the resident (R) in the MICU environment. In this example, the attending physician elaborates on a possible state of the patient as support for the procedure implemented by the resident.

R: I left the pressure support on and put it back up at 10, which was just below what he was doing.

AP: Uhm, that sort of changes my idea about SIMV. I knew you were more negative than that ... but the pressure supported ... they are not working as hard to take a breath, so they have no appetite, so they become more tolerant, in a form, but it doesn't manage, it is not as important to him as if you don't have control over what he's getting out of it. You can't regulate as much if you don't know what it is. It is more variable ... You're giving him 10 breaths ... you're controlling a big chunk of ... what he's doing inside ...

In summary, both the nature of the task and the team structure constrain team decision making and problem solving. In the MICU the task is aimed at both (a) understanding the patient condition and (b) implementing treatment for patient stabilization; in the SICU, the task mostly consists of implementing a course of action for stabilizing the patient. These different tasks lead to differences in the team members' reasoning process. If we assume that the nature of the cases in SICU are more routine than the cases in the MICU—as long as there are no major uncertainties regarding postoperative complications—it is not surprising the SICU team members' use of forward-driven reasoning. In contrast, the nature of the problem is typically known in the SICU. The task usually consists of keeping the patient sufficiently stable that he or she can be sent back to the general ward.

Decision making in the SICU can be viewed in terms of the recognition-primed decision (RPD) model (Klein, 1989; 1993). As long there are no major uncertainties in any complication, decision making takes place by simple pattern matching. The "solution" is seen almost immediately, triggered by observation, especially in urgent situations where other strategies may not be possible (Saito, Rumelhart, & Shigemasu, 1998). However, if complications arise, the solution consists of deciding what to do, that is, determining what procedure to follow, which may involve running mental simulations of potential courses of action to select the most effective treatment and management plan. This is always done dynamically, as any selected plan can be changed rather opportunistically.

In the MICU, the patients are often undiagnosed, so the task becomes one of determining what the problem is (i.e., the diagnosis) to stabilize the patient.

Although the patient is treated and managed regardless of whether a diagnosis is known, management becomes more efficient and effective if the patient problem has been identified. This leads to great effort in problem solving. Solving and managing such complex uncertain problems requires generating backward-driven inferences. This often requires redefining the problem in search of an adequate explanation of the patient problem. This is in keeping with the complex RPD strategy described by the recognition-primed decision making model (Klein, 1989; 1993), which may involve mental simulations of the possible mechanisms underlying the disorder.

Given the complexity of cases seen in the MICU, the major team members (i.e., the decision makers), such as attending physicians and residents, make most of the decisions, with the former determining the critical decisions. The nurses, in this setting, typically make day-to-day decisions, although any critical information is communicated to the residents who, in turn, communicate with the attending physician who makes the final decision.

Although knowledge plays an important role in patient care in both ICU settings, the type of knowledge required is not always explicit, verbalizable knowledge. Team members in the ICU settings often make decisions that they are not able to explicitly state. The next section attempts to discuss the role of tacit knowledge in the ICU.

Tacit Knowledge and Decision Making in the ICU

Medical work in the ICU is often unpredictable. For instance, many patients are put on mechanical ventilators to help them breath. Although this can often save a patient's life, it may also cause complications. The patient may develop stress ulcers and be given drugs to combat this problem, which in turn may change the acid base of the stomach, causing the patient to aspirate bacteria, which may result in pneumonia. Overcoming these difficulties requires rapid actions, which can only be acquired through day-to-day exposure to problems commonly occurring in the ICU setting. Just like the solo expert solves routine problems by deploying forward-directed reasoning (Patel & Groen, 1986; Patel, Groen, & Arocha, 1990), the ICU expert must recognize the problem even when an explanation is not available.

This suggests that tacit knowledge (e.g., knowledge not easily verbalizable) may play an important role in effective patient management. As an illustration, one MICU resident (RES) had made the correct decision of not treating a patient's pneumonia-like symptoms with antibiotics, because he hypothesized that the patient disorder was likely caused by medication (i.e., chemical pneumonitis). If the patient's respiratory difficulties are a result of medication, then antibiotics would have no effect. When questioned by the attending physician

(EXP) about how one discriminates between a chemical pneumonitis and systemic pneumonia, the resident had difficulty articulating the differences. Through the following series of probe questions, the expert physician acting as a teacher challenged the resident's understanding until he was able to describe the pattern of cues that discriminate between these two conditions:

EXP: So how do you decide it is pneumonia or just a new chemical pneumonitis?
RES: I'll wait until he strikes a fever in terms of his ...
EXP: So, could he have a fever with pneumonitis?
RES: He could even have a fever with pneumonitis
EXP: So what else do you need?
RES: White count
EXP: You could have a rise in white count with that too, although that would be more suggestive, and finally, what is going to be the most important thing?
RES: chills, shaking, whether he coughs sputum
EXP: He could have aspirated, some ugly looking sputum coming out is really expanding in the lungs. When he gets the initial pneumonitis, now it gets worse over the next few days and that is a different story, it should start to clear, fairly quickly
RES: chemical pneumonitis on its own doesn't need the antibiotics
EXP: That's right, that's right, you can make him worse if you do that, so for now I would just watch him, and that may be all it is we'll see how he does. O.K., so it's pulmonary, cardiovascular, we have done ... uhhm ... were you happy with his cardiovascular?

The probing generated by the attending physician as a teacher can be likened to the critical decision method developed by Klein, Calderwood, and McGregor (1989), in the sense that it too attempted to identify the critical cues that allowed the resident to distinguish between the two types of pneumonitis (but that the resident could not originally verbalize).

Although the resident's actions were timely and appropriate, the process of supporting those actions in explanation was effortful and reflective. Thus, although the resident made the right decision, he could not spontaneously support it. This is consistent with findings in naturalistic decision making showing that decision makers often make decisions based on a rapid recognition of the situation (cf. Klein, 1993). However, they may require a method to elicit the knowledge that they use in making such decisions (such as the critical decision method presented in Klein, Calderwood, and McGregor, 1989).

This kind of postfacto evaluation was also observed in a weekly session where attending physicians and staff met to discuss the problems of a MICU patient who had suffered a cardiac arrest. On his arrival to the emergency room, the patient had been treated with a blood-thinning agent called streptokinase. As is typical of hospital weekly discussions, the case was summarized by a physician (one cardiologist) followed by a discussion concerning the interpretation of the results of the electrocardiogram (ECG), which suggested that the patient had suffered a myocardial infarction (heart attack). The treatment was the focal issue of a discussion by cardiologists, respirologists, residents, and students. The respirologists argued that the patient had received the appropriate therapy, whereas the cardiologists argued to the contrary. The positive evidence in favor of the use of streptokinase was that it was the usual treatment strategy for patients showing signs of heart attack and that previous research had suggested that streptokinase could reduce morbidity and mortality. The critical question was whether the ECG provided conclusive evidence for heart attack. The cardiologists argued that the ECG did not show a pattern totally consistent with myocardial infarction. Although the ECG pattern was not typical, the respirologists argued that delay in testing may have complicated this measure and could be misleading. One cardiologist argued (supported by a second cardiologist with specific statistics) that only a small percentage of people benefit from this type of treatment, thus questioning the relevance of the decision for the patient.

The patient was stabilized when he was treated with streptokinase but also suffered subsequent bleeding, which is a common side effect of this medication. In the ensuing discussion, they collectively constructed the sequence of events, debated over the interpretation of specific evidence such as the results of the ECG, and discussed the statistical basis of clinical trial research and its interpretation in this therapeutic context.

Although the discussion was not conclusive, evaluation sessions such as this one serve a valuable pedagogical role in that they help to articulate assumptions that would not normally be brought up during clinical rounds. Because decisions are typically made with minimal justification, especially in urgent situations, deliberation is necessary for constructing explanations and possibly learning from them. Evaluation sessions accomplish another function. In day-to-day practice, there is little time available for engaging in extended causal reasoning. However, underlying causal models are often critical to support real-time decision making. By fostering discussion and explanation, this type of environment helps in the development of such models.

The organization of the ICU promotes the distribution of decision making in a manner that is similar to the way in which individuals make decisions. That is, the epistemological model described presents a stratified medical knowledge base (observations, findings, facets, diagnoses). This stratification can be seen in the ICU team. Nurses are chiefly responsible for gathering *observations* and reporting on selected *findings*. Residents synthesize and organize observations into findings and generate a set of organ-system specific *facets* (e.g., gastrointestinal bleed), whereas the attending expert must consider the patient as a whole, provide a *diagnosis,* and generate appropriate plans. In the medical knowledge base, each level contains the levels below: observations are clustered into findings, which are summarized into facets, which are clustered into diagnoses. In the expert, knowledge stratification together with the distribution of tasks leads to a reduction in information management, because the expert does not need to process all levels of information to make decisions. Nurses take care of recording all observations, whereas the resident "filters" that information by focusing on significant benchmark values (e.g., cardiac index) and gathering anomalous findings. Although the expert is chiefly responsible for synthesizing information and planning future courses of action, he or she needs only attend to the most significant information.

In sum, it appears that even with the increased resources of the team, decisions in real life settings, such as ICUs, are performed using a strategy similar to that of the individual expert problem solver. It also appears that the whole team is not acquiring information in the same way. That is, they pick up and use particular pieces of information, depending on task allocation. This task allocation—and the division of labor that results from the way the ICU is organized—allows the team to function more efficiently.

Nurses in the ICU play an important role in the day-to-day running of the unit. However, their decisions are often limited to procedural matters, such as taking the patient's temperature or in making sure that the patient's airways are clear. Difficult and high-level decisions involving selection of treatment are made by the physicians in the team (residents and attending physician). There are settings, however, where nurses play a more critical role, such as in telephone triage calls, where they can make decisions that have life or death consequences. Although nurses receive support from expert physicians here too, often the critical decision making relies solely on the nurses' judgment. The next section summarizes a study conducted to examine nurses' decision making in 911-telephone triage.

LEVELS OF URGENCY AND DECISION MAKING IN 911 TELEPHONE TRIAGE CALLS

Management of urgent situations demands immediate attention and prompt action. Such situations are often complex and uncertain, possibly involving many different decisions and courses of action. Frequently, they also involve high risk. The actions of the decision maker may result in unintended and often unpredictable consequences. Instances of such cases include decision making by anesthesiologists (Gaba, 1992) and dispatchers of emergency medical services (Leprohon & Patel, 1995), where individual variables, such as knowledge structures and skills, interact with modulating variables, such as stress, time pressure, and fatigue, as well as with communication patterns among team members.

Leprohon and Patel (1995) investigated the decision-making processes used by nurses in real emergency telephone triage. In emergency telephone triage, there is a sense of time urgency—decisions often have to be made in seconds and on the basis of partial and frequently unreliable or ambiguous information. The study was based on an analysis of transcripts of 50 nurse–patient telephone conversations conducted by 34 nurses. The data analysis was based on propositional analysis of the tape-recorded conversations. Aside from the conversations, patient records in the hospital were also used to follow up the cases—for those cases where the patient was transported by ambulance to the hospital (i.e., those perceived by the nurse as being either high or moderate urgency). Telephone calls to the patient 10 days after the original call were also used for cases of perceived low urgency (where no ambulance was dispatched).

The decisions made during these conversations were of different levels of urgency: low, medium, and high. *High urgency* refers to the situations where resources should have been sent immediately; *moderate urgency*, to situations where an ambulance should have been dispatched within 20 minutes of the call; and *low urgency*, situations where sending an ambulance before 45 minutes was not necessary or when there was no need to send an ambulance and required only referral to a physician or online advice.

In the analysis of transcripts, errors in judgment were coded as either false positives or false negatives. False positives represented interventions beyond the patient's needs (e.g. sending an ambulance in a nonemergency situation). False negatives corresponded to interventions that were insufficient to meet the patient's needs and may have compromised the patient's health. The nurses' explanations for their decisions were also obtained immediately after their conversations. An independent expert team evaluated the cases, gave their opinion on each of the calls studied, and finally reached consensus. This consensus was used by the researchers to assess the appropriateness of the nurses' decisions.

The results showed three patterns of decision making, reflecting the perceived urgency and ambiguity of the situations. The first pattern corresponds to immediate response behavior evoked in situations of high urgency. In these circumstances, decisions were made rapidly and actions were "triggered" in a forward-directed fashion (59% of the decisions made by all nurses vs. 20% based on backward-driven reasoning). The nurses in this study responded with perfect accuracy in these situations. The second pattern involves limited problem solving in situations of moderate urgency with moderately complex cases. In this case, forward-driven reasoning was 58% and backward-driven reasoning was 20%. The third pattern involves deliberate problem solving and planning and is elicited in response to low urgency situations. In these circumstances, nurses evaluate the whole situation and explore possible solutions, such as identifying the basic needs of a patient and referring the patient to an appropriate clinic. Forward-driven decisions were 82%, but backward-driven strategies increased to 60%.

The number and the quality of errors were also examined. Nurses made no errors in situations of high urgency (i.e., 100% accuracy) and made more errors in situations of moderate urgency than in situations of low urgency (50% and 26%, respectively).

These results are consistent with three patterns of decision making that reflect the perceived urgency of a situation and that can be mapped to the RPD models, as proposed by Klein (1989; 1993). The first pattern corresponds to *immediate response behavior* as reflected in situations of high urgency. In these circumstances, decisions are made with great rapidity. Actions are typically triggered by either the perception of critical patient symptoms or the unknown urgency level of the situation, in a forward-directed manner, which is in keeping with RPD model of simple pattern matching. The nurses in this study responded with perfect accuracy in these situations. The second pattern involves *limited problem solving* and typically corresponds to a situation of moderate urgency and to cases that are of some complexity. The behavior is characterized by information seeking and clarification exchanges over a more extended period of time. These circumstances resulted in the highest percentage of decision errors (50%, of which 43% were false positives). In these instances, a simple RPD strategy of pattern matching cannot be applied and, because of time constraints, a deliberate and detailed mental simulation is not possible. The third pattern was *deliberate problem solving and planning*, which was more frequent in low urgency situations. This involved evaluating the whole situation and exploring alternative solutions (e.g., identifying the basic needs of a patient, referring the patient to an appropriate clinic), which may involve the running of mental simulations to select the most appropriate decision. In this situation, nurses made fewer errors (26%) than in situations of moderate urgency and more errors than in situations requiring immediate response behavior (no errors were made).

Decision-making accuracy was significantly higher in nurses with 10 years or more of experience (65% accuracy) than in nurses with less than 10 years of experience (26% accuracy), which is consistent with what we know about acquisition of expertise in other domains. Furthermore, when the domain of the case was directly relevant to the specific experience of the nurse, the accuracy was 100%.

Decisions in routine triage are often made with minimal reflection and with the focus on patient symptoms rather than on the generation of hypotheses or on the detailed assessment of the situation. Of the total of forward-driven reasoning events (44% of all conversations), only 10% were based on detailed assessment of the situation or on a hypothesis. In many cases, most notably high urgency situations, patterns of symptoms were sufficient to trigger the accurate decisions and, hence, appropriate actions (e.g., sending an ambulance with a physician to treat a potentially life-threatening condition).

Furthermore, an analysis of selected explanations provided by some of the nurses showed that they were unable to articulate the reasons for their appropriate decisions. When probed to do so, they provided erroneous explanations. Hence, these nurses had the ability to act effectively on knowledge, but they were, nonetheless, unable to articulate accurately the reasons for acting. With an increase in problem complexity, causal explanations were used by these nurses, and the decisions were very often inaccurate. In these decisions, contextual knowledge of the situations (e.g., the age of the patient, whether the patient was alone or with others) was exploited to identify the needs of the patients and to negotiate the best plan of action to meet these needs. The dissociation between accuracy of decision and knowledge suggests that tacit knowledge that seems to be independent of deliberate decision making was used.

In summary, the results indicate that in naturalistic decision making situations, such as 911 calls, nurses make decisions mostly by rapidly assessing the urgency of the presented disorder (especially by attending to the case findings). As in other areas of decision making in naturalistic environments, decisions are not made based on the selection of alternatives, although some alternatives were considered by the nurses in the low urgency situations.

Support from this comes from the findings on the reasoning processes used by the nurses and from the dissociation between action and explanation. It may be that through experience the nurses are highly adept at perceiving the urgency of a situation but lack the medical knowledge to articulate a reasonable explanation of the underlying disorder. A more interesting explanation, however, is that the decoupling of action and explanation reflects tacit knowledge that cannot be easily articulated.

CONCLUSION

The studies described in this chapter show how contextual constraints affect the processes of reasoning and decision making by individual team members in health care environments. One of the major constraints comes from the tasks people in such environments have to perform. Task constraints are of utmost importance in the communicative process as these determine the structure of communication among team members. In tasks of little complexity, such as the implementation of a procedure (as in SICU), team members interact with one another in a more equal footing (greater communication among more members). In contrast, with tasks of greater complexity, where problem solving is involved (as in MICU), the structure of the team is more linear, with a more specific role played by each team participant and with communication going from nurse to resident to attending physician.

The reasoning and decision-making processes we discussed here reveal important similarities (as pointed out earlier) with other findings from research on naturalistic decision making and also from investigations of problem solving in laboratory and real-world settings (e.g., recognition-primed decision making, Simon's "satisficing" principle, the relationship between these and the development of expertise). Thus, there is a great deal of data gathered and models proposed (such as Klein's RPD models) that should suggest further attempts to generate a synthesis that gives coherence to seemingly disparate issues and problems (e.g., decision making and problem solving in various settings). We do not try to propose a model as it is a task beyond the scope of this chapter. However, continuing efforts to supply coherent models should be a goal of researchers in naturalistic decision making and problem solving.

Our aim has been to focus on the individual team members' decisions rather than on the functioning of the group as a team. Research on team decision making has examined how teams work together and has pointed out the importance extra-individual issues, such as that of shared mental models for effective decision making (e.g., Orasanu & Salas, 1993). We did not examine the ICU members' shared models, but we believe that these issues are crucial for understanding team learning and performance. For the moment, we suggest that the types of constraints discussed in this chapter need to be taking into serious consideration when developing models of decision making in dynamic naturalistic settings. These constraints cannot be simply taken as add-ons to the models of individual decision making; instead, they should be viewed as integral parts of such models.

To finalize our chapter, we comment on learning in team environments. ICU teams accomplish other functions aside from patient care, one of which is the teaching of medical residents and students. Residents go through conventional learning activities (e.g., seminars and clinical rounds) while devoting considerable time to reviewing clinical research literature and medical references. However, most of the learning takes place in the context of clinical practice—clear examples of "situated" learning—where residents are closely guided by expert role models. Here, correct procedures, ways of thinking about particular problems, and challenges to counterproductive reasoning strategies are demonstrated. This knowledge is often acquired in tacit manner. The problem is that because the two settings are different and impose different types of constraints on reasoning and decision making, it may be possible that residents are not equally qualified to work in dissimilar environments. ICU residents may acquire the reasoning patterns specific to one setting but not the other. If this is so, then it is very important to consider matching board certification to the particular experiences the residents have during their residency training (e.g., in surgical or medical ICU; emergency).

A final aspect concerns the role that tacit knowledge plays in the process of learning medicine. We presented an example of an expert guiding a resident to generate the conditions of applicability for discriminating certain decisions that had remained below conscious control and an example of mismatch between the 911 nurses' explanations and their decisions. In these examples, tacit knowledge seems to play an important role in guiding action in uncertain and urgent settings. This knowledge, however, seems to be decoupled from the explicit declarative knowledge used to explicate the actions. Further research to deepen our understanding of such tacit knowledge may require more controlled studies where one could vary levels of urgency or uncertainty in order to determine more precisely the conditions under which tacit knowledge plays a decisive role and where explicit declarative knowledge is necessary. Investigating how such tacit knowledge is acquired and the constraints that affect such acquisition is an important step in contributing to a general theory of expertise.

ACKNOWLEDGMENTS

The research reported in this chapter was supported in part by a grant from the Social Sciences Research and Humanities Council of Canada (A00-B14-98) to Vimla Patel. We would like to thank the subjects who participated in the studies reported here. Finally, our thanks go to Robert Hoffman, Eduardo Salas, and Gary Klein for their invaluable comments of the chapter.

REFERENCES

Arocha, J. F., Patel, V. L., & Patel, Y. C. (1993). Hypothesis generation and the coordination of theory and evidence in novice diagnostic reasoning. *Medical Decision Making, 13*, 198–211.

Arocha, J. F., & Patel, V. L. (1995). Novice diagnostic reasoning in medicine: Accounting for clinical evidence. *Journal of the Learning Sciences, 4*(4), 355–384.

Chi, M. T. H., Feltovitch, P. J., & Glaser, R. (1981). Categorization and representation of physics problems by experts and novices. *Cognitive Science, 5,* 121–152.

Dunbar, K. (1995). How scientists really reason: Scientific reasoning in real-world laboratories. In R. J. Sternberg & J. Davidson (Eds.), *The nature of insight* (pp. 365–395). Cambridge, MA: MIT Press.

Evans, D. A., & Gadd, C. S. (1989). Managing coherence and context in medical problem solving discourse. In D. A. Evans & V. L. Patel (Eds.), *Cognitive science in medicine: Biomedical modeling* (pp. 211–255). Cambridge, MA: MIT Press.

Gaba, D. (1992). Dynamic decision making in anesthesiology: Cognitive models and training approaches. In D. A. Evans & V. L. Patel (Eds.), *Advanced models of cognition for medical training and practice* (pp. 123–148). Heidelberg, Germany: Springer-Verlag.

Hinsley, D. A., Hayes, J. R., & Simon, H. A. (1977). From words to equations: Meaning and representation in algebra word problem. In M. A. Just & P. A. Carpenter (Eds.), *Cognitive processes in comprehension* (pp. 89–108). Hillsdale, NJ: Lawrence Erlbaum Associates.

Joseph, G.-M., & Patel, V. L. (1990). Domain knowledge and hypothesis generation in diagnostic reasoning. *Medical Decision Making, 10*, 31–46.

Klein, G. A. (1989). Recognition-primed decisions. In W. B. Rouse (Ed.), *Advances in human-machine system design* Vol. 5 (pp. 42–92). Greenwich, CT: JAI.

Klein, G.A. (1993). A recognition-primed decision (RPD) model of rapid decision making. In G. A. Klein, J. Orasanu, R. Calderwood, C. E. Zsambok (Eds.), *Decision making in action: Models and methods* (pp. 138–147). Norwood, NJ: Ablex.

Klein, G. A., Calderwood, R., & McGregor, D. (1989). Critical decision method for eliciting knowledge. *IEEE Systems, Man, and Cybernetics, 19(3)*, 462–472.

Larkin, J. H., McDermott, J., Simon, H. A., & Simon, D. P. (1980). Expert and novice performances in solving physics problems. *Science, 208*, 1335–1342.

Leccisi, M. S. G., & Patel, V. L. (1997, April). *Worlds apart: Decision making strategies in medical and surgical intensive care environments.* Paper presented at the 1997 American Educational Research Association, Chicago, IL.

Leprohon, J., & Patel, V. L. (1995). Decision making strategies for telephone triage in emergency medical services. *Medical Decision Making, 15,* 240–253.

Orasanu, J., & Connolly, T. (1993). The reinvention of decision making. In G. A. Klein, J. Orasanu, R. Calderwood, & C. E. Zsambok (Eds.), *Decision making in action: Models and methods* (pp. 3–20). Norwood, NJ: Ablex.

Orasanu, J., & Salas, E. (1993). Team decision making in complex environments. In H. J. Klein, J. Orasanu, R. Calderwood, & C. E. Zsambok (Eds.), *Decision making in action: Models and methods* (pp. 327–345). Norwood, NJ: Ablex.

Patel, V. L., Arocha, J. F., & Kaufman, D. K. (1999). Medical cognition. In F. T. Durso (Ed.), *Handbook of applied cognition* (pp. 663–693). Chichester, England: Wiley.

Patel, V. L., Arocha, J. F., & Kaufman, D. R. (1994). Diagnostic reasoning and expertise. *The Psychology of Learning and Motivation, 31,* 137–252.

Patel, V. L., Evans, D. A., & Kaufman, D. R. (1989). Cognitive framework for doctor-patient interaction. In D. A. Evans & V. L. Patel (Eds.), *Cognitive science in medicine: Biomedical modeling* (pp. 253–308). Cambridge, MA: MIT Press.

Patel, V. L., & Groen, G. J. (1986). Knowledge-based solution strategies in medical reasoning. *Cognitive Science, 10,* 91–116.

Patel, V. L., & Groen, G. J. (1991). Developmental accounts of the transition from medical students to doctor: Some problems and suggestions. *Medical Education, 25,* 527–535.

Patel, V. L., Groen, G. J., & Arocha, J. F. (1990). Medical expertise as a function of task difficulty. *Memory & Cognition, 18(4),* 394–406.

Patel, V. L., Kaufman, D. R., & Arocha, J. F. (2000). Conceptual change in the biomedical and health sciences domain. In R. Glaser (Ed.), *Advances in Instructional Psychology, Vol. 5: Advances in Instructional Psychology: Educational design and cognitive science* (pp. 329–392). Mahwah, NJ: Lawrence Erlbaum Associates.

Patel, V. L., Kaufman, D. R., & Magder, S. A. (1996). The acquisition of medical expertise in complex dynamic environments. In K. A. Ericsson (Ed.), *The road to excellence* (pp. 127–165). Mahwah, NJ: Lawrence Erlbaum Associates.

Saito, K., Rumelhart, D. E., & Shigemasu, K. (1998). *Decision making under time pressure.* Unpublished manuscript, Stanford University, Stanford, CA.

24

Tactical Mission Analysis by Means of Naturalistic Decision Making and Cognitive Systems Engineering

Arne Worm
National Defense College
Stockholm, Sweden

In this chapter, I describe a broad Command and Control (C^2) research perspective and how I applied a combined theorist's and practitioner's approach to the problem of modeling C^2 processes of military and emergency response units and to modeling the mission itself. I performed case studies, field studies, and experiments using a combined control theory, Naturalistic Decision Making and Cognitive Systems Engineering framework. A composite approach was necessary to fulfill this endeavor. In this chapter, I outline the work on development of theories and models intended to sustain analysis, evaluation, and assessment of military and emergency response units performing complex, high-stake tactical operations. I also tested these concepts in several simulated tactical operations, and finally, validated the concepts in a number of full-scale exercises.

THEORETICAL FRAMEWORK

Control Theory: Powerful Science and Useful Metaphors

An approach based on control theory and dynamic systems can facilitate structuring and understanding of the command and control problem. The mathematical stringency and powerful formalism of control theory makes it possible to describe and treat systems as diverse as technical, organizational, economic, and biological dynamic systems in basically the same manner: as processes, or clusters of processes, with a built-in adherent or assigned control system. The concepts of control theory can be used as metaphors in research on decision making, especially in multiple player, dynamic contexts. The notion that decision making constitutes the regulatory function in command and control processes (Orhaug, 1995) strongly supports the control theory approach. This notion also supports the fact that the hierarchical command structures of military and emergency response organizations are strongly coupled to both centralized and distributed decision-making principles (Brehmer, 1988). Control theory was used by Annett (1997) to investigate team skills. This hints at the use of a control theory framework for analysis and evaluation of command and control in tactical operations. There are four fundamental requirements to be met (Conant & Ashby, 1970; Glad & Ljung, 1989; Brehmer, 1992) if control theory is to be used in analysis and synthesis of dynamic systems:

1. There must be a goal (*the goal condition*).
2. It must be possible to ascertain the state of the system (*the observability condition*).
3. It must be possible to affect the state of the system (*the controllability condition*).
4. There must be a model of the system (*the model condition*).

Distributed Dynamic Decision Making

Brehmer (1992) suggested the use of control theory as a framework for research in distributed, dynamic decision making. The conventional view of decision making, supported by normative theories, reduces decision making to selecting an appropriate action from a closed, predefined action set and to resolving conflicts of choice. As a consequence, the analysis of decision tasks focuses on the generation of alternatives and the evaluation of these alternatives as in Multi-Attribute Utility (MAU) analysis (Kleindorfer, Kunhreuther, & Schoemaker, 1993). Research in dynamic decision making has been based on analysis of several applied scenarios, for example, military decision making, operator tasks in industrial processes, emergency management, and intensive care (Brehmer, 1988, 1992). Two things were clarified in these analyses:

1. The decision making was never the primary task. It was always directed toward some goal.
2. The dynamic character of the assigned tasks became apparent in the study of the applied contexts.

These results are coherent with earlier descriptions by Edwards (1962), Rapoport (1975), and Hogarth (1981) of dynamic decision making, which Brehmer (1992, pp. 212–213) summarized as follows:

1. A series of decisions is required to reach the goal. To achieve and maintain control is a continuous activity requiring many decisions, each of which can be understood only in the context of the other decisions.
2. The decisions are mutually dependent. Later decisions are constrained by earlier decisions and, in turn, constrain those that come after them.
3. The state of the decision problem changes both autonomously and as a consequence of the decision maker's actions.
4. The decisions have to be made in real time. This finding has several significant implications, which are elaborated on in the next section.

The real time properties of dynamic decision making cause special problems:

1. Decision makers are not free to make decisions when they feel ready to do so. Instead, the environment requires decisions, and the decision maker, ready or not, has to make these decisions on demand. This causes stress in dynamic decision-making tasks. In order to cope with this stress, decision makers have to develop strategies for control of the assigned dynamic tasks and for keeping their own workload at an acceptable level.
2. Both the system that is to be controlled and the procedures and resources the decision maker uses to control the system have to be seen and treated as processes. Dynamic decision-making tasks can be characterised as finding a way to use one process to control another process.
3. The different time scales involved in dynamic decision-making tasks have to be monitored and taken into consideration. In most situations the active agents in a dynamic system, such as the directly involved operators and their closest commander or squad leader, operate in a time scale of seconds to minutes. Their commanders and their command and control systems operate in time scales of hours to days.

An application of this approach in studies of distributed decision making in dynamic environments such as firefighting and rescue missions was described by Brehmer and Svenmarck (1995).

Naturalistic Approaches to Decision Making

Zachary and Ryder (1997) reviewed decision-making research during the past decades and elaborated on the major paradigm shift in decision theory. The shift is from analytic, normative decision-making procedures described in Kleindorfer et al. (1993) to Naturalistic Decision Making (NDM) developed and described by Klein (1993a; 1993b), Zsambok and Klein (1997) as well as by Klein and Woods (1993). NDM applies to many dynamic and potentially dangerous areas of activity, such as military missions, air traffic control, fire fighting, emergency response, and medical care. The essentials of this paradigm are condensed below:

- Human decision making should be studied in its natural context.
- The underlying task and situation of a problem is critical for successful framing.
- Actions and decisions are highly interrelated.
- Experts apply their experience and knowledge nonanalytically by identifying and effecting the most appropriate action in an intuitive manner.

Cannon-Bowers, Salas, and Pruitt (1996) reviewed, commented, and related the NDM approach to the extensive research on distributed and dynamic decision making described in an earlier section. They argued that this was how to overcome the limitations of the notions of the classic normative research paradigm in decision making. A fundamental application of NDM, the Recognition-Primed Decision (RPD) model, was presented in detail in Klein (1993a) and was applied to complex command and control environments in Kaempf, Klein, Thardsen, and Wolf (1996).

COGNITIVE ASPECTS ON MISSION COMMAND AND CONTROL ISSUES

The area of Cognitive Systems Engineering (CSE) has grown steadily since the first significant contributions were published in the 1980s by Rasmussen (1983, 1986), who introduced the concept of skill-based, rule-based, and knowledge-based behavior for modeling different levels of human performance. Endsley (1995) developed a comprehensive theory of individual operator, commander, and team situation awareness in dynamic systems. Danielsson and Ohlsson (1996) studied information needs and information quality in emergency management decision making. This work also applies to the military context. Woods and Roth (1988) comprehensively reviewed the CSE domain. Hollnagel and Woods (1983) significantly contributed to this field by their definition of a cognitive system (CS) as a man-machine system (MMS) whose behavior is goal-oriented, based on symbol manipulation, and uses heuristic knowledge of its surrounding environment for guidance. A cognitive system operates using knowledge about itself and the environment to plan and modify its actions based on that knowledge. In complex systems this is indisputable. For example, in C^2 tasks in military missions, a multitude of sensor systems, communication systems, training programs, personnel, and procedures are all elements of the total operational system. Viewing this system as a cognitive system permits the integration of all existing control resources—operators and commanders, technological facilities, procedures, and training—into a coordinated system that accomplishes a mission safely and efficiently. The use of CSE to model, analyze, and describe C^2 in hazardous, real-time, high-stake activities is a powerful approach, given a sufficient understanding by the investigator of the interdependencies and linkages between other research areas, such as those elaborated on earlier, and the CSE field. In this work, I tried to coherently integrate several of these research areas.

METHODOLOGY AND APPROACH

Experiences from the author's initial work on military missions and emergency response (Worm, 1996; 1997; 1998c) made it possible to integrate a dynamic system model with CSE and associated process control concepts.

I developed a set of methods and tools for modeling, analysis, and evaluation of ground forces and their abilities at the battalion and lower levels: Mission-critical skills of individual operators and teams, commander mission resource management, and overall unit performance. The central point of this project was the integration into a multidiscipline mission and unit evaluation and assessment technique of a number of methods and tools used in trade and industry as well as in military systems development. To facilitate this integration, a set of concepts was introduced in order to analyze and evaluate accomplishments and shortcomings of various military units in an unambiguous and comprehensive way. The concepts were:

- Conceptual modeling of dynamic, complex tactical systems and processes, of their states and state transitions, based on CSE and control theory.
- Identification of mission and unit state variables and of different action and decision-making mechanisms as a process regulator.
- Mission Efficiency Analysis (MEA) of fully manned and equipped company units executing missions in an authentic environment against a realistic opposing force.
- Applying the Cognitive Reliability and Error Analysis Method (CREAM), developed by Hollnagel (1998), for detailed and comprehensive investigation of probable causes of mission failure or system malfunction.

MODELS

The striking properties of tactical forces performing hazardous, time-critical operations can be characterized in brief as improved mobility and lethality, increased risks and resource requirements, and complex decision-making and action selection situations (Worm, 1997). These dynamic properties raise a demand for increased personal and equipment performance requirements and an escalating need for personal protection. Military commanders of today, and to an even greater extent in the future, face dynamic and nonlinear C^2 problems. Implementing modern C^2 principles requires advanced human, organizational, and technical resources with very high information processing capabilities. Modern C^2 systems demonstrate true real-time properties at all levels: the individual soldier and weapon system as well as where the systems are integrated into higher order structures, such as joint operations forces. This calls for unique and innovative approaches to the mission C^2 problem. Improving operator and commander abilities to train, assess, evaluate, and master these belligerent dynamics will have decisive influence on all decisions and selections of action, mission course of events, logistics, the number of casualties, and many other vital components of emergency response or other kinds of severe crisis (Worm, 1998c). However, the specific skills and properties that commanders and operators have to possess in order to yield optimal mission performance in such critical and uncertain situations are not easily identified, and hence, they are difficult to improve.

A major problem in addressing a topic area with such a vast scope is to fulfill a number of diverse and many times conflicting requirements for supporting C^2. Three major conclusions by Serfaty and Entin (1997, p. 172) concerned the properties and abilities of teams performing tactical, hazardous operations:

1. The team structure adapts to changes in the task environment.
2. The team maintains open and flexible communication lines. This is important in situations where lower levels in a command hierarchy have access to critical information not available to the higher command levels.
3. Team members are extremely sensitive to the workload and performance of other members in high-tempo situations.

Using experiences from studies performed during military instrumented force-on force exercises (Worm, 1997), together with two studies in the emergency response domain (Worm, Jenvald, & Morin, 1998a, 1998b), My colleagues and I began trying to define the systems studied utilizing a common frame of reference. Based on these experiences, we defined the requirements imposed on forces performing hazardous, time-critical operations as follows:

- Capable of rapid and reliable information acquisition and processing, both manual and automated.
- Able to execute distributed team decision making in dynamic environments.
- Access to high bandwidth, jamming-resistant communications between and within the units engaged in a mission.

- Highly efficient, agile, robust, and adaptable to a multitude of missions and tasks.
- Constantly exposed to risk for own and others' lives and property.
- Striving to meet ever-increasing performance requirements by means of advanced training.
- Capable of coping with complex and ambiguous decision and action selection situations.

These requirements call for unique and innovative approaches to the problem of modeling the dynamics of tactical missions. Similar demands are also imposed on the modeling and analysis of the units performing such missions. The identification and specification of optimal performance requirements in such critical and uncertain situations are cumbersome tasks that are often subject to misinterpretation.

The Main System Model: The Tactical Joint Cognitive System

By the term *dynamic system* is meant, in control theory, an object, driven by external input signals $u(t)$ for every t and as a response produces a set of output signals $y(t)$ for every t. From the work of Conant and Ashby (1970) and Brehmer (1992), it is well known that most complex systems have real-time, dynamic properties; the system output at a given time depends not only on the input value at this specific time, but also on earlier input values. It is also known that a good regulator of a system has to implement a model of the system that is to be controlled. Put otherwise, Ashby's law of requisite variety (Ashby, 1956) states that the variety of a controller of a dynamic system has to be equal to or greater than the variety of the system itself. This implies the tactical robustness and adaptability characteristics identified earlier. The basic principles of human-centered automation (Billings, 1996) provided a thorough and clear-cut set of requirements to make the mission and unit modeling comprehensive and explicit. The principles are based on a premise: Some supervisory human operators bear ultimate responsibility for the achievement of operational goals. This premise is the followed by an axiom that states: These supervisory human operators must be in command. To bear responsibility, and hence, to be in control of a tactical operation, operators and commanders involved must be in command. This basic axiom supports this notion by the following corollaries:

- To remain in command and to execute command effectively within one's range of responsibility, the human operator must be actively involved.
- To remain involved within one's range of responsibility, the human operator must be appropriately informed.

- The human operator in command must be able to monitor the automated technological systems or other assisting subordinate agents.
- The automated technological systems or other assisting subordinate agents must also monitor the human operator in command.
- The human operator must be able to comprehend and predict the behavior and performance of such agents.
- The human operator and every other intelligent system element must be able to communicate its intent to other system elements, and to track and acquire the intent of other system elements.

Consequently, my colleagues and I identified the main system in a tactical mission process as the system

- to which a mission is assigned.
- to which the operational command of the mission is commissioned.
- to which the responsibility for effecting the mission is authorized.
- to which the resources needed for performing the mission is allocated.

We designated this main system as the Tactical Unit, an aggregate consisting of one or several instances of four principal subsystem classes:

Technological Systems, for example vehicles, intelligence acquisition systems, communication systems, sensor systems, life support systems, and other kinds of mission-specific equipment, including the system operators.

Command and Control Systems, consisting of an information exchange and command framework built up by technological systems and directly involved decision makers, arranged in a hierarchical or quasihierarchical system (Brehmer, 1988).

Support Systems, comprising staff functions, logistic functions, decision support functions, organizational structures, and other aiding and assisting services.

Tactical Teams, for which we chose to use the concepts of Salas et al. (1992, p. 4) to define a team as follows:

Two or more people who interact, dynamically, interdependently, and adaptively toward a common and valued goal/objective/mission, who have been assigned specific roles or functions to perform, and who have a limited life-span of membership.

The primary issue of the definition was that task completion required support of the following processes:

- A dynamic exchange of information and resources among team members,
- Coordination of task activities (for example active communication, back-up behaviors),
- Constant adjustment to task demands, and some organizational structuring of members.

The terminology used in CSE proved to be quite adequate for describing and modeling technological systems and human operators, as Artificial Cognitive Systems (ACSs), and Natural Cognitive Systems (NCSs), respectively (Hollnagel & Woods, 1983). The complete system, where the technology and the operator together perform a complex task, could correspondingly be described and modeled as a Joint Cognitive System (JCS). Also in the other cases, the CSE framework yielded a powerful ability to describe and model these systems as JCS. This designation implies that novel solutions to the problems of performing efficient tactical operations must be focused at investigating the nature of the entire array of cognitive systems and processes together with analysis and synthesis of different means to support and control them. As a result of the work on studying, modeling, and evaluating fully operational tactical units and tactical teams, my colleagues and I concluded that such systems could be described making use of a common CSE framework, and therefore we define these as Tactical Joint Cognitive Systems (TJCSs).

The Contextual Control Model: Control, Cognition, and Context

Earlier in my work, I had not been able to accurately describe the intricate mechanisms and processes of controlling a process such as a tactical mission. The need for building a conceptual framework for time-critical, complex, and situation-dependent command and control of dynamic high-stake missions was satisfactorily met by the concepts of the cybernetics-influenced Contextual Control Model (COCOM; Hollnagel, 1998). Also in this area of context-dependent control, Ashby's law of requisite variety (Ashby, 1956) has some important implications. The requisite variety of mental modeling was investigated by Hollnagel (1992), who described the importance of the law of requisite variety to mental model development and minimization. A mental model can be seen as the basis for generating input to a system, in order to keep the variety of the system within given limits. A mental model—although difficult to study and measure, particularly mental models shared within a team (Rouse, Cannon-Bowers, & Salas, 1992)—synthesizes the steps of a process and organizes them as a unit (Allen, 1997). In CSE, cognition and control are always embedded in a context. The context includes demands and resources, tasks, goals, organization, and social and physical environments. When modeling cognitive processes, such as command, control, and intelligence processes, one must account for how cognition depends on the overall context rather than on the input. Procedural prototype control models assume a characteristic sequence of actions, whose ordering is determined by the control prototype (Hollnagel, 1998). COCOMs, however, focus on how the choice of next action is determined mainly by the current context of the mission. COCOMs describe action sequences as constructed rather than previously defined. The choice of action is controlled by the context, and actions can be both reactive and proactive. COCOMs distinguish between competence and control, in that competence—or in other words, capability—describes what the operator and commander is able to do, and control describes how he achieves it.

Models and Modes of Control. Control modes are theoretical constructs. Control will vary along a continuum instead of shifting between discrete modes. It is, however, important to make the main characteristics of different modes of control distinctly separable in order to observe, identify, and classify control. The four main control modes defined by Hollnagel are:

1. Strategic control,
2. Tactical control,
3. Opportunistic control, and
4. Scrambled control.

Each control mode has certain distinct characteristics regarding:

1. Operator's subjectively available time to execute control in order to meet task objectives, emphasizing the impact of time pressure.
2. Operator's familiarity of encountered situation, emphasizing the importance of expertise.
3. Operator's level of attention, emphasizing the influence of cognitive capabilities.
4. Operator's number of concurrently active operational goals, emphasizing the importance of ability to monitor and control parallel processes.
5. Operator's choice of next action, emphasizing the influence of abilities to predict future system states and to identify alternative future courses of action.
6. Operator's evaluation of performed missions and tasks, emphasizing the significance of adaptability in process control.

Control modes are theoretical constructs. Control will vary along a continuum instead of shifting between discrete modes. It is, however, important to make the main characteristics of different modes of control distinctly separable in order to observe, identify, and classify control.

Important model issues concerning the use and utility of the COCOM are:

- Transition between control modes. What causes control to change from one mode to another, whether it is lost or gained?
- Performance in a control mode. What is the characteristic performance for a given control model?
- Interaction between competence and control. Higher competence makes it more likely that control is maintained.

The Mission Model: Controlling Joint Systems and Processes

The combined view of control theory in technical as well in behavioral domains is crucial for success in this research area. When a function is implemented at one level of abstraction, represented at a second level of abstraction, and controlled at a third level of abstraction, the requirement for timely and complete information varies accordingly. On the other hand, it is not important whether a function or mission is carried out by an operator or by an automated system under higher order supervision; the operators and the supervisory controllers still need to maintain an adequate situation understanding—or situation awareness.

If it is not feasible to reliably and timely observe and measure the system output, and situation understanding cannot be based on the information supplied by the system; it must be based on the current process knowledge and understanding of the situation. Operators and controllers must compensate by means of accurate system performance prediction. This prediction ability is based on the axiom that a cognitive system must be able to think ahead in time and anticipate the dynamics of the process. To accomplish this, a cognitive system must rely solely on exact model knowledge of the system input's influence on the system output. This is normally referred to as *open-loop control*. Open-loop control can be a cumbersome and arduous task, especially when the system environment and the mission context are highly dynamic and the system process is unstable and nonlinear, that is, small changes or state transitions in the process can generate out-of-proportion, unpredictable, or even chaotic system behavior. In some cases the disturbances can be measured. It is then possible to almost entirely eliminate the influence of those disturbances by using feedforward control. However, this requires extremely good system knowledge of the process that one wishes to control. Feedforward control is also sensitive to variability in the system dynamics. The main advantage of feedforward control is the possibility to counteract the effects of disturbances before they are visible as an undesired deviation from the reference. Control theory has proven that although feedforward control can be considered the perfect mode of control, it is often only achievable for a limited amount of time due to, among other things, the time constants of the process. However, if the system output can be used to determine the system state, there is only a limited need for detailed knowledge of system dynamics, and feedback control can be executed. The necessary adjustments can be made by constantly measuring the deviation of the system output from the reference value. The joint cognitive system is unstable without feedback, and thereby feedback will be needed to correct deviations and compensate for the incompleteness and inadequacy of the internal system model. The importance of balance between feedback (reactive) control and feedforward (proactive) control is crucial to achieve optimal C^2 performance in a tactical mission. Feedforward control is often combined with feedback control because of its practical reliability limitations. Missions can be described as aggregates of joint cognitive systems and processes, arranged in a dynamic process cluster, depicted in Fig. 24.1.

THE MISSION EFFICIENCY ANALYSIS TECHNIQUE

Worm et al. (1998b) developed and improved the applicability of these concepts from the military domain (Worm, 1997) into a generic *Mission Efficiency Analysis (MEA)* technique. This was achieved mainly by extending the conceptual framework to cover larger, more diverse, and temporarily organized forces, and by making use of a more general terminology (Worm, 1998a). The mission efficiency definition, its determinants, and their relations to the mission efficiency measure are illustrated in Fig. 24.2.

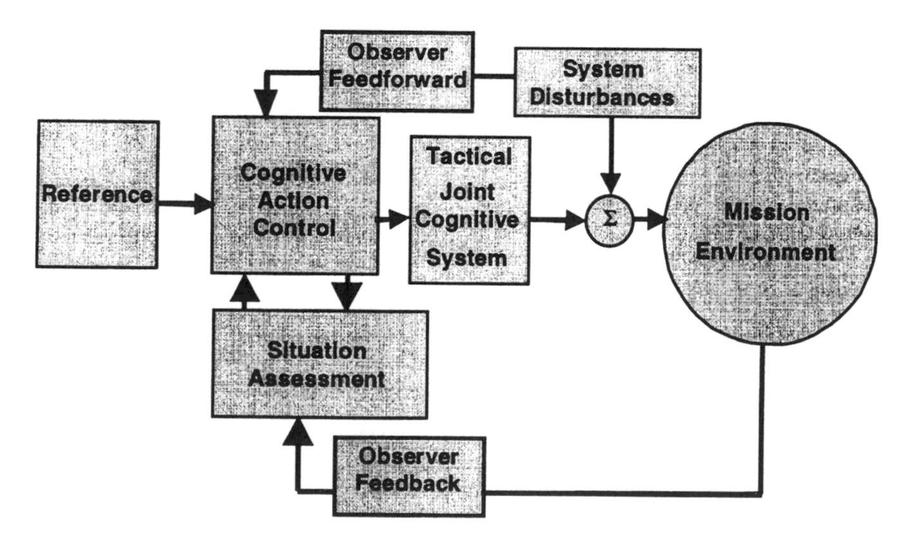

FIG. 24.1. The mission model.

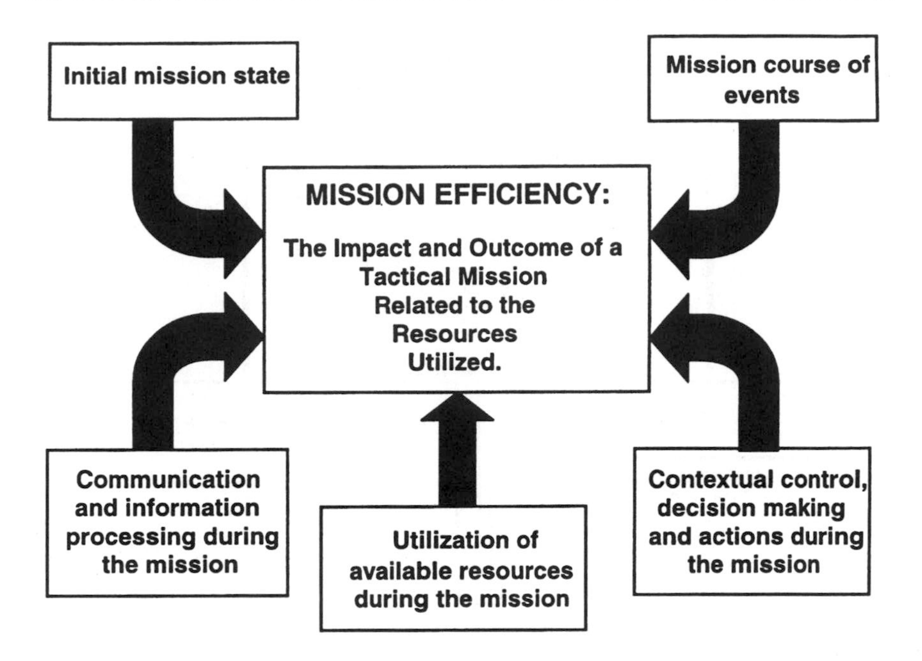

FIG. 24.2. The definition and determinants of the mission efficiency measure.

Determining Initial Mission State and Mission Course of Events

The term *Mission State* means at a given time a set of information, that is, variable values that makes it possible to determine future output if future input is known.

Before the mission, the mission has an *initial state.* When the course of events of the mission, or any activity, function, status, or mission objective of the unit changes, due to any cause, a *mission state transition* takes place. The initial mission state was determined primarily by the factors described in Fig. 24.3.

Communication and Information Processing During the Mission

The ability of the mission commander and staff to manage and process the information available at the time of mission execution was analyzed by studying:

- Information acquisition performance.
- Information processing performance.
- Information exchange and distribution effectiveness.

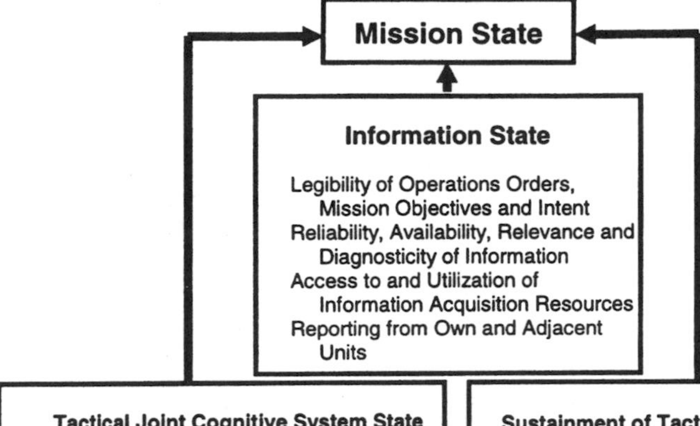

FIG. 24.3. A schematic description of mission state determinants.

- Time relations when critical orders and reports are transmitted in the studied scenarios.

Utilization of Available Resources During the Mission

The management and utilization of available resources were analyzed by studying:

- Resource depletion, replenishment and allocation .
- Unit recovery and reorganization.
- Time relations at critical re-deployment moments in the studied scenarios.

Contextual Control, Decision Making, and Actions During the Mission

Decisions and actions of the mission commander and staff were documented, along with the information related to the mission that was available at the decision points. Crucial properties and topics were:

- Information reliability (How do I know that this information is true?).
- Information availability (How do I access this information, and what are my sources?).
- Information relevance to the actual decision situation (What is the significance and use of this information?).
- Information diagnosticity (Does this information represent what I want to know?).
- Information complexity (Can this information be integrated into a comprehensible entirety?).
- Time relations at critical decision points in the studied scenarios.
- Actual and predicted control level of commanders, staff members and operators in mission-critical moments in the scenarios.

Data Collection Techniques

The course of events and the mission environment were simulated, registered, and reconstructed in an integrated simulation and data collection system for instrumented mission training, currently used by the Swedish National Defence Research Establishment. The system incorporates networks for supporting simulation and registration of unit activities during an exercise. The flow of information in these networks is controlled by a central unit, which includes both automatic procedures for simulation and registration and operator interfaces for exercise control and for replay of recorded events. The central unit also compiles and stores exercise data in a database for subsequent analysis and evaluation. The system has three different networks that are used in studies and exercises:

- The *instrumentation network* is the main simulation and registration network in the system. It consists of a number of mobile nodes with the necessary hardware and software to handle simulation and registration tasks at the level of a single vehicle. Each node is equipped with an off-the-shelf Global Positioning System (GPS) receiver with internal memory and is capable of recording the position of the vehicle at regular intervals.

- The *observer network* connects the expert observers that follow each rescue team to the central using cellular telephones. The observers are responsible for making observations that complement the automatically recorded data from the instrumentation network. To support this task, we provide a set of protocols for structured reporting (Thorstensson, 1997). By restricting the form of the reports, we ensure that they can be automatically interpreted in the central, which saves valuable time. Furthermore, the use of structured reports improves data quality which, in turn, facilitates the analyzability of the recorded information.
- The *tactical radio communication network* comprises the ordinary communication channels used by the units participating in the exercise. The tactical radio communication provides valuable information about what orders were given in different situations and for information exchange analysis. Therefore, these networks are monitored by equipment in the central. Each transmission is recorded digitally and time stamped. In this way, the radio communication from a particular situation is immediately available for replay.

The Mission Efficiency Analyses were based on aggregation and evaluation of expert observations, registered mission events, and After Action Reviews (Rankin, Gentner, & Crissey, 1995) of mission training sessions. All available information was presented synchronized in time and put into its proper context in order to create an elaborate description of a tactical scenario. During the chemical disaster response study, a total of some 9,000 events were recorded by the instrumentation system. During the military study, a total of some 5,000 events were recorded. These numbers include both automatically recorded position reports from GPS receivers and manual subject-matter expert reports entered using the structured report tool. This comprehensive material provided a solid basis for after-action reviews and for subsequent Mission Efficiency Analyses. Information lost due to technical failure was not taken into consideration, but judging from the consistency of the results, this fact did not affect the data compilation and analysis in any decisive way.

THE EXPERIMENTAL PROGRAM: TWO CASES

Mission: Attack Enemy Airborne Assault

The study in Case 1 (Worm, 1997) was performed in March 1996. The unit studied was a reduced mechanized infantry company, transported in armored personnel carriers, and equipped with man-portable antitank missile systems as its main armament. The unit also had access to indirect fire support, such as artillery, mortars, and antitank mines. The mission of the mechanized infantry company was to locate and attack enemy airborne landings in any of three geographical sectors. The opposing force in this particular situation was a special force, composed of uniquely equipped and trained airborne and mechanized units, with tactics, mobility, and firepower equal to a hypothetical, modern, enemy air assault unit. The mission of the air assault forces was to seize and defend a number of vital crossing points in order to support following airborne landings. In the following after-action review, grave inconsistencies and misconceptions in the commanders decisions were revealed. Several technical malfunctions and inability to use back-up procedures also contributed to the unit's failure to execute its mission.

Mission Course of Events

The outline of the mission, the preparations, the opponent's approach to the landing zone, and the acquired intelligence prior to engagement are briefly depicted in the mission timeline in Table 24.1. From the time of landing and engagement of the air assault forces, the course of events were monitored and reported by expert observers and controllers, most of which were experienced mission training officers in active duty. Battle course of events was intense and expeditious. The combat activity culminated approximately 60 minutes from the landing of the assault forces.

TABLE 24.1
Mission Timeline Depicting the Main Events During the Military Air Assault Exercise

Time	*Own unit events*	*Opposing force events*	*Time*
1900	Session start. Tactical operations orders issued: Prepare location and attack of enemy airborne assault in sector A, B, or C.	Session start. Tactical operations orders issued: Prepare airborne assault in sector A, B, or C.	1900
0600	Company operations orders issued: Locate and attack enemy airborne assault in sector A, B, or C.	Company operations orders issued: Commence airborne assault in sector A. Bomber sorties launched.	0500
0630	Air defense warning issued: Air attacks east Skövde city, sector A and B.	Assault preparations. Air strikes in sector A and B. Assault units boarding.	
	Air Defence warning issued: Air attacks Karlsborg city westward, sector A and B.	Last bomber sorties launched. Assault force launched.	
	Air Defence warning issued: Air assault Karlsborg city heading Skövde city.	Assault force approach landing zones in sector A.	

(Continued)

TABLE 24.1 *(CONTINUED)*
Mission Timeline Depicting the Main Events During the Military Air Assault Exercise

Time	*Own unit events*	*Opposing force events*	*Time*
0725	Battlefield intelligence reports: Air assault units landing in sector A.	Time T: Assault units landing.	0725
	Indirect fire support authorized.	Assembly of units after landing.	0745
0800	Maneuvering towards enemy location. Visual contact. Engagement.	Time T + 35 minutes: Advance towards target zone. Engagement.	0800
1030	Session stops. Preparations for After Action Review.	Session stops. Preparations for After Action Review.	1030

Mission: Search and Rescue Chemical Warfare Victims

The study in Case 2 (Worm et al., 1998b) was performed in September 1997, when a full-scale Chemical Warfare (CW) exercise was conducted in the northern part of Sweden. The scenario assumed there were hostilities between countries in the vicinity of Sweden. Intelligence reports indicated that chemical weapons were used. The Swedish Armed Forces were in a state of alert. The local rescue brigades prepared themselves to handle a chemical attack against specific high-risk targets, including the air base 15 kilometers northwest of a small town. In the morning of September 25, enemy aircraft attacked the air base. The air base area became the target of an air raid with high explosive, fragmentation bombs, followed by a CW attack. Vehicles, buildings, and other parts of the infrastructure were severely damaged. The air raid inflicted some 50 casualties, who needed immediate rescue. Fire brigades from the whole county were engaged, along with large police forces and extensive medical expertise from the county general hospital. Outside the contaminated area, an emergency management command post was deployed, composed of a communications central and a mission control room. In the exercise debriefing, insufficient intelligence support and communications difficulties were found to constitute the major part of the inconsistencies in the unit's shared situation awareness and the inaccuracies of the commander's decisions. The commander and his staff did not realize the need for more optimal use of the gas indication team, which showed later to be of crucial importance to the fulfillment of the mission. This led to feedback delays in decision making and to actions that caused severe consequences for the units' abilities to operate in a coordinated and timely manner. The outline of the mission, preparations, and the main events from the exercise mission are briefly described in chronological order in the mission timeline in Table 24.2. Approximately two hours from the time of the air strike alarm at 0903 hours, the mission activity culminated.

TABLE 24.2
Mission Timeline Depicting the Main Events During the Emergency Response Exercise

Time	*Event*
08.30	The exercise starts
09.03	Alarm to the Rescue Commander in Piteå from the SOS emergency communication centre: Enemy air strike against the Piteå air base. Possible use of nerve-gas.
09.03	The rescue brigade in Piteå sends the first rescue team to the air base
09.04	The rescue commander requests support from the rescue brigades in Luleå, Boden and Älvsbyn.
09.04	The Police force receives the alarm from the SOS emergency communication centre.
09.05	The rescue commander alerts the support team.
09.07	The five medical teams and the ambulance groups receive their first orders: Assemble at meeting point.
09.08	The first weather report. Wind: West to West-North-West 3 to 10 meters per second. Temperature: 10 degrees Celsius.
09.27	First gas indication report, no indication.
09.46	The police helicopter reaches the disaster area. The Rescue Commander gets his first observation report from the air base.
09.48	The first rescue team arrives at the impact area at the air base. Gas indication show the use of VX-gas.
09.59	Order from the rescue commander to the support unit: Deploy decontamination station at the northern end of the airstrip.
10.03	The first victims found at the air base. The rescue commander receives a report on 20–25 casualties.
10.08	The second rescue team arrives at the air base from Luleå.
10.12	The Police have cordoned off the gas-covered area.
10.17	Order from the rescue commander to the support unit: Deploy the area of medical attendance at Skravelbäcken.
10.22	The special indication vehicle detects the boundary of the area contaminated by VX-fluid.
10.38	Weather report. West to West-North-West 3 to 5 meters per second. Temperature 14 degrees Celsius.

(Continued)

TABLE 24.2 ***(CONTINUED)***
Mission Timeline Depicting the Main Events During the Emergency Response Exercise

Time	*Event*
10.39	First transport of injured persons to the decontamination station.
10.40	The local radio station broadcasts an Important Emergency Message. The people in the area around the air base are requested to stay inside their houses and to put their gas masks on.
10.43	The Police helicopter uses a loudspeaker system to inform the inhabitants of the village of Böle that the Police prepare an evacuation of the villagers.
10.50	The decontamination station is fully deployed.
11.00	The area of medical attendance at Skravelbäcken is deployed.
11.01	The inhabitants of the village Kyrkbyn are evacuated.
11.18	The inhabitants of the village Böle are evacuated.
11.40	The south boundary of the area affected by gas is established.
12.12	The medical teams at the decontamination station are exhausted and have to be relieved by a back-up team.
13.00	The exercise ends.

RESULTS

The researchers applied the CREAM analysis technique based on facts revealed in the Mission Efficiency Analysis and used the same verbal and timeline scenario description. The results from this modeling, analysis, and evaluation endeavor in these two cases were consistent.

The predominant error modes were:

- Timing of movement and of tactical unit engagement.
- Speed of movement or maneuver, which is especially important in the initial phase of engagement.
- Selection of wrong object. The environments of ground warfare or emergencies offer many opportunities for choosing wrong objects, in navigation, in engagements, or in visual contact.

The main causes of partial or total mission failure seemed to be interpretation and communication failures, caused by:

- Slow organizational response.
- Ambiguous or missing information.
- Equipment malfunction, for example, power failure or projectile/missile impact.
- Personal factors, for example, lack of team training and experience.

From the performed MEA and CREAM analyses, researchers found that in these particular cases the following factors constrained the ability of the units to execute their respective mission:

- Lack of expertise and specific skills needed by all unit members, commanders and soldiers alike, to rapidly and accurately identify and locate enemy targets and own units, to evaluate the battlefield terrain, and to apply the right procedure in the right moment.
- Insufficient abilities to rapidly and accurately build and sustain individual and team situation understanding, causing difficulties in selecting alternative actions when the situation and the course of events changed in an unanticipated way.
- Lack of abilities to explicitly formulate resource needs coupled to the mission at hand and to allocate adequate and available resources to facilitate optimal mission accomplishment.
- Limited access to and use of a mission information structure that supports and improves real-time information and intelligence acquisition and permits mission-relevant information and intelligence to penetrate the organizational hierarchy to reach the intended decision maker in a secure and timely manner.
- Limited access to and limited experience in using a robust, wide-band communication system that permit fast and accurate transfer of data and speech.

CONCLUSIONS

My colleagues and I integrated various research domains, such as control theory, naturalistic decision making, quality control, operations research, and statistics, and applied them together with extensive utilization of the MEA and CREAM techniques. This approach seemed effective in achieving several goals. Applied in a context of tactical mission training, combined with advanced medium- and high-fidelity mission simulations and the systematic use of domain expert knowledge, this novel approach facilitated comprehensive mission and system evaluation and assessment.

Applicability of the Technique

To be able to objectively and reliably identify the limiting factors of a specific unit, system, or operating procedure, and to assess the magnitude of influence of these factors on unit's overall mission performance, a series of within-mission mission efficiency analyses will have to be performed in each typical case. In mission training situations, most factors, except the ones studied, can be held constant to the greatest extent possible, which makes the validation of results easier. Also actual emergency response missions can be analyzed in the same way. However, the extreme risk exposure and the reduced reliability, diagnosticity, and availability of information obtained in such situations make validation more cumbersome. This conclusion, together with the findings of Rouse et al. (1992), implies that individual team members must develop task skills and knowledge in a team-oriented environment and that the team must build and refine team-specific competence by practicing together. This requirement calls for dedicated, scenario-based training in a realistic setting. However, evaluating teamwork skills and providing meaningful performance feedback in real time, or near real time, are complex and demanding tasks. The mission efficiency analysis technique provides the means for performing these tasks. Applying it extensively will have decisive influence on tactical performance and, most likely, on the outcome of missions, too.

Who Can (and Should) Contribute?

One of the most important management decisions is what kind of personnel to assign to a project such as this. My colleagues and I identify three factors that have been crucial to ensure overall project progress and to accomplish the underlying goals of the experimental battle training center. First, it is essential that both the operators and the investigators have solid theoretical and practical competence. By this we mean that the people involved have to be able both to grasp the big picture, including the overall objectives of the project, and to focus on microscopic details when that is called for. Second, all who work in the project need experience from the actual work domain, that is, they should have some domain-specific training, preferably as an operator or commander. This requirement ensures that the investigators have the necessary domain understanding. It also aids the cooperation with the planners and officers who are responsible for the actual studies and exercises. Third, the investigators and the planners/officers should work closely together. The investigators have to be ready to modify their methods and tools and adapt their systems when new lessons are learned during the studies. Conversely, the planners/officers have to be fast learners to handle software and hardware that change frequently, as a result of those adaptations.

CONTRIBUTIONS TO GUIDELINES AND PRINCIPLES

The main objective of this research effort was to actively contribute to the science of command and control. The contention was that integrating relevant and effective methods and tools for analysis, synthesis, and development is of the utmost importance to achieve successful improvement of command and control procedures as well as to successfully design and operate future command and control systems. Furthermore, the science of command and control has a number of lessons to learn from its more specialized applications in different areas, such as military operations and intelligence, emergency response operations, and air traffic control. This work was performed in an attempt to understand how those applications can help in unifying the science of command and control. Science should look on the applications as tools that can be used for tasks possibly extended much further than what they were originally designed for. Based on the findings described, there are a few principles that I would like to communicate to other researchers interested in the NDM paradigm:

1. Participation of Subject Matter Experts (SMEs) is critical when studying and evaluating dynamic, complex missions. High-stake, multiple-player missions performed by highly professional and motivated operators can only be analyzed reliably with full participation of these operators.
2. To be able to use the concepts of NDM and CSE in system evaluation, assessment, and design, a model of cognition that supports prediction of system behavior and system reliability must complete the descriptive models of the NDM paradigm.

REFERENCES

Allen, R. B., (1997). Mental models and user models. In M. Helander, T. K. Landauer, & P. Prabhu (Eds.), *Handbook of human-computer interaction* (2nd ed., pp. 49–63). New York: Elsevier.

Annett, J. (1997). Analysing team skills. In R. Flin, E. Salas, M. Strub & L. Martin (Eds.), *Decision making under stress: Emerging themes and applications* (pp. 315–325). Aldershot: Ashgate.

Ashby, W. R. (1956). *An introduction to cybernetics*. London: Chapman & Hall.

Billings, C. E. (1996). *Aviation automation: The search for a human-centered approach*. Mahwah, NJ: Lawrence Erlbaum Associates.

Brehmer, B. (1992). Dynamic decision making: Human control of complex systems. *Acta Psychologica 81,* 211–241.

Brehmer, B., & Svenmarck, P. (1995). Distributed decision making in dynamic environments: Time scales and architectures of decision making. In J.-P. Caverni, M. Bar-Hillel, F. H. Barron, & H. Jungermann (Eds.), *Contributions to decision making* Vol. 1, (pp. 155–174). New York: Elsevier.

Cannon-Bowers, J. A., Salas, E., & Pruitt, J. S. (1996). Establishing the boundaries of a paradigm for decision-making research. *Human Factors, 38,* 193–205.

Conant, R. C., & Ashby, W. R. (1970). Every good regulator of a system must be a model of that system. *International Journal of System Science, 1,* 89–97.

Danielsson, M., & Ohlsson, K. (1996, October). Models of decision making in emergency management. *Proceedings of the 1st International Conference on Engineering Psychology and Cognitive Ergonomics*, Cranfield: Cranfield University.

Edwards, W. (1962). Dynamic decision theory and probabilistic information processing. *Human Factors, 4,* 59–73.

Endsley, M. R. (1995). Towards a theory for situation awareness in dynamic systems. *Human Factors, 37,* 32–64.

Glad, T. & Ljung, L. (1989). *Reglerteknik. Grundläggande teori* [Automatic Control. Basic Theory]. Lund, Sweden: Studentlitteratur.

Hogarth, R. M. (1981). Beyond discrete biases: Functional and dysfunctional aspects of judgmental heuristics. *Psychological Bulletin, 90,* 197–317.

Hollangel, E. (1998). Context cognition and control. In Y. Waern (Ed.), *Co-operative process management* pp. 27–52. London: Taylor & Francis.

Hollnagel, E., (1992, March). *Coping, coupling, and control: The modeling of muddling through.* Invited presentation for "Mental Models and Everyday Activities," The 2nd Interdisciplinary Workshop on Mental Models, Cambridge, England.

Hollnagel, E., & Woods, D. D. (1983). Cognitive systems engineering: New wine in new bottles. *International Journal of Man-Machine Studies, 18,* 583–600.

Kaempf, G. L,. Klein, G. A., Thordsen, M. L., & Wolf, S. (1996). Decision making in complex command-and-control environments. *Human Factors, 38,* 220–231.

Klein, G. A. (1993a). *Naturalistic decision making—Implications for design.* Dayton, OH: Wright-Patterson Air Force Base, Crew Systems Ergonomics Information Analysis Center.

Klein, G. A. (1993b). A recognition-primed decision (RPD) model of rapid decision making. In G. A. Klein, J. Orasanu, R. Calderwood, & C. E. Zsambok (Eds.), *Decision making in action: Models and methods* (pp. 138–147). Norwood, NJ: Ablex.

Klein, G. A., & Woods, D. D. (1993). Conclusions: Decision making in action. In G. A. Klein, J. Orasanu, R. Calderwood, & C. E. Zsambok (Eds.), *Decision making in action: Models and methods* (pp. 404–411). Norwood, NJ: Ablex.

Kleindorfer, P. R., Kunhreuther, H. C., & Schoemaker, P. J (1993). *Decision sciences: An integrative perspective.* Cambridge, England: Cambridge University Press.

Orhaug, T. (1995). *Ledningsvetenskap - En diskussion av beskrivningsmässiga och teoretiska problem inom ledningsområdet* [Science of command and control - A discussion of descriptive and theoretical problems within the domain of command and control]. Stockholm: The Swedish National Defence College.

Rapoport, A. (1975). Research paradigms for the study of dynamic decision behavior. In D. Wendt & C. Vlek (Eds.), *Utility, probability and human decision making* (pp. 349–375). Reidel, The Netherlands: Dordrecht.

Rasmussen, J. (1983). Skills, rules, and knowledge: Signals, signs, and symbols, and other distinctions in human performance models. *IEEE Transactions on Systems, Man, and Cybernetics, SMC-13,* 257–266.

Rasmussen, J. (1986). *Information processing and human-machine interaction: An approach to cognitive engineering.* New York: North-Holland.

Rankin, W. J., Gentner, F. C., & Crissey, M. J. (1995, November). After Action Review and Debriefing Methods: Technique and Technology. In *Proceedings of the 17th Interservice / Industry Training Systems and Education Conference* (I/ITSEC) (pp. 252–261). Arlington, VA: National Defense Industrial Association.

Rouse, W. B., Cannon-Bowers, J. A., & Salas, E. (1992). The role of mental models in team performance in complex systems. *IEEE Transactions on Systems, Man, and Cybernetics, 22,* 1296–1308.

Serfaty, D., & Entin, E. (1997). Team adaptation and co-ordination training. In R. Flin, E. Salas, M. Strub, & L. Martin, (Eds.), *Decision making under stress: Emerging themes and applications* (pp. 170–184). Aldershot: Ashgate.

Thorstensson, M. (1997). *Structured reports supporting manual observations in instrumented mission training*. Unpublished Master's thesis, Linköping University, Linköping, Sweden.

Woods, D. D., & Roth, E. M. (1988). Cognitive engineering: Human problem solving with tools. *Human Factors, 30*, 415–430.

Worm, A. (1996). *Metoder och verktyg för värdering och utveckling av krigsförband* [Methods and Tools for Evaluation and Assessment of Military Units]. Unpublished Master's thesis, Linköping University, Linköping, Sweden.

Worm, A. (1997). Simulation-supported analysis and development of tactical and lower level ground battle units. In *Proceedings of the Human Factors and Ergonomics Society Europe Chapter 1997 Annual Conference* (pp. 164–174). Bochum, Germany: The Human Factors and Ergonomics Society Europe Chapter.

Worm, A. (1998a). *Command and control science: Theory and tactical applications*. (Linköping studies in science and technology, Thesis No. 714). Linköping University. Linköping, Sweden.

Worm, A. (1998c). Modeling tactical joint cognitive systems performing dynamic, time-critical operations in distributed environments. In *Proceedings of The 4th International Symposium on Command and Control Research and Technology* (pp. 361–370).

Worm, A., Jenvald, J., & Morin, M. (1998a). Mastering the dynamics of crisis: Improving situation awareness in high-risk, time-critical emergency operations. In *Proceedings of the International Emergency Management Society 1998 Annual Conference* (pp. 527–539). Washington, DC: The George Washington University.

Worm, A., Jenvald, J., & Morin, M. (1998b). Mission efficiency analysis: Evaluating and improving tactical mission performance in high-risk, time-critical operations. *Safety Science, 30*, 79–98.

Zachary, W. W., & Ryder, J. M. (1997). Decision support systems: Integrating decision aiding and decision training. In M. Helander, T. K. Landauer, & P. Prabhu (Eds.), *Handbook of human-computer interaction*, (2nd ed., pp. 1235–1258). New York: Elsevier.

Zsambok, C. E., & Klein, G. A. (Eds.). (1997). Naturalistic decision making. Mahwah, NJ: Lawrence Erlbaum Associates.

Author Index

H

I

J

K

N

O

P

Subject Index